Quantum Mechanics
with Applications

David B. Beard and George B. Beard

New foreword by
Dr. Dian Curran and Dr. Kevin B. Beard

D0141342

Dover Publications, Inc.
Mineola, New York

Bibliographical Note

This Dover edition, first published in 2014, is an unabridged republication of the work originally published by Allyn and Bacon, Inc., Boston, in 1970. The Dover edition adds a new Foreword by Dr. Dian Curran and Dr. Kevin B. Beard, and corrects several errors in the original text.

Beard, David B.
    Quantum mechanics with applications / David B. Beard and George B. Beard. — Dover edition.
        pages cm
    Originally published: Boston : Allyn and Bacon, 1970.
    Includes bibliographical references.
    ISBN-13: 978-0-486-77990-4
    ISBN-10: 0-486-77990-4
    1. Quantum theory. I. Beard, George B. (George Breckenridge), 1924– joint author. II. Title.
QC174.1.B37 2014
530.12—dc23

                                                                    2014005919

Manufactured in the United States by Courier Corporation
77990401   2014
www.doverpublications.com

FOREWORD

Dr. David Breed Beard and Dr. George Breckenridge Beard were orphaned brothers, children of the Great Depression. After serving their country in WWII (David in the Navy and George in the Army Air Forces) they finished their educations. George attended Harvard University with the aid of scholarships and went on to the University of Michigan to earn his doctorate in experimental nuclear physics. David attended Hamilton University, also with the aid of scholarships, and received his Ph.D. in nuclear physics at Cornell University, later shifting his focus to theoretical space physics.

After conducting research and teaching for half a decade at Michigan State University, George moved to Wayne State University in Detroit, Michigan, where, except for a few sabbaticals and many summers at Argonne National Lab, he spent the rest of his career as a respected and popular professor, researcher, and physics department chair. He did extensive work in nuclear resonance fluorescence and low-energy nuclear reactions.

David accepted a professorship at the University of California, Davis, and later moved to the University of Kansas to serve as department chair. Many of his sabbaticals were spent collaborating with researchers at Imperial College in London. Among his most notable accomplishments were his derivation of a model of planetary magnetospheres and his study of comets.

In 1970, the brothers coauthored this book, *Quantum Mechanics with Applications,* an updated version of David's earlier book *(Quantum Mechanics)* including new material.

Both men had the satisfaction of having one of their children follow them into the field and later work with them. George and his son, Dr. Kevin Breckenridge Beard, together conducted experiments with the decay of ^{152m}Eu in doped calcium fluoride crystals. Recoil from a neutrino in the decay sometimes compensates for the subsequent recoil from the gamma ray, giving rise to both temperature and orientation-sensitive Mössbauer-like effects. Likewise, David's daughter, Dr. Dian Curran, worked with her father on modeling the interaction of the solar wind with Mercury. Although Mercury has no intrinsic magnetic field, its high iron content does interact with the solar wind, creating a weak magnetosphere surrounding the planet. Both brothers are now gone, but their good work, of which this book is a part, remains.

Dr. Dian Curran
Dr. Kevin B. Beard
September 2013

CONTENTS

PREFACE

This book is based on *Quantum Mechanics* by David Breed Beard which was published by Allyn and Bacon in 1963. It differs from the earlier book in the simplification of the more advanced topics and in the inclusion of much new material on applications stressing atomic and molecular physics and other fields of physics.

The first chapters are designed to familiarize the reader with the mathematical treatment and particular properties of ordinary wave motion which apply also to particle motion and have crucial importance to the subsequent development of the subject. The close relation of quantum theory to physical optics is stressed. By use of Feynman's derivation of quantum mechanics, the wave theory for particles is made to appear as inevitable and necessary as Huygen's wave theory for light.

The student has seen wave motion, understands it, and knows that the only reason for accepting the wave theory of light in preference to the ray theory is that diffraction and interference of light can be explained by the wave theory. He has worked with phase differences of permissible light paths; for example, when using Fraunhofer diffraction theory to analyze a diffraction grating. Therefore, once he is shown in detail how a ray theory of particles breaks down for electrons transmitted through a crystal, he should be quite prepared to accept a simple Fraunhofer diffraction theory for particles identical to that for light. This argument was first presented by Feynman in 1948. While at least as rigorous as the initial arguments by Schrödinger and Heisenberg in 1926, the argument focuses on the wave function directly. Thus the student who has previously worked simple problems in physical optics should have little difficulty realizing what the wave function is supposed to represent, what its properties are, and how it behaves in an homogeneous or inhomogeneous medium (*i.e.*, how it depends on the light or particle path in a medium of constant or changing index of refraction or particle potential energy).

Following this introduction, the text stresses the physical consequences of a wave theory of material particles by examining the extreme, non-classical characteristics of abrupt changes in potential energy as a function of space. Classical mechanics is later shown to arise as an approximation valid when lengths, momentum, energy or time are not specified with a precision greater than that permitted by the Uncertainty Principle. This semiclassical approximation, standard example, and other approximation methods are discussed in order to develop the student's confidence and his ability to apply the theory to diverse physical problems.

Chapters Eleven through Thirteen contain rather extensive introductory applications of quantum mechanics to atomic structure and diatomic molecules. It is anticipated that the reader will have attained a solid foundation in both quantum mechanics and atomic physics at this point. The physical applications stressed in the remaining chapters are introduced to help prepare the student in topics he is likely to encounter early in further studies, particularly studies of nuclear and solid state physics.

David B. Beard
George B. Beard

Quantum Mechanics
with Applications

I

BLACK-BODY
RADIATION

Historically, the quantum theory first arose from studies of radiation, and it is useful to introduce the subject along historical lines in which revelant aspects of electromagnetic waves may be reviewed. Quantum mechanics has, with good cause, also been called wave mechanics since it emphasizes and treats wavelike behavior of particles. Hence, it is appropriate to begin our study by an examination of familiar classical radiation theory. By this means we will be able to investigate relevant consequences of wave motion and develop for later advantage the mathematical tools frequently used in classical treatments of electromagnetic waves. As an introduction to waves in three dimensions the reader may wish to review the theory of waves confined to one dimension in a string, presented in Appendix A.

1.1 DERIVATION OF THE ELECTROMAGNETIC
WAVES IN A CAVITY

Suppose a light wave to be incident on a small hole in the side of a box which is otherwise completely light-tight so that radiation enters or leaves the box only through the small hole. Such a box is called a *black body* since the

1

hole completely absorbs all light incident on it. The incident light wave will be lost inside the box after repeated reflections from the walls and will come into thermal equilibrium with the walls. Thus a light wave escaping outside the box through the hole will have been in thermal equilibrium with the interior of the box. The radiation in equilibrium within the box consists of electromagnetic waves or, in the parlance of classical radiation theory, oscillations of the ether, the vibrating medium within the box by which the waves are propagated and sustained. *In order to estimate the radiation to be expected from the pinhole, it is necessary first to calculate the possible radiation frequencies and the energy of each separate electromagnetic wave.* The energy or light intensity emitted from the opening at a particular frequency is proportional to the total energy of all the waves having that frequency contained within the box. Hence, the intensity of the electromagnetic radiation within the box must first be estimated by means of Maxwell's equations for the electric and magnetic field vectors of the waves in the empty interior of the box. Expressed in Gaussian units, Maxwell's equations are

$$\nabla \times \mathbf{H} = \frac{1}{c}\frac{\partial \mathbf{E}}{\partial t} \tag{1.1}$$

$$\nabla \times \mathbf{E} = -\frac{1}{c}\frac{\partial \mathbf{H}}{\partial t} \tag{1.2}$$

$$\nabla \cdot \mathbf{H} = 0 \tag{1.3}$$

$$\nabla \cdot \mathbf{E} = 0 \tag{1.4}$$

Taking the curl of both sides of Eq. (1.2), interchanging the order of the space and time derivatives, and substituting Eqs. (1.1, 1.2) in the result, by the the rules of vector analysis we obtain three equations for the three components of the electric field vector, summarized in

$$\nabla^2 \mathbf{E} = +\frac{1}{c^2}\frac{\partial^2 \mathbf{E}}{\partial t^2} \tag{1.5}$$

This is a very familiar differential equation in physics, occurring for such diverse motions as a vibrating string, pressure variations in an organ pipe, flow of electricity in a cable, or the waves on a drum head. It solution describes wave motion, and it is called the *wave equation.*

In Cartesian coordinates Eq. (1.5) has the particular form or *representation*

$$\frac{\partial^2 \mathbf{E}}{\partial x^2} + \frac{\partial^2 \mathbf{E}}{\partial y^2} + \frac{\partial^2 \mathbf{E}}{\partial z^2} = \frac{1}{c^2}\frac{\partial^2 \mathbf{E}}{\partial t^2} \tag{1.5a}$$

Equation (1.5a) may be solved for one component E_x by assuming E_x to be written as a product of functions of one variable each. (This is known as assuming the variables separable.) Thus, let $E_x = X(x)Y(y)Z(z)T(t)$. This procedure will be justified if a solution is obtained; not all partial differential equations have solutions which can be obtained in this way. Dividing Eq. (1.5a) for the x component of the field by E_x, one finds in Cartesian coordinates

$$\frac{1}{X}\frac{\partial^2 X}{\partial x^2} + \frac{1}{Y}\frac{\partial^2 Y}{\partial y^2} + \frac{1}{Z}\frac{\partial^2 Z}{\partial z^2} = \frac{1}{Tc^2}\frac{\partial^2 T}{\partial t^2} \tag{1.6}$$

The left-hand side is completely independent of the time, while the right-hand side is independent of space. Therefore, both sides of the equation are equal to a constant, independent of both time and space. Let the constant be written as $-\omega^2/c^2$. Then the time-dependent part of (1.6) becomes

$$\frac{1}{Tc^2}\frac{\partial^2 T}{\partial t^2} = -\frac{\omega^2}{c^2} \tag{1.7}$$

whose solution is

$$T = D_1 e^{i\omega t} + D_2 e^{-i\omega t} \tag{1.8}$$

where D_1 and D_2 are arbitrary constants for the time-dependent part of E_x.

Similarly,

$$\frac{1}{X}\frac{\partial^2 X}{\partial x^2}$$

is equal to a constant, $-k'^2$, and thus the solution to the space-dependent part of (1.6) for E_x is

$$
\begin{aligned}
X &= A_{1k'x} \sin k'x + A_{2k'x} \cos k'x \\
Y &= B_{1l'x} \sin l'y + B_{2l'x} \cos l'y \\
Z &= C_{1m'x} \sin m'z + C_{2m'x} \cos m'z
\end{aligned}
\tag{1.9}
$$

where l' and m' are additional constants introduced for the Y, Z terms. Identical solutions may be obtained for E_y and E_z.

Since the tangential components of $\mathbf{E}$ must be continuous at the walls, $\mathbf{E}$ at the sides of the box cannot be arbitrary but is fixed by the physical properties of the wall material. In a box with infinitely conducting walls, for example, the tangential components of $\mathbf{E}$ at the walls must be zero (that is, E_x must be zero in the four walls of the box perpendicular to the y or z axes). Hence, in a rectangular infinitely conducting box of dimensions a, b and d, $Y(0) = 0$ and therefore $B_{2l'x} = 0$, and $Y(b) = 0$ and therefore $\sin l'b = 0$, which means that $l'b = \pi l$ where l is an integer. Thus, two of Eqs. (1.9) become

$$Y = B_{1l'x} \sin \frac{\pi l y}{b}; \qquad Z = C_{1m'x} \sin \frac{\pi m z}{d}$$

Similar results are obtained for the other two components of $\mathbf{E}$. Let $B_{1l'x}, C_{1m'x}, A_{1k''y}, C_{1m'y}, A_{1k'''z}$, and $B_{1l''z}$ equal unity, which may be done with no loss of generality by readjusting the values of the remaining unde-

termined constants. Then the total solution for the space-dependent part of
E may be written as

$$E = (A_{1k'_x} \sin k'x + A_{2k'_x} \cos k'x) \sin l'y \sin m'z \hat{\mathbf{i}}$$
$$+ (B_{1l''_y} \sin l''y + B_{2l''_y} \cos l''y) \sin k''x \sin m''z \hat{\mathbf{j}}$$
$$+ (C_{1m'''_z} \sin m'''z + C_{2m'''_z} \cos m'''z) \sin k'''x \sin l'''y \hat{\mathbf{k}}$$

The total solution for **E** must also satisfy Eq. (1.4), from which we obtain
(after dividing out the time-dependent part)

$$A_{1k'_x}k' \cos k'x \sin l'y \sin m'z + B_{1l''_y}l'' \sin k''x \cos l''y \sin m''z$$
$$+ C_{1m'''_z}m''' \sin k'''x \sin l'''y \cos m'''z$$
$$- A_{2k'_x}k' \sin k'x \sin l'y \sin m'z$$
$$- B_{2l''_y} l'' \sin k''x \sin l''y \sin m''z$$
$$- C_{2m'''_z}m''' \sin k'''x \sin l'''y \sin m'''z = 0$$

Since this equation must hold for any values of x, y, and z, the coefficients
of each term must each individually be equal to zero, and therefore only the
$A_{2k'_x}k'$, $B_{2l''_y}l''$, and $C_{2m'''_z}m'''$ can be nonzero (although their sum must
be zero) provided that $k''' = k'' = k'$, $l''' = l'' = l'$, and $m''' = m'' = m'$.
Thus the spatial dependence of **E** is given by

$$E_x = A_{klm} \cos \frac{k\pi x}{a} \sin \frac{l\pi y}{b} \sin \frac{m\pi z}{d}$$

$$E_y = B_{klm} \sin \frac{k\pi x}{a} \cos \frac{l\pi y}{b} \sin \frac{m\pi z}{d} \qquad (1.10)$$

$$E_z = C_{klm} \sin \frac{k\pi x}{a} \sin \frac{l\pi y}{b} \cos \frac{m\pi z}{d}$$

where

$$k = ak'/\pi, \quad l = bl'/\pi, \quad m = dm'/\pi,$$

$$A_{klm} = A_{2k'_x}B_{1l'_x}C_{1m'_x}, \quad B_{klm} = A_{1k'_y}B_{2l''_y}C_{1m''_y}$$

and
$$C_{klm} = A_{1k'''_z}B_{1l'''_z}C_{2m'''_z}$$

Equations (1.8, 1.10) together describe a sinusoidal wave of angular frequency
ω ($\omega = 2\pi\nu$ where ν is the number of cycles per second and ω the number of
radians per second), and velocity c. Equations (1.10) alone would describe
a standing wave such as the particular example illustrated in Figure 1.1
for $C_{klm} = C_{410}$,

$$E_z = C_{410} \sin \frac{4\pi x}{a} \sin \frac{\pi y}{b}$$

By substituting Eqs. (1.7) and (1.10) into Eq. (1.5), one finds that

$$\left(\frac{k\pi}{a}\right)^2 + \left(\frac{l\pi}{b}\right)^2 + \left(\frac{m\pi}{d}\right)^2 = \frac{\omega^2}{c^2} \qquad (1.11)$$

For a particular frequency ω, all integral values of k, l, and m are possible (k, l, m) which satisfy Eq. (1.11). For a cube, for example, besides the mode $(4, 1, 0)$ for E_z illustrated in Figure 1.1, there are four other modes for E_z, $(1, 4, 0)$, $(3, 2, 2)$, $(2, 3, 2)$, and $(2, 2, 3)$ with frequency $\omega = \sqrt{17}(\pi/a)c$. (The modes $(4, 0, 1)$, $(1, 0, 4)$, $(0, 4, 1)$, and $(0, 1, 4)$ are not included since $E_z \equiv 0$ for these modes.)

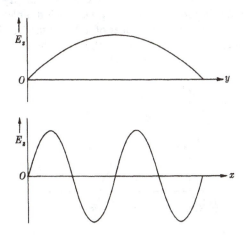

FIGURE 1.1   Graphs of E_z as function of x and y for the mode $k = 4$, $l = 1$, and $m = 0$.

1.2 NUMBER OF MODES HAVING THE SAME FREQUENCY

In order to obtain the total energy of radiation of a particular frequency ω contained within and emitted from the box, it is necessary to find the number of different modes of oscillation all having the same frequency ω. The number of modes having a frequency equal to or less than ω is found by summing all integers l, k, and m which make the left-hand side of Eq. (1.11) less than or equal to the right-hand side. For large values of ω such that the wavelength $2\pi c/\omega$ is orders of magnitude smaller than the box dimensions a, b, and d, the summation may be replaced by a triple integral over k, l, and m, that is, a volume integral over a k, l, m coordinate space. Equation (1.11) is the equation for an ellipsoid in which k, l, and m are the coordinates, and the semi-axes are $(a\omega/\pi c)$, $(b\omega/\pi c)$, and $(d\omega/\pi c)$ respectively (Fig. 1.2). The number of modes having a frequency equal to or less than ω is the number of different combinations of positive integers k, l, and m such that the left-hand side of Eq. (1.11) is less than or equal to ω^2/c^2. This number is given by the volume of the positive octant of the ellipsoid,

$$(\pi/6)abd(\omega/\pi c)^3$$

not the total volume of the ellipsoid since k, l, and m must all be positive integers. The three amplitudes A_{klm}, B_{klm}, C_{klm} may have any value provided that $A_{klm}k' + B_{klm}l' + C_{klm}m' = 0$. Therefore there are two independent directions of vibration for each set of k, l, m, which doubles the number of possible modes. Hence, the total number of different modes of oscillation up to and including a frequency ω will be $V\omega^3/3\pi^2c^3$, where V

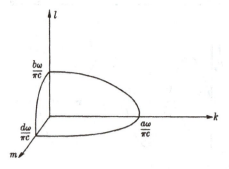

FIGURE 1.2 Illustration of the octant of the ellipsoid expressed in Eq. (1.11) containing all those values of k, l, m which make the left-hand side of Eq. (1.11) less than or equal to the right-hand side. Each mode of oscillation is represented by a single point on the diagram at an integral value of k, l, and m. The dimensions of the graph have been so chosen that each point is equivalent to a unit volume of the ellipsoid.

is the volume of the cavity. The total possible number of distinct electromagnetic waves or modes of oscillation per unit volume within the cavity, whose frequency is less than or equal to ω, is therefore $\omega^3/3\pi^2c^3$. The number of vibrational modes per unit volume having a frequency between ω and $\omega + d\omega$ is found by differentiating that expression:

$$N(\omega)d\omega = \left(\frac{\omega^2}{\pi^2c^3}\right)d\omega \qquad (1.12)$$

or, in cycles per second,

$$N(\nu)\,d\nu = (8\pi\nu^2/c^3)\,d\nu$$

As long as the total energy in the box is conserved, two of the amplitudes A_{klm}, B_{klm}, C_{klm} of each single oscillation are in no way restricted. The validity of this important assumption will be questioned later in this chapter.

1.3 THE RADIATION ENERGY DENSITY

The energy in each mode of oscillation is proportional to the square of the amplitude of the particular oscillation. To obtain the energy density inside the box as a function of frequency, Eq. (1.12) must be multiplied by the average or mean energy of each mode of oscillation. To find the average energy of a wave we may note that the electromagnetic vibrations are analogous to mechanical oscillations. By introducing normal coordinates to replace each separate degree of freedom or mode of oscillation one may identify each wave with a corresponding mechanical oscillator having the same frequency. Statistical mechanics is unfortunately outside the scope of this course, but from his classic studies of heat Maxwell was able to show that any particular part of a classical system, such as a single oscillator in a system of oscillators, has a probability of possessing an energy E between E and $E + dE$ proportional to $e^{-E/kT}dE$, where k is the experimentally determined Boltzmann constant, $1.38 \cdot 10^{-16}$ erg/°K, and T is the absolute temperature of the system. The constant of proportionality is determined by requiring that the total probability that this part of the system has any energy at all must be unity. Hence the probability that any given mode of oscillation will have an energy E between E and $E + dE$, $p(E)dE$, will be given by

$$p(E)dE = \frac{e^{-E/kT}\,dE}{\int_0^\infty e^{-E/kT}\,dE} = \frac{e^{-E/kT}dE}{kT} \qquad (1.13)$$

The average or expected energy of each mode of oscillation is obtained by averaging the energy over the probability of having that particular energy:

$$\langle E \rangle = \int_0^\infty Ep(E)\,dE = \int_0^\infty \frac{E\,e^{-E/kT}\,dE}{kT} = kT \qquad (1.14)$$

Hence the energy density $u(\omega)$ of the modes of oscillation having a frequency between ω and $\omega + d\omega$ is the product of Eqs. (1.12) and (1.14):

$$u(\omega)d\omega = \langle E \rangle N(\omega)d\omega = kT\left(\frac{\omega^2}{\pi^2 c^3}\right)d\omega \qquad (1.15)$$

Equation (1.15) was first derived by Rayleigh and Jeans and is called the *Rayleigh-Jeans law*. Unfortunately, as ω increases, the energy density also increases without limit to infinity, leading to an absurd result for infinite frequencies. As shown in Figure 1.3, Eq. (1.15) does indeed fit experimental observations made for very small frequencies or long wavelengths. Observations at high frequencies, however, were empirically described by Wien to conform to

$$u(\omega)\,d\omega = \omega^3 e^{-\hbar\omega/kT}\,d\omega \qquad (1.16)$$

where $\hbar$ is an experimentally determined constant ($1.05 \cdot 10^{-27}$ erg-sec).

Every step of the derivation of Eq. (1.15), the Rayleigh-Jeans law, resulted from what at the time were well-understood and experimentally confirmed physical principles. Yet it did not fit all of the experimental observations. The trouble lay in assuming that all the modes of oscillation have the average

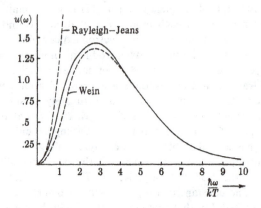

FIGURE 1.3 Energy density of the modes of oscillation within a box as a function of frequency. Experimental observations are given by the solid line, the low and high frequency approximations by dotted lines.

energy kT. Actually, somehow the high-frequency modes are frozen out so that their average energy is zero. The solution thus lay in questioning the assumption that the amplitude (and therefore the energy) of each mode of oscillation was unrestricted. If we assume that only discrete values of wave energies are possible, then the average energy of each wave or mechanical oscillator will be different from the constant classical value kT, as we will now show. This is as though one were to assume, for example, that ocean waves enclosed in a breakwater could have heights of the square root of one, two, or three feet but not of one and a half or two and three quarters feet or any other in-between value. If the energy of each wave can be only some integral multiple of $\hbar\omega$, that is,

$$E = n\hbar\omega \qquad (1.17)$$

then Eq. (1.13) must be replaced by a different expression for the probability that any given mode of oscillation will have a unique energy E. First note that

$$\sum_{n=0}^{\infty} e^{-n\hbar\omega/kT} = \sum_{n=0}^{\infty} y^n = \frac{1}{1-y} = \frac{1}{1-e^{-\hbar\omega/kT}} \qquad (1.18)$$

where $x = \hbar\omega/kT$ and $y = e^{-x}$. Then in place of Eq. (1.13) we have for the number of discrete states $p(E)$ having an energy $E = n\hbar\omega$,

$$p(E) = \frac{e^{-n\hbar\omega/kT}}{\sum_{n=0}^{\infty} e^{-n\hbar\omega/kT}} = (1 - e^{-\hbar\omega/kT})e^{-n\hbar\omega/kT} \qquad (1.19)$$

The average energy of each mode of oscillation becomes

$$\langle E \rangle = \sum_{n=0}^{\infty} Ep(E) = (1 - e^{-\hbar\omega/kT}) \sum_{n=0}^{\infty} n\hbar\omega e^{-n\hbar\omega/kT}$$

$$= (1 - e^{-\hbar\omega/kT})\hbar\omega \frac{e^{-\hbar\omega/kT}}{(1 - e^{-\hbar\omega/kT})^2}$$

$$= \frac{\hbar\omega}{e^{\hbar\omega/kT} - 1} \qquad (1.20)$$

This result for the summation can be easily seen if we take the derivative with respect to x of both sides of Eq. (1.18).

For $\hbar\omega \gg kT$, $\langle E \rangle$ is seen to be approximately $\hbar\omega e^{-\hbar\omega/kT}$, which does approach zero as ω increases. This high-frequency result for the average energy is readily understood from the first two terms in the summation, $0 + \hbar\omega e^{-\hbar\omega/kT}$, which illustrate that if the first level above $n = 0$ has energy much greater than kT it will not be occupied on the average because $e^{-\hbar\omega/kT} \ll 1$. Multiplying Eq. (1.12), based on Maxwell's theory of electricity and magnetism, by our newly adopted value for the average energy of each mode of oscillation, Eq. (1.20), we find for the energy density of waves having a frequency ω within a range $d\omega$,

$$u(\omega)d\omega = \frac{(\hbar\omega^3/\pi^2 c^3)}{e^{\hbar\omega/kT} - 1}d\omega \qquad (1.21)$$

which is the Planck formula for the energy distribution for the modes of oscillation within a black body. Throughout this book a bar through a letter means that the physical constant or quantity the letter represents has been divided by 2π; thus $\hbar = h/2\pi$ where h is a natural constant, Planck's constant, experimentally observed to be $6.6252 \pm 0.0005 \times 10^{-27}$ erg-sec. If one expresses the frequency in cycles per second ν instead of in radians per second ω, the energy of a wave would be written $E = nh\nu$ instead of as in Eq. (1.17).

At low frequencies $e^{\hbar\omega/kT}$ is approximately equal to $1 + \hbar\omega/kT$, so that the right-hand side of Eq. (1.20) becomes simply the classical result kT. At high frequencies the one in the denominator of Eq. (1.21) may be neglected. Thus we should expect the Rayleigh-Jeans law to be valid at low frequencies and Wien's law to be valid at high frequencies, which is in fact the case. Intermediate frequencies can be described only by the exact formula, Eq. (1.21).

It should be emphasized that Planck's use of $E = n\hbar\omega$ (Eq. 1.17) was a theoretical attempt to justify an empirical equation (1.21) which fitted the experimental observations remarkably well; there was no theoretical justification for using (1.17) other than that the experimental observations could be interpreted as requiring (1.17) to be satisfied by the individual electromagnetic waves in the box, or by the fictitious mechanical oscillators in normal coordinates corresponding to the electromagnetic waves in the box. It is easiest and most straightforward, however, to conclude for the present, subject to further examination later, that the energy transitions of all waves or oscillators occur in quantum jumps. Nondiscrete energy changes for any given frequency or oscillation are somehow forbidden.

In deriving our results for the radiation inside a black body we assumed metallic walls. But it would have made no difference if the walls had been made of any other material. Black-body radiation is completely independent of the wall materials, and is observed and predicted to be always the same function of temperature provided only that all radiation incident on a pinhole in one wall is either lost through multiple reflections inside the box or, if incident on the pinhole from within the box, is transmitted freely to the exterior.

1.4 SIGNIFICANT MATHEMATICAL PROCESSES USED IN THE DERIVATION

There are several results incidental to the derivation of Eqs. (1.10) which should be emphasized because of their significance and utility to later developments:

1. If Eq. (1.7) is substituted into Eq. (1.5) there results

$$\nabla^2 \mathbf{E} = -\frac{\omega^2}{c^2} \mathbf{E} \qquad (1.22)$$

∇^2 defines an *operation* to be carried out on $\mathbf{E}$; the particular operation stated in Eq. (1.22) consists of taking partial derivatives of $\mathbf{E}$ stated explicitly in Eq. (1.6) in terms of Cartesian coordinates, but in other cases the operation may be multiplication by a constant or by a function of space or time, for example. ∇^2 is an *operator* whose operation on the wave amplitude or wave function $\mathbf{E}$ results, in this instance, in a constant (namely ω^2/c^2) times the wave function.

2. Since the waves were confined in a box, the boundary conditions (i.e., conditions imposed on the wave function at the walls, which required the tangential component of $\mathbf{E}$ to be zero in the case of metallic walls) allowed only those solutions for which ω satisfies Eq. (1.11) where k, l, and m must be integers. This result is a phenomenon common to all confined waves; in vibrating violin strings or organ pipes, for example, it also happens that only those frequencies which satisfy the boundary conditions are permitted.

3. The values of ω^2/c^2 which satisfy the boundary conditions are called characteristic values or more commonly *eigenvalues* of the system. The solutions for $\mathbf{E}$ of Eq. (1.22) for particular eigenvalues ω^2/c^2 are called characteristic functions or more commonly *eigenfunctions*. Thus, the operator ∇^2 operating on the particular eigenfunction E_z illustrated in Figure 1.1 has the eigenvalue $17(\pi/a)^2$ in the case of the cube of side a. This particular eigenvalue

$17(\pi/a)^2$ has five different nonzero eigenfunctions corresponding to the different k, l, and m consistent with this eigenvalue. An eigenvalue of $16.7(\pi/a)^2$ with its associated eigenfunctions, for example, is not possible for such a system since no combination of integral k, l, and m can give such an eigenvalue. (If we include the E_x and the E_y solutions, as we must, a total of nine different eigenfunctions are possible for $\omega^2/c^2 = 17(\pi/a)^2$.)

Problem 1.1 Find the four lowest eigenvalues and their corresponding eigenfunctions of a vibrating violin string 10 cm long if the tension on the string T is 10 newtons and the string has a mass per unit length μ of 0.2 kg/m. The wave equation for transverse waves in a string is

$$\frac{\partial^2 S}{\partial x^2} = \frac{\mu}{T} \frac{\partial^2 S}{\partial t^2}$$

where S is the perpendicular displacement of the string from equilibrium, x is the distance along the string, and t is the time. (Include the possibility that S may be taken in each of two perpendicular directions.)

Problem 1.2 Graph the Rayleigh-Jeans law and the Planck blackbody distribution as a function of frequency for the surface of the sun, whose temperature is 6000°K. Be careful in labeling the ordinate to give the units and magnitude of what is observed from a black-body surface. On the same graph paper plot the average energy of the electromagnetic waves on the solar surface as a function of their frequency. (Surface emission per unit area is proportional to the velocity of light c times the energy density per unit volume $u(\omega)$.)

Problem 1.3 What are the minimum energies electromagnetic waves of wavelengths 5000 Å and 10 m may have?

Problem 1.4 What are the eigenvalues and eigenfunctions of the operator $(i \, d/dx)^2$ if the eigenfunctions are required to be zero when $x = 0$ and 2?

BIBLIOGRAPHY

Suggested books for further reading:

Bergmann, P. G. *Basic Theories of Physics*, Vol. II. Englewood Cliffs, N. J., Prentice-Hall Inc., 1951. This book contains an excellent and detailed treatment of black-body radiation and the early quantum theory, on which much of this chapter is based. The author is particularly gifted in presenting apt examples and models in his lucid presentation.

Slater, J. C., *Modern Physics*. New York: McGraw-Hill Book Co., 1955. This book is notable for its exceptionally clear and interesting exposition.

II

FOURIER
ANALYSIS

In describing the electric field vector everywhere within the box (the black-body cavity) at a given instant of time through Eq. (1.10), we obtained an alternative way of specifying the condition of the radiation field within the box. It is equally possible to specify the electric field vector as an explicit function of the spatial coordinates x, y, and z, or alternatively to specify the amplitude of each standing wave, as is suggested by Eq. (1.10). Either description serves to *represent* completely the physical *state* of the interior of the box. The method of representing a state of radiation in terms of its various modes of oscillation is developed more generally in this chapter.

The customary representation in terms of spatial coordinates is replaced by alternative representations in terms of modes, which are of great utility in computing and understanding quantum mechanical behavior. It will be shown, by representing a wave in terms of its frequency dependence as well as in terms of its time dependence, that *one cannot simultaneously measure the frequency of a wave and the time at which the wave is absorbed in a detector with unrestricted precision.* This imprecision or *uncertainty* is a necessary consequence of all wave motion and is of central importance to the quantum theory. Indeed, Heisenberg originally derived quantum mechanics by postulating this uncertainty as fundamental to all physical measurement. In later chapters it will transpire that a description of any physical system can be made only in terms of pairs of *complementary* physical variables, both of which cannot be measured simultaneously with infinite precision. In this chapter it will be shown that it is impossible to limit the range of frequency or wave number *and* the temporal duration or spatial position of a wave; in the following chapter, in addition to discussing the complementary variables of time and frequency or energy, we will take up the physical impossibility of simultaneously measuring position and momentum with infinite precision.

Sines and cosines are good examples of functions whose product with any other function of the same class gives zero when integrated over all ranges of the variable, unless the two multiplied functions are identical. Such sets of functions are said to be orthogonal. For example, the integrals over a period of the fundamental frequency ω_0 of the products $\sin m\omega_0 t$ $\sin n\omega_0 t$, $\sin m\omega_0 t \cos n\omega_0 t$, and $\cos m\omega_0 t \cos n\omega_0 t$, where m and n are integers, are

$$\frac{\omega_0}{2\pi} \int_0^{2\pi/\omega_0} \sin m\omega_0 t \sin n\omega_0 t \, dt$$

$$= \frac{\omega_0}{2\pi} \left[\int_0^{2\pi/\omega_0} \frac{\sin (m-n)\omega_0 t}{2(m-n)\omega_0} - \frac{\sin (m+n)\omega_0 t}{2(m+n)\omega_0} \right] = 0 \qquad \text{if } m \neq n$$

$$= \frac{\omega_0}{2\pi} \int_0^{2\pi/\omega_0} \sin^2 m\omega_0 t \, dt = \frac{\omega_0}{2\pi} \int_0^{2\pi/\omega_0} \left(\frac{1}{2} - \frac{1}{2} \cos 2m\omega_0 t \right) dt = \frac{1}{2}$$

$$\text{if } m = n$$

$$\frac{\omega_0}{2\pi} \int_0^{2\pi/\omega_0} \sin m\omega_0 t \cos n\omega_0 t \, dt$$

$$= -\frac{\omega_0}{2\pi} \left[\int_0^{2\pi/\omega_0} \frac{\cos (m-n)\omega_0 t}{2(m-n)\omega_0} + \frac{\cos (m+n)\omega_0 t}{2(m+n)\omega_0} \right] = 0 \qquad \text{if } m^2 \neq n^2$$

$$= \pm \frac{1}{2} \frac{1}{m\omega_0} \frac{\omega_0}{2\pi} \left[\int_0^{2\pi/\omega_0} \sin^2 m\omega_0 t \right] = 0 \qquad \text{if } m^2 = n^2$$

and similarly

$$\frac{\omega_0}{2\pi} \int_0^{2\pi/\omega_0} \cos m\omega_0 t \cos n w_0 t \, dt = \frac{1}{2} \delta_{mn} \qquad\qquad n \neq 0$$

$$= \delta_{m0} \qquad\qquad n = 0$$

where δ_{mn}, called a *Kronecker delta*, is zero for $m \neq n$ and unity for $m = n$. Besides sines and cosines, other examples of orthogonal functions that we shall encounter are Bessel functions and Legendre polynomials. The coefficients of such functions may frequently be used to describe a given function of coordinates or some other variable, so that such functions furnish a new and frequently more convenient kind of "space" in which to describe the given function.

2.1 FOURIER SERIES

Suppose that a given oscillation or periodic function of time $f(t)$, with period $2\pi/\omega_0$, is to be described by a sum or superposition of many waves, specifically an infinite sine and cosine series, called a *Fourier series*:

$$f(t) = \sum_{n=1}^{\infty} A_n \sin n\omega_0 t + B_n \cos n\omega_0 t + B_0 \qquad (2.1)$$

where the n's are integers, ω_0 is a given fundamental frequency, and the A's and B's are the amplitudes of each component frequency. Multiplying both sides of Eq. (2.1) by $\sin m\omega_0 t$ and integrating over the fundamental period, we find that

$$\int_0^{2\pi/\omega_0} f(t) \sin m\omega_0 t \, dt \qquad (2.2)$$

$$= \sum_{n=1}^{\infty} \left(A_n \int_0^{2\pi/\omega_0} \sin n\omega_0 t \sin m\omega_0 t \, dt + B_n \int_0^{2\pi/\omega_0} \cos n\omega_0 t \sin m\omega_0 t \, dt \right)$$

The sines and cosines are orthogonal functions, and all the terms of the infinite series in n on the right-hand side of Eq. (2.2) give zero except for the one term in the sine series for which $n = m$, which gives

$$\int_0^{2\pi/\omega_0} f(t) \sin m\omega_0 t \, dt = \frac{\pi}{\omega_0} A_m \qquad (2.3)$$

Similarly by multiplying (2.1) by $\cos m\omega_0 t$,

$$B_m = \frac{\omega_0}{2\pi} \int_0^{2\pi/\omega_0} f(t) \, dt \qquad m = 0$$

$$= \frac{\omega_0}{\pi} \int_0^{2\pi/\omega_0} f(t) \cos m\omega_0 t \, dt \qquad m \neq 0 \qquad (2.4)$$

As in the preceding, our function may be given as an explicit function of time, $f(t)$; alternatively, however, we may compute the A's and B's from a knowledge of $f(t)$ and use them to describe our function:

$$f(t) = \sum_{n=1}^{\infty} \left\{ \frac{\omega_0}{\pi} \left[\int_0^{2\pi/\omega_0} f(t) \sin n\omega_0 t \, dt \right] \sin n\omega_0 t \right\}$$

$$+ \frac{\omega_0}{\pi} \left[\int_0^{2\pi/\omega_0} f(t) \cos n\omega_0 t \, dt \right] \cos n\omega_0 t \right\} + \frac{\omega_0}{2\pi} \int_0^{2\pi/\omega_0} f(t) dt \qquad (2.5)$$

where the quantities in brackets are functions only of frequency $n\omega_0$, since the time will have been integrated out. In this way $f(t)$ may be described, for example, as being made up of oscillations having frequencies $n\omega_0$, each of whose relative amplitudes A_n and B_n represent the quantity just as completely as $f(t)$ does. There is one essential difference, however, in that while all values of t are possible in the time representation, only certain values of frequency, integral multiples of ω_0, are possible in the frequency representation. If $f(t)$ is not periodic this difference disappears, as will be shown in the following section.

As an example, let us consider a 60-cycle alternating current of unit amplitude $f(t) = \sin 2\pi 60t$. If we attempt to find a series to describe it using the basic frequency $\omega_0 = 2\pi 60$, we immediately find the trivial result $A_1 = 1$. All other coefficients are zero. For graphs of alternative representations of 60-cycle ac current see Figure 2.1. Note that in this example the Fourier series consists of only one term, because we implicitly assumed that the wave started at $t = -\infty$ and continued to $t = +\infty$. Actually, every time an electric switch is turned on or off, higher frequencies called *transients* are introduced (see Problem 2.2). These are frequently picked up by household radios as sharp audible clicks.

For a second example, suppose that at

$$0 < t < t_0, \qquad f(t) = 0$$
$$t_0 < t < 2t_0, \qquad f(t) = 1$$

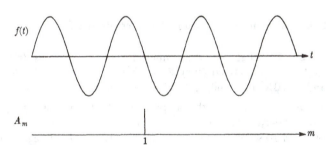

FIGURE 2.1 Alternative representations of 60-cycle ac; the second graph illustrates the fact that only one frequency is represented.

and then the function is repeated for larger and smaller integral multiples of t_0. Obviously, a convenient fundamental frequency will be π/t_0:

$$A_n = \frac{1}{t_0} \int_0^{t_0} (0) \sin n\frac{\pi}{t_0} t \, dt + \frac{1}{t_0} \int_{t_0}^{2t_0} (1) \sin n\frac{\pi}{t_0} t \, dt$$

$$= 0 - \frac{1}{t_0} \frac{t_0}{n\pi} \left[\cos n\frac{\pi}{t_0} 2t_0 - \cos n\frac{\pi}{t_0} t_0 \right] \qquad \textbf{(2.6)}$$

$$= -\frac{2}{\pi n} \quad \text{for odd } n; \ = 0 \text{ for even } n.$$

Similarly,

$$B_n = 0 \quad \text{for} \quad n > 0, \quad B_0 = 1/2$$

and hence

$$f(t) = \frac{1}{2} - \sum_{m=0}^{\infty} \frac{2}{\pi(2m+1)} \sin{(2m+1)} \frac{\pi}{t_0} t \qquad (2.7)$$

Alternate representations of this system are presented in Figure 2.2.

Problem 2.1 What are the amplitudes of the various wavelengths present in a vibrating violin string of length l plucked initially to form an isoceles triangle?

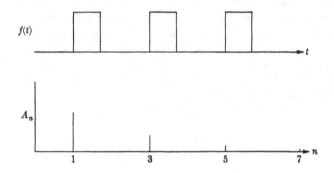

FIGURE 2.2 Alternative representations of a periodic step function pulse.

Problem 2.2 What are the amplitudes of the various frequencies present in half-wave rectified 60-cycle ac, that is, $0 < t < 1/120, f(t) = \sin 2\pi 60 t$, and $1/120 < t < 1/60, f(t) = 0$, and then the function repeats?

Problem 2.3 What are the amplitudes of the various frequencies present when 60-cycle ac is repeatedly turned on for $1/10$ second and then turned off for $4/10$ second? Do not bother to carry out the integration.

Problem 2.4 Find the two Fourier series for $f(x) = x$ and $|x|$ for $-l < x < l$ ($f(x)$ repeats for other values of x); then compare the series for $0 < x < l$.

Problem 2.5 Find the Fourier series for $f(x) = \cos \mu x$ for $0 < \mu < 1$, $-\pi < x < \pi$ ($f(x)$ repeats for other values of x).

Problem 2.6 Find the two Fourier series for $f(x) = x^2$ and x^3, over $-l < x < l$ ($f(x)$ repeats for other values of x).

2.2 FOURIER TRANSFORMS

The preceding discussion was devoted to alternative representations of functions varying periodically with time. The periodicity resulted in a discrete frequency representation, called a *Fourier series*. A Fourier series is composed of harmonic terms whose frequency is some integral multiple of the fundamental frequency. In order to represent the larger class of functions which are not necessarily periodic, we now discuss the case in which the fundamental period approaches infinity and instead of a discrete set of frequencies we have a continuum of oscillation frequencies. We need a new and very important concept, that of the *Dirac delta function* which is zero everywhere except for one point where it is infinite. Consider the integral of the function $(\sin bu)/u$. It is a standard definite integral whose value is π when integrated over u from $-\infty$ to ∞. For large values of b, however, the sine function oscillates very rapidly as a function of u, as shown in Figure 2.3;

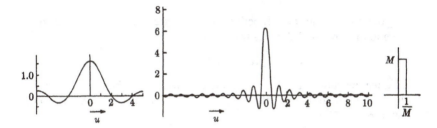

FIGURE 2.3 Illustration of the function $(\sin bu)/u$. The first figure is for $b = \pi/2$; the second for $b = 2\pi$; the third function approaches the Dirac delta function as M approaches infinity.

hence the integral is close to zero between limits of large values of bu of the same sign. As $b \to \infty$ the integral approaches zero, unless the origin is included in the integration interval in which case the value of the integral remains π, and as the maximum of this function increases its width decreases in such a way that the total area under the curve is a constant. That is, the integral of this function has a nonzero value only at the origin; such a function is called a Dirac delta function, and is represented by the symbol $\delta(t)$. By definition, $\delta(t) = 0$ except at the point $t = 0$ where it is infinite in such a way that its integral is unity. It may be thought of as a function of infinitesimal width and infinite height such that the area under its curve is one, as is illustrated in the third diagram in Figure 2.3. If the integrand is multiplied by any more slowly varying function of u, say $f(t + u)$, the value of this function is only important at the origin where the integral has a nonzero value. That is:

$$\int_{-a}^{a} f(t + u)\frac{\sin bu}{u}\,du \xrightarrow[b \to \infty]{\lim} \pi f(t) \tag{2.8}$$

and since

$$\frac{\sin bu}{u} = \int_{0}^{b} \cos u\omega \, d\omega \tag{2.9}$$

$$f(t) \underset{b \to \infty}{\text{lim}} \frac{1}{\pi} \int_{-a}^{a} f(t+u) \, du \int_{0}^{b} \cos u\omega \, d\omega \qquad (2.10)$$

where $\underset{b \to \infty}{\text{lim}}$ is read "equals in the limit as $b \to \infty$." Interchanging the order of integration and letting first a then b go to infinity, we obtain

$$f(t) = \frac{1}{\pi} \int_{0}^{\infty} d\omega \int_{-\infty}^{\infty} f(t+u) \cos u\omega \, du$$

Let
$$u = v - t \qquad (2.11)$$

then
$$f(t) = \frac{1}{\pi} \int_{0}^{\infty} d\omega \int_{-\infty}^{\infty} f(v) \cos \omega(v-t) \, dv$$

Since the cosine is an even function, we may extend the integral over ω to $-\infty$ and multiply by $1/2$. Since the sine is an odd function,

$$f(v)i \sin \omega(v-t)$$

may then be added to the above integral over v without changing the value of the integral, since the integral over ω of this added term is zero. Therefore,

$$f(t) = \frac{1}{2\pi} \int_{-\infty}^{\infty} d\omega \int_{-\infty}^{\infty} f(v)e^{-i\omega(t-v)} \, dv \qquad (2.12)$$

Thus
$$f(t) = \frac{1}{\sqrt{2\pi}} \int_{-\infty}^{\infty} F(\omega)e^{-i\omega t} \, d\omega \qquad (2.13)$$

where
$$F(\omega) = \frac{1}{\sqrt{2\pi}} \int_{-\infty}^{\infty} f(v)e^{i\omega v} \, dv \qquad (2.14)$$

$F(\omega)$ is called the Fourier transform of $f(t)$, and is obtained from $f(t)$ by Eq. (2.14), where v is a dummy variable of integration. Conversely, $f(t)$ may be obtained from $F(\omega)$ by the inverse Fourier transform, Eq. (2.13). Hence, if t were to represent time and ω frequency, a particular oscillation or wave could be equally well represented as a function of time $f(t)$, or as a function of frequency $F(\omega)$. Similarly in later chapters we will frequently find it convenient to transform from a spatial coordinate representation in ordinary space to a wave number or momentum representation in which momentum forms an equally valid space or set of continuous variables in terms of which the system may be explicitly described.

Equations (2.13) and (2.14) define the Fourier exponential transforms. The analogous Fourier cosine and sine transforms may be obtained from a similar derivation. Alternative spherical harmonic or Bessel integral transform representations may be obtained for spherical or cylindrical waves, for example, but our discussion in this chapter will be limited to Fourier exponential transforms.

2.3 RELATION BETWEEN THE WIDTH OF A FUNCTION AND ITS TRANSFORM

For our trivial 60-cycle ac current of unit amplitude for example, we have $F(\omega) = \sqrt{2\pi}\delta(\omega - 2\pi 60)$, since only $\omega = 2\pi 60$ is present, as may be readily confirmed by evaluating Eq. (2.14) for $f(t) = e^{-i2\pi 60 t}$. Conversely, from Eq. (2.13), $f(t) = e^{-i2\pi 60 t}$ for $F(\omega) = \sqrt{2\pi}\delta(\omega - 2\pi 60)$. However, if the wave were not of infinite duration (for instance if a light switch were rapidly flicked on and off), $F(\omega)$ would no longer be a delta function consisting of only one frequency. Suppose the switch were flicked on at time $-t_0$ and off at time $+t_0$; then from Eq. (2.14) we find

$$F(\omega) = \frac{1}{\sqrt{2\pi}} \int_{-t_0}^{t_0} e^{-i2\pi 60 t} e^{i\omega t}\, dt$$

$$= \frac{1}{\sqrt{2\pi}} \frac{e^{i(\omega - 2\pi 60)t_0} - e^{-i(\omega - 2\pi 60)t_0}}{i(\omega - 2\pi 60)}$$

$$= \frac{2}{\sqrt{2\pi}} \frac{\sin(\omega - 2\pi 60)t_0}{\omega - 2\pi 60} \tag{2.15}$$

which we recognize from our earlier discussion to be a delta function times $\sqrt{2\pi}$ as t_0 increases without limit. For finite t_0, however, $F(\omega)$ is not a delta function but contains frequencies other than $2\pi 60$, as is shown in Figure 2.4.

From this example one can see that the range of frequencies present in a finite wave train is, at least in this instance, inversely related to the duration of the wave train. This is not at all a mathematical trick; an apparatus such as a spectroscope employed to detect any of these frequencies in such a finite wave train would yield a positive result.

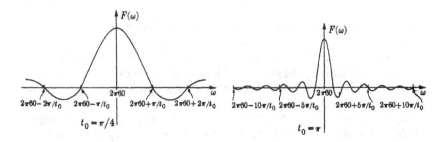

FIGURE 2.4 The Fourier transform of a wave train of frequency $2\pi 60$ which is turned on for a time $2t_0$. Note that the frequency $2\pi 60 + (5\pi/25t_0)$, for example, is present with amplitude as much as 13% of the amplitudes of $2\pi 60$ itself. The figure on the left is for $t_0 = \pi/4$; the figure on the right is for larger $t_0 = \pi$. The ordinate and abscissa scales of the two figures are the same.

The Fourier transform of a Gaussian distribution function in t is a Gaussian distribution function in ω. Specifically, if

$$f(t) = \frac{A}{\sqrt{2}} e^{-t^2/2(\Delta t)^2} \tag{2.16}$$

is inserted into Eq. (2.14) there results

$$F(\omega) = \frac{A}{\sqrt{4\pi}} \int_{-\infty}^{\infty} e^{-t^2/2(\Delta t)^2 + i\omega t} \, dt$$

$$= \frac{1}{\sqrt{2}} A\Delta t e^{-\omega^2(\Delta t)^2/2} \tag{2.17}$$

Thus a Gaussian distribution in time with a root-mean-square deviation in time of Δt leads to a Gaussian distribution in frequency with a root-mean-square deviation in frequency of $\Delta\omega = 1/\Delta t$. The inverse relationship between spread in time of a wave and the spread in frequency is a very general one, as may be inferred from Eqs. (2.13) or (2.14). If $f(v)$ is not oscillatory and peaks at $v = 0$, the integrand in Eq. (2.14) represents a sum of waves $e^{i\omega v}$ of differing v with weight $f(v)$. The waves will interfere constructively only if their phase differs by less than π. Hence, a broad maximum in v would result in destructive cancellation except near $\omega = 0$, whereas a sharply peaked distribution in v would result in little destructive interference over a large spread in $\omega \neq 0$.

We must conclude that the product of the spread in frequency of a pulse of radiation $\Delta\omega$ and the duration of the pulse Δt is quite generally of the order of unity. That is,

$$\Delta\omega\Delta t \sim 1 \tag{2.18}$$

or, remembering the relation between the least unit of radiation energy and frequency stated in the previous chapter (Eq. 1.17), $E = \hbar\omega$,

$$\Delta E\Delta t \sim \hbar \tag{2.19}$$

The relationship between frequency and time expressed in Eq. (2.18) is a fundamental characteristic of all wave motion.

A wave train whose duration is limited is necessarily comprised of a range of frequencies. A single cycle of high-frequency music has a very short duration and requires a large range of AM broadcast frequency to carry it. The higher the audio frequency with which the amplitude of the broadcast frequency is modulated, the more rapid must be the changes in this amplitude. For really high fidelity broadcasting this rapid modification requires the availability of a finite range of frequencies above and below the assigned broadcast frequency. Hence, Eq. (2.18) requires broader frequency assignments for higher fidelity in AM broadcasting, which may be represented approximately by $\Delta v/\Omega \gtrsim 1$ where Δv is the assigned spread in broadcast frequency (ordinarily 10 kc) and Ω is the highest reproducible audio frequency for the assigned AM band width. Of course, the AM receiver should then not be too selective of broadcast frequency, as superheterodyne receivers are, in order to make full use of the high-fidelity transmission.

Problem 2.7 What is the frequency spectrum of the intensity of a flash of lightning which might be represented as a sudden magnetic field decreasing exponentially with time as e^{-t/t_0}? That is, compute $F(\omega)$ and then the intensity, $I = F^*(\omega)F(\omega)$, where $F^*(\omega)$ is the complex conjugate of $F(\omega)$. What is the frequency spread in which the intensity falls to one-half its peak value if $t_0 = 10^{-2}$ sec? if $t_0 = 1$ sec? (These frequencies show up as whistles in unselective AM radio receivers, due to the difference in propagation velocity of the various frequencies in the ionosphere. Thus a lightning stroke in South America produces an electromagnetic disturbance which travels along a magnetic field line out into the ionosphere and back into the United States. The higher frequencies arrive first, the low last, and one hears a high note which drops rapidly in pitch, a whistle in the receiver.)

2.4 WAVE PACKETS

A plane wave propagated along the x axis may be represented by the expression

$$E = E_0 e^{i(k_0 x - \omega t)} \tag{2.20}$$

where k_0 is the wave number, defined in terms of the wavelength,

$$k_0 = 1/(\lambda_0/2\pi) = 1/\lambda_0$$

The phase $k_0 x - \omega t$ remains constant for $x = (\omega/k_0)t = \omega \lambda_0 t$. Hence, the phase of the wave propagates with a velocity $\omega \lambda_0$ and this velocity may be written as c/n. Therefore, if the wave represents light propagating in a medium, ω may be written as $c/n\lambda_0$ where c is the phase velocity of the wave *in vacuo* ($2.998 \cdot 10^8$ m/sec for electromagnetic radiation) and n is the ratio of the phase velocity *in vacuo* to the phase velocity in the propagating medium. In general n is some function of the wavelength of the radiation and ω is not a linear function of k.

The previous discussion concerning the complementary variables ω and t may be taken over in entirety for the one-dimensional variables k and x, which are also complementary. The formal mathematical treatment is identical; by taking the Fourier transform of $f(x)$ one may go to k space, just as in the previous discussion one transformed from the variable t to ω:

$$F(k) = \frac{1}{\sqrt{2\pi}} \int_{-\infty}^{\infty} f(x)e^{ikx} \, dx \tag{2.21}$$

If one wishes to represent a pulse of light, that is, a bundle of radiation or a *wave packet* which propagates through the medium, a monochromatic wave having wave number k_0 such as is represented by Eq. (2.20) will not do. Instead, one requires

$$E(x, t) = f(x, t)e^{ik_0 x} \qquad (2.22)$$

Assuming the wave packet centered at $x = 0$ when $t = 0$, $f(x,0)$ is a function which is large near $x = 0$ and small at x far from the origin. Alternatively $E(x,t)$ can be expressed in terms of $F(k',t)$, the x-space Fourier transform of $f(x,t)$,

$$F(k', t) = \frac{1}{\sqrt{2\pi}} \int_{-\infty}^{\infty} f(x, t)e^{ik_0 x} e^{ik'x} \, dx$$

$$= \frac{1}{\sqrt{2\pi}} \int_{-\infty}^{\infty} f(x, t)e^{ikx} \, dx \equiv F(k - k_0, t)$$

where $k = k' + k_0$. Thus $F(k - k_0, t)$ is a distribution in wave number, rather than a delta function in k representing a monochromatic wave (whose inverse transform would give Eq. (2.20) instead of the desired Eq. (2.22)).

We illustrate the behavior of a wave packet, using a Gaussian distribution function again because of the convenience of its simple transform properties. Suppose that

$$F(k - k_0, t) = Be^{-(k-k_0)^2/2(\Delta k)^2} e^{-i\omega t} \qquad (2.23)$$

Equation (2.23) is the weighting function of a group of waves whose integral over k (the Fourier transform) will yield $E(x,t)$:

$$E(x, t) = \frac{1}{\sqrt{2\pi}} \int_{-\infty}^{\infty} Be^{-(k-k_0)^2/2(\Delta k)^2} e^{i(kx-\omega t)} \, dk \qquad (2.24)$$

Since ω is a function of k it must be included in the integral, and hence the integral can be solved exactly only for specific examples of $\omega(k)$. Fortunately, if $F(k - k_0)$ is a reasonably narrow distribution function, one may solve (2.24) in terms of the derivatives of ω to good approximation. In this case ω is expanded in a Taylor's series about $k = k_0$,

$$\omega(k) = \omega(k_0) + \left(\frac{\partial \omega}{\partial k}\right)_{k=k_0} (k - k_0) + \frac{1}{2}\left(\frac{\partial^2 \omega}{\partial k^2}\right)_{k=k_0} (k - k_0)^2 \qquad (2.25)$$

The first derivative with respect to k has a particularly useful and simple physical interpretation. In a dispersive medium, n and therefore the phase velocity of a wave depend on the wavelength, and one cannot define a single phase velocity for the entire group of waves or wave packet given by Eq. (2.24). In detecting a wave packet one observes the intensity of the wave. A maximum signal is received when the maximum amplitude of the wave arrives. Under these circumstances one does better to find the velocity of the maximum of the wave packet and call it the *group velocity*. The maximum of the wave packet in momentum space will occur for $k = k_0$ (because $F(k - k_0)$ is a maximum at $k = k_0$). The wave packet will have a maximum in coordinate space as well, at that position where the least rapid phase variation with k occurs, yielding maximal constructive interference in the integral given in Eq. (2.24) for the wave amplitude. That is, the extremum in the phase of the integrand in Eq. (2.24) must be sought:

$$\frac{\partial}{\partial k}(kx - \omega t) = x - \frac{\partial \omega}{\partial k}t = 0 \tag{2.26}$$

The *position* of maximal constructive interference is seen therefore to propagate with a velocity $\partial \omega / \partial k$, called the *group velocity* v_g.

The difference between phase and group velocity of a wave packet may be illustrated by a rotating barber pole with helical stripes moving with a translational velocity parallel to its axis. Even when the "group velocity" (translational velocity) of the pole is zero, a colored portion of the pole is observed to "move" if the pole rotates, with a "phase velocity" given by sv where s is the pitch of the stripes (distance along the pole that a stripe moves in one rotation) and v is the number of rotations per second. The phase velocity would be different if the pole moved with a velocity v_g, namely, it would be $v_g + sv$.

Another frequently cited example of differing phase and group velocities is that of wavelets caused by a stone dropped in water. The outermost ring of ripples propagates outward with a group velocity v_g, but ripples within this outer ring are observed to propagate with a higher velocity and to disappear when they reach the outer, slower-moving edge of the ripples.

Labeling $\partial \omega / \partial k$ as v_g and $\partial^2 \omega / \partial k^2$ as a, substituting Eq. (2.25) into Eq. (2.24), and letting $k' = k - k_0$, we obtain

$$
\begin{aligned}
E(x, t) &= \frac{B}{\sqrt{2\pi}} \exp\{i[k_0 x - \omega(k_0)t]\} \exp\left\{-\frac{1}{2}\frac{(x - v_g t)^2}{(\Delta k)^{-2} + iat}\right\} \\
&\quad \times \int_{-\infty}^{\infty} \exp\left\{-\frac{1}{2}[(\Delta k)^{-2} + iat]\left[k' - \frac{i(x - v_g t)}{(\Delta k)^{-2} + iat}\right]^2\right\} dk' \\
&= \frac{B}{\sqrt{(\Delta k)^{-2} + iat}} \exp\{i[k_0 x - \omega(k_0)t]\} \exp\left\{-\frac{1}{2}\frac{(x - v_g t)^2}{(\Delta k)^{-2} + iat}\right\} \\
&= B\left[\frac{(\Delta k)^{-2} - iat}{(\Delta k)^{-4} + a^2 t^2}\right]^{1/2} \exp\left\{i\left[k_0 x - \omega(k_0)t + \frac{at(x - v_g t)^2}{2[(\Delta k)^{-4} + a^2 t^2]}\right]\right\} \\
&\quad \times \exp\left\{-\frac{1}{2}\frac{(x - v_g t)^2 (\Delta k)^{-2}}{(\Delta k)^{-4} + a^2 t^2}\right\} \tag{2.27}
\end{aligned}
$$

which is illustrated in Figure 2.5.

The observed intensity of the wave packet is the square of the absolute value of the wave amplitude $E(x,t)$,

$$I = E(x, t)E^*(x, t) = \frac{B^2}{\sqrt{(\Delta k)^{-4} + a^2 t^2}} \exp\left\{-\frac{(x - v_g t)^2 (\Delta k)^{-2}}{(\Delta k)^{-4} + a^2 t^2}\right\} \tag{2.28}$$

The expression on the right-hand side of Eq. (2.28) is simply a Gaussian

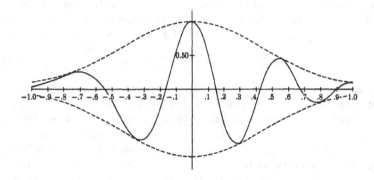

FIGURE 2.5   The real part of the amplitude of a wave packet as a function of spatial coordinate x, which has a Gaussian distribution in k and x. The dashed line is the Gaussian envelope of the packet.

function in x about a mean value of $v_g t$ whose root-mean-square deviation in x, Δx, is given by

$$\Delta x = \sqrt{\frac{(\Delta k)^{-4} + a^2 t^2}{(\Delta k)^{-2}}} = \frac{1}{\Delta k}\sqrt{1 + a^2 t^2 (\Delta k)^4} \qquad (2.29)$$

Hence a wave packet in a dispersive medium ($a \neq 0$) spreads out in time, as would be expected since then different elements of the wave packet (waves of differing frequency) propagate with different velocities.

Problem 2.8   Radio waves in the ionosphere propagate with an index of refraction given approximately by

$$n = 1 + \frac{\omega_p^2/2}{\omega_p^2 - \omega^2 + [\gamma^2/(\omega_p^2 - \omega^2)]}$$

where ω_p is the natural frequency of electron oscillations, equal to a maximum of 5×10^7 radians/sec in the F layer where the electron density is a maximum. If $\gamma = 10^{15}$ (radians/sec)^2, what are the phase and group velocities of a wave of frequency 3×10^7 radians/sec? 7×10^7 radians/sec? Within how great a distance will a pulse 10^4 meters long, with $\omega_0 = 3 \times 10^7$ radians/ sec, have lengthened 10%? (A pulse of this length would be required to broadcast high audio frequencies and a change in the length would result in distortion.) (Hint: First find the *time* in which the pulse lengthens by 10% and note that the frequency of the radio wave is constant, but that

$$\omega = ck/n(\omega)$$

$$\frac{\partial \omega}{\partial k} = \frac{c}{n}\left[1 + \frac{ck}{n^2}\frac{\partial n}{\partial \omega}\right]^{-1}$$

and similarly for $\partial^2\omega/\partial k^2$.)

Problem 2.9 The velocity of crests of gravitational waves on water is related to their wavelength λ as follows:

$$v = \sqrt{g\lambda}$$

where g is 980 cm/sec². What is the group velocity?

Problem 2.10 If the group velocity of a wave is given by $v_g = 3v_p$ where v_p is the phase velocity, how does v_p depend on ω?

BIBLIOGRAPHY

Bohm, David, *Quantum Theory*. Englewood Cliffs, N. J.: Prentice-Hall, Inc., 1951. This book contains an exceptionally fine and detailed discussion of wave packets and their transform properties. See Chapter 3.

Jenkins, F. and H. White, *Fundamentals of Optics*. New York: McGraw-Hill Book Co., 1950.

Rossi, Bruno, *Optics*. Reading, Mass.: Addison-Wesley, Inc., 1957. See Chapter 5 for a discussion of the velocity of propagation of waves.

III

DIFFRACTION AND INTERFERENCE OF PHOTONS AND PARTICLES

From our study of black-body radiation we saw that Planck obtained a satisfactory theoretical prediction of black-body radiation by assuming, instead of an arbitrary energy of oscillation, that the modes of oscillation in a black body could possess only discrete amounts of energy. The basis for this assumption was no more or less than that the experimentally observed spectrum may be interpreted in this way. This discrete, quantized behavior of electromagnetic radiation strongly suggests that light waves are transmitted as packets or quanta of energy. This chapter will be devoted to exploring further particle-like characteristics of electromagnetic waves and the wave-like characteristics of particles.

Atoms emit and absorb light energy in discrete amounts or *quanta*, as was initially pointed out by Einstein in his interpretation of the photo-electric effect and has subsequently been confirmed by many diverse experimental observations. The light is emitted and absorbed, in the volume of a single atom, as light quanta called *photons*. The position and motion of the photons are determined by means of wave packets localized in space and time but much less precisely localized than the atoms themselves. Our discussion of diffraction and interference phenomena in terms of wave packets will prove very suggestive and illuminating for an understanding of diffraction and interference phenomena of particles, which phenomena in turn require an inherent space-time localization as the essence of the "common sense" model of a particle.

This chapter will further review those aspects of particle behavior which cannot be explained by a model of particle behavior requiring infinitely precise space-time localization. The uncertainty or indeterminacy in complementary variables, such as frequency (or energy) and time, which is a fundamental characteristic of wave motion, will be shown to characterize particle behavior as well. The discreteness or noncontinuity of available values of normally continuous physical observables of radiation (for example, the frequency of a confined wave) is observed also for closely confined particles (for example, the energy of electrons in an atom). Finally, diffraction and interference phenomena, which are the sole basis for the credibility of the wave theory of light, will be discussed for particle experiments where they are also realized.

3.1 THE PHOTOELECTRIC EFFECT

When light is incident on the atoms of a gas, a liquid, or a solid, electrons are instantaneously emitted. The remarkable feature about this phenomenon, called the *photoelectric effect*, is that the energy of the electrons is observed to be unaffected by changes in the *intensity* of the incident radiation. An increase in light intensity only increases the number of electrons released. However, the kinetic energy of the electrons is observed to decrease when the *frequency* of the incident radiation is lowered. Indeed, no photoelectrons whatever are observed if the frequency is less than some critical value unique for each substance.

It is impossible to explain these observations by means of the wave theory of light. The wave theory of light leads one to anticipate that a long radio wave incident on an atom could cause enough energy to be absorbed (over sufficiently long radiation times) for an electron to be released. Moreover, when electrons are emitted, an increase in radiated power for a particular wave should cause an emitted electron to have more kinetic energy rather than more electrons of the same average energy to be emitted. One would further expect a rate of electron emission which would increase during the radiation time, being greatest at the end of the radiation pulse, rather than the observed electron emission rate which is constant, beginning instantaneously at the moment the radiation is turned on. Also the wave theory does not explain how the energy of radiation having a wavelength of thousands of atomic diameters may be concentrated in a single atomic electron.

Einstein theorized that the radiation itself consists of discrete quantities of energy, called quanta, which are localized in space and time. Photon is the specific name given to a light quantum. If the radiation is quantized, then the least possible energy of an electromagnetic wave will be $\hbar\omega$. The energy of n photons associated with a particular mode of oscillation in a black body, i.e., the energy of this mode of oscillation, is $n\hbar\omega$. The intensity of electromagnetic radiation is given equally well by either the absolute square of the electric field intensity times the velocity of light, or the number of photons arriving per cm^2-sec times the energy of each photon $\hbar\omega$. This is very different from Planck's initial interpretation, which suggested that the material oscillators in the walls of the box had quantized energies, rather than the radiation field itself.

Conservation of energy requires that, if no energy is lost due to collisions with atoms in the metal, an electron ejected from a metal surface by the absorption of an electromagnetic wave of frequency ω will have a kinetic energy determined from Einstein's simple equation

$$E = \hbar\omega - W \qquad (3.1)$$

where W is the work function or energy required to free an electron from the attractive force field of the metal. Millikan and others quickly confirmed experimentally this relationship between electron energy and radiation frequency, and observed that no matter how intense the incident radiation, no electrons were released by incident waves of frequency less than $W/\hbar$.

Conversely, Franck and Hertz observed that in a gaseous discharge tube atoms bombarded by an electron beam remain unexcited until the electron kinetic energy is made greater than $\hbar\omega$ for the atomic spectra of the excited atoms. Until excitation, the electron beam is not absorbed or deflected from the detector, showing that the electrons behave as though in a vacuum and the gas atoms were not there. But as soon as $\hbar\omega$ of a spectral line is exceeded, the electron beam disappears and light of this frequency is produced, showing that electron kinetic energy can be accepted by the atoms only if it produces a quantum jump, and that de-excitation occurs with the creation of a quantum of radiant energy. (The atoms could not be excited by successive smaller nudges, for example.) Therefore, the interaction of radiation with matter occurs by the loss or gain of discrete amounts of radiation energy called photons. Electromagnetic waves behave in this regard like little lumps or packets of energy.

Problem 3.1: Mercury vapor has two strong absorption lines, at $\lambda = 2537$ Å and $\lambda = 1849$ Å. At what voltages would you expect a current drop in a Franck-Hertz experiment on mercury vapor?

3.2 THE COMPTON EFFECT

A. H. Compton observed experimentally that monochromatic X rays incident on a scattering target are scattered at all angles from the atoms and from the atomic electrons in two frequency components; the component being scattered from the entire atom has the original frequency of the radiation and the other has a slightly lower frequency, decreasing with larger observed scattering angles. He successfully interpreted his result in terms of scattering by electrons in which momentum and energy were conserved.

As a consequence of his electromagnetic theory of light, Maxwell showed that a beam of radiation has momentum equal to the energy transported by the beam divided by the velocity of light. This is well confirmed, for example, by observing the spinning of a radiometer vane mounted in a vacuum due to radiation reflected from its arms, or by a more recent example, the orbital changes of the artificial earth satellite Echo I brought about by solar radiation pressure. The minimum altitude of Echo I above the earth decreases and increases periodically by hundreds of kilometers due to solar radiation pressure. Hence, each photon should have a momentum whose magnitude is given by the quantum energy divided by its velocity, $\hbar\omega/c$. If a photon were scattered by an electron in a target material, momentum would be exchanged, and since the electron is essentially stationary, the photon would leave the scattering volume with the speed of light, but with less momentum and energy.

If a photon is deflected through an angle θ by collision with an electron at rest as illustrated in Figure 3.1, then conservation of momentum and energy require

$$\left.\begin{aligned}
\frac{\hbar\omega}{c} &= \frac{\hbar\omega'}{c}\cos\theta + \frac{mv}{\sqrt{1-(v/c)^2}}\cos\varphi \\
\frac{\hbar\omega'}{c}\sin\theta &= \frac{mv}{\sqrt{1-(v/c)^2}}\sin\varphi \\
\hbar\omega &= \hbar\omega' + mc^2\left(\frac{1}{\sqrt{1-(v/c)^2}} - 1\right)
\end{aligned}\right\} \qquad (3.2)$$

three simultaneous equations in the unknown ω', v, and φ. Solving for ω' we find

$$\frac{c}{\hbar\omega'} - \frac{c}{\hbar\omega} = \frac{1}{mc}(1-\cos\theta)$$

or using the wavelength $c/\omega = \lambda$,

$$\lambda' - \lambda = \frac{h}{mc}(1-\cos\theta) \qquad (3.3)$$

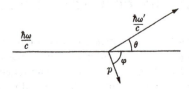

FIGURE 3.1 Illustration of a collision between a photon and an electron initially at rest, showing momenta exchanged.

Thus, in the scattering, the wavelength of the photon (or quantum of light energy) is lengthened while the frequency or energy is lessened. h/mc has the dimensions of length and is called the Compton wavelength for the particle of mass m. The Compton wavelength of the electron is 2.43×10^{-10} cm. The recoil electrons have been observed in cloud chambers, and the momentum and energy of the recoil electrons have been measured and found to agree with Compton's predictions. It is especially important to note here that the precise accounting of the photon frequency and wavelength implicit in Eqs. (3.2) and (3.3) is possible only for long times ($t > |\omega - \omega'|^{-1}$) and over large volumes ($v > |\lambda' - \lambda|^3$).

Problem 3.2: What is the average pressure on an interplanetary solar sail at the earth's orbit where 1.374×10^6 ergs/cm²-sec of solar radiation power is incident?

Problem 3.3: What is the Compton wavelength of a piece of chalk having a mass of 100 gm? a uranium-238 nucleus? a proton? a π meson whose mass is 264 electron masses? How does this last value compare with the range of nuclear forces, 1.7×10^{-13} cm?

3.3 RECONCILIATION OF THE PHOTON AND WAVE THEORIES OF LIGHT

The preceding discussion makes it apparent that light frequently exhibits particle-like behavior. That is, it travels and interacts with matter as a small bundle of energy and momentum well localized in space and time. Indeed, in any medium where the absorption and index of refraction change negligibly within a radiation wavelength, the wave nature of light is not apparent and ray optics becomes an excellent approximation. Newton, as is well known, was able to explain many of his observations by a corpuscular theory of light.

Yet when the pertinent dimensions of an obstruction to a photon become sufficiently small, diffraction and interference phenomena are observed which have been successfully interpreted only on the basis of a wave theory (see Figure 3.2). We here undertake a discussion of interference in terms of photons whose average position and motion are determined by means of wave packets extending over large volumes of space and long times. In the wave theory of light, the light intensity or light energy/cm²-sec, which is necessarily always the observed quantity, is computed by taking the square of the absolute value of a complex quantity called the *wave amplitude*. (In Maxwell's electromagnetic theory of light, the wave amplitude is the electric or magnetic field vector of the wave.) First the phase of any particular part of the total wave disturbance at a given point is computed along the wave path up to the given point. The total wave amplitude at any particular point

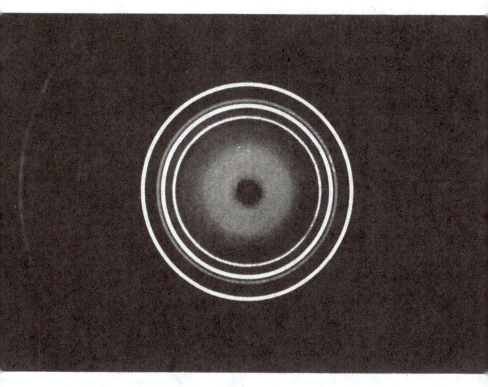

FIGURE 3.2 X-ray diffraction pattern of polycrystalline aluminum. (Courtesy of Mrs. M. H. Read, Bell Telephone Laboratories, Murray Hill, New Jersey.)

is then computed by summing or integrating all the contributions from all possible wave paths. The square of the absolute value of this quantity is the intensity. (See the texts on optics listed in the bibliography of this chapter.)

The frequency spectrum of any particular atomic bright-line radiation (which can be observed, incidentally, only in experiments involving the emission of many photons) shows a bright center with monotonically decreasing intensity as a function of frequency difference with the frequency of the center of the line. The finite frequency breadth of the line causes the Fourier transform of the frequency spectrum to represent (as it must) a sharp pulse in time of very short duration $\Delta t = (\Delta\omega)^{-1}$ where $\Delta\omega$ represents the difference in frequencies at which the intensity is at a maximum and at half maximum value ($\sim 10^{-8}$ sec for ordinary atomic spectra). The spatial extent of the radiation, or rather the length within which there is an appreciable probability of finding a photon, would be the velocity of light times the time duration of the probability of emission (~ 3 meters for atomic spectra in a vacuum). A representative wavelength of the radiation might be 5×10^{-7} m (i.e., about 5000 atomic diameters).

The light intensity may be expressed alternatively as the energy of a photon (whose position is represented by a wave packet) times the number of photons of that frequency arriving per cm²-sec. The average number of photons per cm²-sec or probability of finding one photon per cm²-sec at any particular point is found, as just stated, by computing the amplitude of the light intensity. Suitably normalized, the amplitude of the light intensity (the electric field vector) may be considered as proportional to a *probability amplitude* (which in general is complex), whose absolute value squared gives the probability of counting a photon within, say, the volume of a silver grain in a photographic plate. A diffraction experiment could be performed, in principle, by a source which emitted one photon per minute. A detector placed behind a diffraction grating would then count one photon per minute, each photon being absorbed by a particular detector atom at a particular position. The average counting rate, or total observed counting rate at any position,

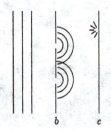

FIGURE 3.3 Illustration of a photon being registered at c (indicated by a star) after the plane wave incident on the slits at b is diffracted and expands in concentric circles beyond the slit openings.

results from the interference of the probability wave of each photon simultaneously transmitted through all the separate openings in the grating. Niels Bohr has given a very nice illustration of this phenomenon, which is depicted in Figure 3.3. A more detailed treatment of diffraction of probability waves will be presented in Chapter IV.

3.4 WAVE CHARACTERISTICS OF PARTICLES

The wave characteristics of particles are intimately connected with the *uncertainty principle* which we shall discuss in this section. In Chapter II we discussed the fundamental relationship between the timing of an electromagnetic signal and its frequency. Equation (2.18) expresses the fact that if we attempt to specify the duration of an electromagnetic wave, we are unable to assign the wave a unique frequency. If a light signal has a finite duration (not infinitesimal), we are unable to know precisely when the light wave is absorbed in a detector. Furthermore, if the light signal has a finite duration (not infinite), the frequencies of the photons which are absorbed are also not precisely known, since the Fourier analyzed frequency components of the signal consists of many frequencies. Rephrased, Eq. (2.18) says that the uncertainty in the timing of a light signal (or change in amplitude of a wave) times the uncertainty in specifying the frequency can never be less than unity. Equation (3.3), expressing the Compton effect in which the scattered light gives an unknown momentum to a particle, enables us to illustrate that that same uncertainty is necessarily with us in any and all measurements on particles. Suppose we attempt with an optical microscope to measure the position and velocity of an electron simultaneously (see Figure 3.4). The position of the electron will be uncertain due to the

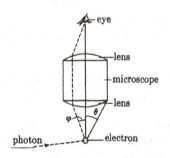

FIGURE 3.4 Illustration of path of photon (dashed line) scattered from an electron into a microscope in the xz plane where z is the axis of the microscope.

diffraction of light. A photon will be reflected from the electron and travel through the microscope to our eye but its point of origin, the electron, will be uncertain due to the wavelike diffraction of the photon by the microscope aperture. The total uncertainty in position, or reciprocal of the resolving power of the microscope, is found from elementary physical optics to be $\Delta x = 1.22\lambda/\sin\theta$ where θ is the angle made by the microscope aperture with the axis of the instrument at the electron. Knowing that the momentum of a photon is $h\nu/c = h/\lambda$, we know that the amount of momentum exchanged by the electron and photon is $h/\lambda \sin\varphi$ where φ is the angle the photon is scattered into by the electron. The maximum angle of scattering is θ, φ is unknown between the limits $\pm\theta$, and therefore the momentum acquired by the electron is equal to or less than $h/\lambda \sin\theta$. Hence, the product of the uncertainty in position and momentum of the electron parallel to a particular direction, say the x axis, after observation would be

$$\Delta x \Delta p_x = (1.22\lambda/\sin\theta)(h\sin\theta/\lambda) \gtrsim h \qquad (3.4)$$

$\Delta x \Delta p_y$, however, can be zero. We may obtain the same relation for photons if Eq. (2.18) is rewritten $[\Delta(h\nu/c)](c\Delta t) \geqslant h$, since $c\Delta t$ is the uncertainty in position of a photon. Many ingenious experiments have been proposed in an attempt to disprove this relation, but none has met with success. This relationship between complementary physical variables, in this case position and momentum, is a well-substantiated property of nature. Similarly, the uncertainty in the energy of a particle and in the length of time it has that energy has been shown to satisfy the identical condition Eq. (2.19). Heisenberg enunciated the *uncertainty principle* and has discussed the intimate relation between it and wave mechanics. By methods beyond the scope of this text, one could state the uncertainty principle as a fundamental postulate and derive quantum mechanics therefrom.

Problem 3.4: What is the approximate expected lifetime of an atomic state whose emission wavelength of ~ 5000 Å can be determined with a precision of one part in 10^3, i.e., $\Delta\lambda/\lambda = 0.001$?

Problem 3.5: If the lifetime of a state is 10^{-9} seconds, what is the line width or uncertainty in the energy of the emitted radiation?

3.5 THE DISCRETENESS OF ATOMIC ENERGY LEVELS

Another characteristic shared by particles and waves confined within a small volume is the discreteness of the energies or frequencies which the system may have. A great many diverse observations, such as the photoelectric or Compton effects or, in general, the emission of electrons by material particles, have well confirmed the fact that atoms contain electrons having mass negligible compared to observed atomic masses, and electromagnetic radiation emitted by atoms has long been identified with the motion of atomic electrons. Zeeman observed that if atoms were placed in a strong magnetic field, the frequency of the emitted radiation was shifted by an amount which is an integral multiple of $eH/2m$, where H is the magnetic field intensity and e/m is the ratio of the charge on the electron to the mass of the electron. Let n be the integer multiplier; then $nheH/2m$ is the magnetic energy, in the applied field H, of a circular electron current. By observing the sign of the shift in frequency and the circular polarization of the radiation parallel to the magnetic field axis, Zeeman and Lorentz determined that the charge, which was the evident source of the radiation, was negative and circulating in the magnetic field. This is known as the *normal* Zeeman effect.

The size of the entire atom is obtained from the density of the material and Avogadro's number: that is, by determining the volume per atom of a macroscopic quantity of a given material. Atomic radii are thus determined to be of the order of 10^{-8} cm. By scattering alpha particles from thin films of metal, however, Rutherford and subsequent experimenters observed unequivocally that virtually all the atomic *mass and positive charge* are concentrated within a radius of the order of 10^{-12} cm. The observed few, large-angle deflections of the alpha particles cannot be interpreted in any other way.

It was natural for physicists of Rutherford's era to attempt to construct a theoretical atomic model resembling the solar system, the analogy resulting from electromagnetic forces acting between orbiting electrons and positive nuclei replacing the gravitational forces acting between the planets and the sun. One of the main difficulties with this theory is that the centripetal acceleration of the electrons would cause them to radiate electromagnetic radiation copiously, which would be observed as a "broad-band" frequency source, not as a single "bright-line" source. Moreover, this theory would predict a rapid energy loss by radiation; thus atoms would not be stable and would lose all their potential energy of charge separation in a small fraction of a second in collapsing to a volume only 10^{-12} cm in radius.

Since atoms are observed to be stable and to emit light of only a few discrete frequencies unique to each atom, one must conclude that atomic electrons can have only certain discrete energies. This suggests that their behavior too is quantized and may be satisfactorily predicted from solutions of a wave equation which satisfy certain boundary conditions resulting from the small confinement of the electrons in the nuclear electrostatic field of force. That is, the potential and kinetic energies of electrons in the atomic electric field are limited somehow to only a few particular values, as are the energies of photons confined in a black body. Semiclassically, one has to conclude that only those electron orbital radii are possible consistent with the energy values theoretically inferred from the observed discrete atomic energy levels. How such limitations on the electron energies could arise was not understood; but the existence of these limitations certainly constitutes wavelike behavior of the electrons.

Problem 3.6: Calculate the classical lifetime of a circularly moving electron in the hydrogen atom on the Rutherford model if the electron is originally at $5 \cdot 10^{-9}$ cm from the nucleus. (Hint: The total energy of a particle in a circular orbit in an attractive central force field with radial dependence $1/r^2$ is equal to minus one-half the absolute value of its potential energy. The power radiated is $\partial E/\partial t = \frac{2}{3}(e^2/c^3)a^2$ in cgs units, where a is the particle acceleration, or $\frac{2}{3}(e^2/c^3)(a^2/4\pi\epsilon_0)$ in mks units. The centripetal acceleration may be obtained nonrelativistically by equating $m\omega^2 r = e^2/4\pi\epsilon_0 r^2$.)

3.6 THE DAVISSON-GERMER EXPERIMENT

The old corpuscular theory of light was suggested by observations that light traveled in straight lines in a homogeneous medium just as particles do when moving in any force-free field. This theory was proven to be incomplete by the fundamental observation that light shows interference effects and experiences diffraction. Only wave motion may result in these phenomena; hence diffraction observations permanently established the wave nature of light.

Just as electromagnetic radiation exhibits corpuscular characteristics, so also do material particles appear wavelike in their behavior. The analogous behavior of particles and photons led de Broglie to suggest that associated with particles there might be a wave whose length is given by

$$\lambda = h/p$$

Here p is the particle momentum corresponding to the momentum of a photon in the Compton effect $p_{(\text{light})}$, which is given by

$$\lambda = c/v = h/(hv/c) = h/p_{(\text{light})}$$

Here λ is the wavelength of the electromagnetic radiation and v its frequency in cycles per second.

Niels Bohr determined theoretically the allowed planetary orbits of the electrons by placing appropriate conditions on the solar system atomic model described in the previous section. De Broglie showed that the path lengths over one period of the Bohr orbits of atomic electrons were all integral multiples of h/p, suggesting that electrons in these allowed orbits were in actuality characterized by standing waves.

Davisson and Germer subsequently studied the reflection of 200-ev electrons from a nickel crystal and observed a diffraction pattern for the reflected electrons identical to the Bragg reflection pattern for X-rays from crystals. (See Figure 3.5 and compare with Figure 3.2 for photon diffraction on a different crystal.) The number of reflected electrons at any point could be

FIGURE 3.5 Electron diffraction pattern of polycrystalline tellurium chloride. (Courtesy of *RCA* Laboratories, Princeton, New Jersey.)

successfully predicted if one assumed the electrons diffracted like X-radiation with a wavelength equal to the de Broglie wavelength of 200-ev electrons (~ 0.027 Å). Many subsequent experiments on the reflection and transmission of atomic particles, protons, and neutrons through crystals have abundantly confirmed that particles experience wave diffraction and interference. These observations are as fundamental to a wave theory of particles as a Young slit experiment or ordinary diffraction-grating experiments are to a wave theory of light. How one may construct a wave theory of particles to correspond in success with the Huygens wave theory of light will be undertaken in the following chapter; the necessity for such a theory is established absolutely by these experiments.

Problem 3.7:   What are the de Broglie wave lengths for a 10-ev electron? 1-mev electron? a 10-ev proton? a 100-g rifle bullet moving 3000 feet per sec?

BIBLIOGRAPHY

Bohr, Niels, *Atomic Physics and Human Knowledge*. New York: John Wiley & Sons, Inc., 1958. A detailed discussion of the uncertainty principle and complementary variables is given in Chapter 4.

There are many excellent books on modern physics, a few of which are:

Weidner, Richard T. and Robert Sells, *Elementary Modern Physics*, Second Ed., Boston: Allyn & Bacon, Inc., 1968.

Leighton, Robert B., *Principles of Modern Physics*. New York: McGraw-Hill Book Co., 1959.

Richtmeyer, F. K., E. H. Kennard, and T. Lauritsen, *Introduction to Modern Physics*. New York: McGraw-Hill Book Co., 1955.

Sproull, Robert L., *Modern Physics*. New York: John Wiley & Sons, Inc., 1956.

Slater, J. C., *Modern Physics*. New York: McGraw-Hill Book Co., 1955.

A few of the finest elementary texts in physical optics are:

Rossi, Bruno, *Optics*. (Cited in Bibliography of Chapter 2.)

Jenkins, F. and White, H., *Fundamentals of Optics*. (Cited in Bibliography of Chapter 2.)

Andrews, Charles, *Optics of the Electromagnetic Spectrum*. Englewood Cliffs, N. J.: Prentice-Hall, Inc., 1960.

IV

THE FEYNMAN
PATH INTEGRALS
AND THE
SCHRÖDINGER
EQUATION

4.1 ELECTRON DIFFRACTION

Since the original Davisson and Germer experiment, electron diffraction has become a very common laboratory phenomenon. Diffraction is peculiar to wave motion; the observation of it is central to our understanding of particle behavior. Figure 4.1 illustrates a simple electron diffraction experiment. A monoergic source of electrons at a emits electrons which pass through two slits at b to be counted on a photographic plate c. The two curves at the right of c represent respectively the actual wavelike result obtained, entailing diffraction and interference effects, and the classically predicted result.

It is a very remarkable result that opening the lower slit causes some spots at c to be inaccessible (the minimum in the diffraction pattern) *which were accessible when the lower slit was shut!* This is interference phenomena recognized much earlier and more easily in studies of the propagation of light. Interference phenomena in the propagation of particles is the foundation rock on which the wave theory for particles is built.

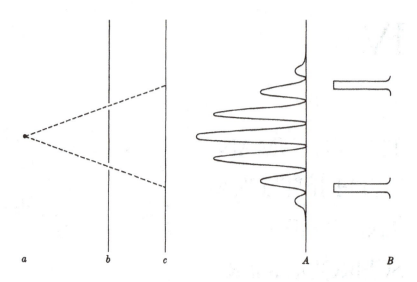

FIGURE 4.1 Symbolic illustration of a simple electron diffraction experiment showing monoergic electron source at a, two slits at b, and photographic plate c. A is a graph representing the actual photographic result obtained on c, showing typical wave diffraction and interference maxima and minima. B represents the classically predicted result.

The dashed curve in Figure 4.2 illustrates the diffraction pattern at c when only the upper slit is open and the lower slit is closed. The solid line curve in Figure 4.2 illustrates the pattern at c when both slits are open. In Figure 4.2 A is a graph illustrating the changes in grain density in the photographic film at c as a function of distance parallel to the plane containing the slit openings. While the electrons arrive at the film one at a time, as could be proved by using a scintillation detector at c, the counting rate or film density at the point c_1 in the figure is as much a function of whether the lower slit is open as of whether the upper slit is open. That is, an electron arriving at c_1, for example, is simultaneously influenced in its motion by *both* slit openings. What is wrong with the classical prediction illustrated by the curve B? It says that the electrons will go through *either* the upper *or* lower slit before being counted at the plate. Those electrons passed by the upper slit will be expected to contribute the upper classical counting peak and those passed by the lower slit, to the lower classical peak. The presence of the two peaks would imply that the electrons went through one slit or the other, the electrons in the upper counting peak, for example, having traveled via the upper slit. But if the two slits should be located a distance less than $h/\Delta p$ apart, where Δp is the uncertainty in the component of the

FIGURE 4.2 The diffraction patterns at c in Fig. 4.1 when only the upper slit is open (dashed curve – – –) and when both slits are open (solid line curve ———).

electron momentum parallel to the slit opening (each slit presumably being less than $h/\Delta p$ wide), then specifying by means of the photographic result which of the two slits the particles went through would mean locating the electrons within a distance less than $h/\Delta p$, in violation of the uncertainty principle. Since we cannot say which of the slits any particular electron or group of electrons went through, the classical description of an electron going through a particular hole is meaningless.

If the experiment were performed with only one slit open, the curve illustrated in Figure 4.3 would result. A broad, almost uniform distribution (A) results if the slit opening Δx is small. Since the electron penetrates the slit its position at the slit is known with a precision of Δx. Therefore, its momentum in the x direction Δp_x can be known to a precision no better than $h/\Delta x$. In traveling a distance $\overline{bc}$ from b to c, therefore, for a time $t = m\overline{bc}/p_y$ it will move in the x direction a distance roughly equal to or less than $\pm \Delta p_x t/m = \pm h\overline{bc}/\Delta x p_y = \pm \lambda(\overline{bc}/\Delta x)$. It is the interference between these broad diffraction distributions from the separate slits which results in the familiar diffraction-grating pattern in ordinary physical optics.

A more elegant description of the classical prediction, which has been discussed by Feynman, is the following: The probability of observing an electron at a particular point c emitted from a source at a, P_{ac}, is given by the probability of observing an electron at one of the slits at b, P_{ab}, times the subsequent chance of finding the electron at c, P_{bc}, and then summing over all the slits at b, i.e.,

$$P_{ac} = \sum_b P_{ab} P_{bc} \tag{4.1}$$

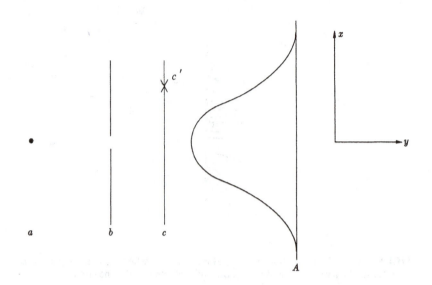

FIGURE 4.3 Illustration of a diffraction experiment (a, b, c) and a graph of the counting rate or film density (A) as a function of distance up the screen at c for only one narrow slit open.

As we know, this expression is incorrect since it treats the electrons as though they travel from the source a to a particular opening b and then from this particular opening b to the spot c, and we know it is wrong to state that the electrons travel via any one particular opening.

4.2 DIFFRACTION OF LIGHT

From previous courses the student is quite familiar with a situation similar to the diffraction of electrons, namely, the diffraction of light. *A correct particle theory must predict an electron diffraction pattern identical to that for light.* We will briefly review the theory for the preceding diffraction experiment, using photons instead of electrons.

In this case, the number of photons arriving at a point c_1 on screen c is proportional to the intensity of the light wave, given by the absolute square of the electric field of the wave. The electric field may be expressed as a complex amplitude $e^{i(kx - \omega t)}$ where $k = 2\pi/\lambda = 1/\lambda$ ($\lambda =$ wavelength). The phase of the wave passed by each slit is given by an integral over the path of the wave. In the case of two slits, the relative number of photons arriving at c_1, or the probability of a photon arriving at c_1, is then proportional to

$$P_{ac_1} = \left| e^{-i\omega t} \left\{ \exp\left(i \int_{b_1} k\, dx \right) + \exp\left(i \int_{b_2} k\, dx \right) \right\} \right|^2 \tag{4.2}$$

where the two integrals are taken over respective paths through the two slits. If the two paths differ by an odd multiple of $\lambda/2$, then $P_{ac_1} = 0$, a familiar condition for destructive interference. For more than two slits, we have

$$P_{ac_1} = \left| e^{-i\omega t} \sum_j \exp\left(i \int_{b_j} k\, dx \right) \right|^2 \tag{4.3}$$

or in the case of finite slit openings, say in the case of a lens opening or an astronomical mirror, the summation for all the path contributions must be replaced by an areal integration,

$$P_{ac_1} = \left| e^{-i\omega t} \int dz \int dy \exp\left(i \int_{b(z,\, y)} k\, dx \right) \right|^2 \tag{4.4}$$

Straight-line paths through points in the lens or mirror are assumed for the integrations, since it can be shown that the contribution to the intensity from the summation over the other paths is zero. In general,

$$P_{ac_1} = \left| e^{-i\omega t} \int_{\substack{\text{all} \\ \text{possible} \\ \text{paths}}} \exp\left(i \int_b k\, dx \right) db \right|^2 \tag{4.5}$$

The phase $\int_b k\, dx$ in (4.5) is computed along a particular path b in order to obtain that path's contribution to the total amplitude of the wave originating at a and arriving at c. One must then obtain the total wave amplitude by integrating the contributions from all the separate paths over the lens, mirror, or slit openings. Let

$$\varphi_{ac_1} = \exp\left(i \int_a^{c_1} k\, dx \right), \quad \varphi_{ab_j} = \exp\left(i \int_a^{b_j} k\, dx \right)$$

and

$$\varphi_{b_j c_1} = \exp\left(i \int_{b_j}^{c_1} k\, dx \right)$$

Since the sum in an exponential argument may be represented as a product of exponentials, Eq. (4.3) may be rewritten

$$P_{ac_1} = |\varphi_{ac_1}|^2 = \left| \sum \varphi_{ab_j} \varphi_{b_j c_1} \right|^2 \tag{4.6}$$

which is *not* equal to the classical equivalent for particles which is

$$P_{ac_1} = \sum_j |\varphi_{ab_j}|^2 |\varphi_{b_j c_1}|^2 = \sum_j P_{ab_j} P_{b_j c_1} \tag{4.7}$$

The phase of our complex probability amplitudes φ_{ac_1} for computing the probability of a photon arriving at c_1, is given by the path integral $\int k\, dx$ where k may be better written for our purpose

$$k = \frac{2\pi}{\lambda} = \frac{2\pi\nu}{c} = \frac{2\pi h\nu}{hc} = \frac{p}{\hbar}$$

Similarly,

$$\omega = \frac{E}{\hbar}$$

Thus if we express the physical optics treatment of light diffraction in terms of physical quantities possessed by particles (i.e., energy or momentum instead of frequency or wavelength) we arrive at the following description of the diffraction of light:

The probability of a photon's arriving to be counted at a particular place c_1 from an origin a is given by the square of the absolute value of the integral, over all contributing paths, of a complex quantity which is a function of the path. The complex quantity of unit modulus has a phase given by $(1/\hbar) \int p \, dx$ where p is the photon momentum and the integral is taken over the path. The complex quantity $\exp \{(i/\hbar)(px - Et)\}$ is a *probability amplitude* or *wave function*, and in customary discussions of diffraction of electromagnetic waves it is given as a factor in the expression for the electric field.

If the polarization of light waves is neglected, the probability of counting a photon on a photographic plate or any other detector is given by the intensity of the wave within the detector volume, which is proportional to $\int |\varphi^2| \, d\tau$, where φ is defined following Eq. (4.5) and $d\tau$ represents a differential element of the detector volume.

4.3 PARTICLE PROBABILITY AMPLITUDES

In constructing a wave theory for the motion of particles too minute to be directly observed in a microscope, we are presented with the great difficulty of having to extrapolate observed macroscopic models of physical behavior to a totally different realm of observation. Our conceptual difficulty is comparable to the problem Huygens might have faced if he had never observed waves in water, strings, or other substances in deriving a wave theory of light. Here we must refrain from applying too closely a physical model or mental picture of submicroscopic particle motion based on macroscopic objects, and refer always only to what are *observed*, those physical consequences of the particle motion which may be detected in laboratory apparatus.

In view of the similar characteristics of light and particles, and particularly since *our objective is to obtain the same diffraction pattern for particles as is obtained for light*, we introduce a theory for particle motion identical to that for electromagnetic wave propagation. That is, we take over complete the methods, functions, and interpretation of the preceding section, substituting particles for photons, with but one modification which concerns the space integral in the argument of the wave function. In optics, a space integral $\int p\,dx$ is taken along the path in order to obtain the phase of the wave function. This integral is well known in advanced classical mechanics, where it is given the name "action." A closely related integral over the time, having the same units and called *Hamilton's principle function*, must be used instead in those physical circumstances where the time-dependent part of the wave function is not independent of the path. Since ω for a light wave is independent of coordinates, and in our discussion of optics we were concerned with equal time intervals for all our paths, $e^{-i\omega t}$ was simply excluded from the integration; but more generally $e^{-iEt/\hbar}$ may be included in the integration, the space integral may be made a time integral by substituting $(dx/dt)\,dt$ for dx, and the entire argument of the wave may be integrated over the time along the wave path. That is,

$$\Psi_{ac} = e^{-(i/\hbar)Et}\exp\left\{\frac{i}{\hbar}\int_{x_a}^{x_c} p\,dx\right\} = \exp\left\{-\frac{i}{\hbar}\int_{t_a}^{t_c} E\,dt\right\}\exp\left\{\frac{i}{\hbar}\int_{t_a}^{t_c} p\dot{x}\,dt\right\}$$

$$= \exp\left\{\frac{i}{\hbar}\int_{t_a}^{t_c} (p\dot{x} - E)\,dt\right\} \tag{4.8}$$

where t_a and t_c are the times at the ends of the path and p and E are here the momentum and total energy of a particle rather than of a photon as in previous section. We will use Ψ for particle wave functions instead of φ which we have been using for optical wave functions. If we had to evaluate the integral in (4.8) for a particular example, we would express the momentum, velocity, and total energy of the particle as a function of time and integrate over the time of the particle motion from point a to point c. For non-relativistic motion in Cartesian coordinates, which we have been using,

$$p\dot{x} - E = 2T - T - V = T - V = L \tag{4.9}$$

where T is the particle kinetic energy, V the potential energy, and the difference L is called the *Lagrangian*. Therefore, the phase of the probability amplitude for the particle taken over a particular path is given by Hamilton's principle function, the time integral of the Lagrangian for the particle taken along the path of the wave. Hence we may write

$$\Psi_{ac} = \exp\left\{\frac{i}{\hbar}\int_{t_a}^{t_c} (p\dot{x} - E)\,dt\right\} = \exp\left\{\frac{i}{\hbar}\int_{t_a}^{t_c} L\,dt\right\} \tag{4.8a}$$

We may summarize the entire procedure by quoting Feynman*: "In quantum mechanics the probability of an event which can happen in several different ways is the absolute square of a sum of complex contributions, one from each alternative way. The probability that a particle will be found to have a path $x(t)$ lying somewhere within a region of space-time is the square of a sum of contributions, one from each path in the region. The contribution from a single path is postulated to be an exponential whose

---

* See bibliography of this chapter.

(imaginary) phase is the classical action (in units of $\hbar$) for the path in question. The total contribution from all paths reaching x,t from the past is the wave function $\Psi(x, t)$ . . . A probability amplitude is associated with an entire motion of a particle as a function of time, rather than simply with a position of the particle at a particular time."

Thus the probability of finding a particle at c which was initially at a is given by the absolute square of the wave function Ψ_{ac} given in Eq. (4.8), where the integral in (4.8) is evaluated along the classical path of the particle between points a and c.

Equation (4.8) assumes that the particle was known to be at a so that the wave function or probability amplitude of finding the particle at a was something like a delta function. At some time t_c later than t_a the wave function or probability amplitude for finding the particle at any point c is given by Eq. (4.8) for Ψ_{ac}. If we wished to know the wave function at some later time t_d in terms of the wave function at c, Eq. (4.6) tells us it would be

$$\Psi_{acd} = \Psi_{ac}\Psi_{cd} = \exp\left\{\frac{i}{\hbar}\int_{t_a}^{t_c}(p\dot{x} - E)\,dt\right\}\exp\left\{\frac{i}{\hbar}\int_{t_c}^{t_d}(p\dot{x} - E)\,dt\right\}$$

$$= \Psi_{ac}\exp\left\{\frac{i}{\hbar}\int_{t_c}^{t_d}(p\dot{x} - E)\,dt\right\} \tag{4.10}$$

Equation (4.10) treats only one of many possible intermediate points c (on the many possible paths) between a and d. If we wish to know the *total* wave function or probability amplitude for finding the particle at d, having known the particle was once at a, we must sum over all intermediate points c given in Eq. (4.6),

$$\Psi_{ad} = \sum_c \Psi_{ac}\exp\left\{\frac{i}{\hbar}\int_{t_c}^{t_d}(p\dot{x} - E)\,dt\right\} \tag{4.11}$$

For a particle "diffraction grating," we would sum over the slits c in the grating. For an opening (like an optical lens) we would integrate over the area c of the opening.

The argument in the phase integral of Eqs. (4.10) and (4.11), $p\dot{x} - E$, is the Lagrangian function given in Eq. (4.9), so that Eq. (4.11) may also be written

$$\Psi_{ad} = \sum_c \Psi_{ac}\exp\left\{\frac{i}{\hbar}\int_{t_c}^{t_d}L\,dt\right\} \tag{4.11a}$$

This expression is similar to the expression inside the absolute value signs in Eqs. (4.2–4.5) for the optical wave function. The chief difference between Eqs. (4.10, 4.11) and the optical wave functions in homogeneous media is that $\omega t = Et/\hbar$ is taken outside the integral over space or time in computing the optical wave functions (4.2–4.5).

In general, our initial wave function is not the special case of $\delta(x - a)$ at $t = t_a$, and we will not be concerned with the special example of a particle "diffraction grating" or "lens opening." Hence, in general, we must integrate over the volume elements $dx\, dy\, dz$ rather than the grating or lens surface elements $dy\, dz$. To simplify the discussion and expressions in the following we will assume a one-dimensional space in x and eliminate the y and z coordinates. We then see that if the probability amplitude for finding the particle at position x_1 at time t_1 is $\Psi(x_1, t_1)$, then the probability amplitude for finding the particle at x_2 at time t_2 is given by

$$\Psi(x_2, t_2) = \frac{1}{A} \int \exp\left\{\frac{i}{\hbar}\int_{t_1}^{t_1}(p\dot{x} - E)\, dt\right\} \Psi(x_1, t_1)\, dx_1$$

$$= \frac{1}{A} \int \exp\left\{\frac{i}{\hbar}\int_{t_1}^{t_1} L\, dt\right\} \Psi(x_1, t_1)\, dx_1 \tag{4.12}$$

which must be integrated over all the values x_1 may assume (i.e., over all one-dimensional x-space). A is a normalization constant, to be determined later, which makes $\Psi(x_2, t_2) = \Psi(x_1, t_1)$ when $x_2 = x_1$ and $t_2 = t_1$.

Problem 4.1: What is the exact expression for the relative number of electrons detected in a photographic plate as a function of distance along the plate in a direction perpendicular to the beam and the crystal axis when a beam of 400-ev electrons bombards a thin crystal of KF N atoms wide, placed perpendicular to the beam, if the plate is also placed perpendicular to the beam one meter beyond the crystal? Assume the crystal cut parallel to the crystal axes. Potassium and fluorine have identical atomic radii of 1.33 Å, and KF forms a simple cubic lattice similar to the NaCl crystal.

Problem 4.2: What is the probability of observing a neutron at an angle of θ radians from the normal to the face of an atomic pile in a plane perpendicular to a long straight crack in the face of the pile, if the neutron is emerging from the crack with energy 0.025 ev and the crack is $2 \cdot 10^{-4}$ m wide? (Hint: In integrating $e^{(i/\hbar)px}$ over the area of the opening ($\int dy$) use the Fraunhofer approximation formula for diffraction; that is, the length x of a path through the point y in the opening is given by $x_0 - y \sin \theta$ where x_0 is the path length of a ray passing through the center of the opening and y is measured from the center of the opening.)

Problem 4.3: Determine where a particle is most likely to be found whose wave function is given by

$$\psi = (1 + ix)/(1 + ix^2)$$

Solution: The probability of finding a particle within a volume V has already been said to be given by the integral of $\int |\psi|^2\, d\tau$ over the volume. The average value of any property of a state, for example the position x, is defined to be the sum or integral of the product of each possible value of the property times the probability that the state has this value. For example, the probability that a particle is at a position between x and $x + dx$ is

$|\psi(x)|^2\ dx$. The average value of x is found by integrating the expression $x|\psi|^2\ dx$ over all values of x, and is indicated by $\langle x \rangle$.

Problem 4.4: If $\psi = C/(a^2 + x^2)$, where C is found by requiring $\int |\psi|^2\ d\tau = 1$, what is the average value of x, $\langle x \rangle$? of $(x - \langle x \rangle)^2$?

Problem 4.5: A linear harmonic oscillator, for example a mass on a spring, oscillates classically so that the probability of finding the particle at x is given by $P(x)\ dx = C/|\cos(\pi x/2L)|$, where L is the amplitude of the vibration and C is found by requiring

$$\int_{-L}^{L} P(x)\ dx = 1$$

Calculate the expectation of x, x^2, and $\sqrt{(x - \langle x \rangle)^2}$.

Problem 4.6: Calculate the expectation, that is, average value of x, x^2, and $(\cos \pi x)/a$ for a plane wave whose wave function is $\psi = C(\sin \pi x)/a$ for $0 < x < a$. C is found by requiring that

$$\int_{0}^{a} \psi^*\psi\ dx = 1$$

4.4 THE SCHRÖDINGER WAVE EQUATION

In the previous section we attained our objective of formulating a particle theory which would predict a wave diffraction pattern for particles identical with that obtained in the wave theory of light. We have not yet arrived, however, at the stage where we can solve or explain one other confusing breakdown in the classical theory of particles. Specifically, the free particle wave function or probability amplitude we used in the previous section to calculate the interference or diffraction patterns of particles incident on crystals or openings will not help us to predict the bright-line spectra from atoms or the experimental consequences of a host of other spherically or cylindrically symmetric, bound-state particle problems in which the particles are pretty well confined in regions less than or of the order of their wavelengths. We could, for example, calculate the wave function

$$\exp \frac{i}{\hbar} \int L\,dt$$

for all the possible electron orbits in an atomic potential subject to appropriate boundary conditions and use this function to predict bright-line spectra and other experimental observations, but in many instances it would be an unnecessarily difficult task. The integral approach expressed by Eq. (4.8), being a recent theoretical development, is not the method by which standard problems have been handled, and the reader will find virtually no examples of this method in the literature on the standard problems. The method he will find in the literature, which is frequently more convenient, involves a differential equation instead. In discussing black-body radiation in Chapter I, where we had electromagnetic radiation bound or confined in a box, we were able to derive the correct wave functions for the photons by using Maxwell's electromagnetic wave equation and finding solutions which satisfied the boundary conditions. We could proceed with particle problems in the same way if we had a similar but not necessarily identical wave equation (i.e., a second-order differential equation) for particles. Because of the simplicity of the differential wave equation and its frequent use in the literature, we will now undertake the transformation of (4.12), involving an integration over the paths, to a differential equation, and in later chapters we will stress its use.

If we know the wave function at time t as a function of position x', then the wave function at a slightly later time $t + \epsilon$ as a function of position x will be given by

$$\Psi(x, t + \epsilon) = \frac{1}{A} \int e^{(i/\hbar)L\epsilon} \Psi(x', t) \, dx' \qquad \text{(4.12a)}$$

where the integral must be taken over all the values x' may assume. In Cartesian coordinates the Lagrangian may be written

$$L = T - V = \frac{1}{2} m \left(\frac{x - x'}{\epsilon} \right)^2 - V(x)$$

Substituting L and

$$\xi = x - x', \qquad d\xi = -dx'$$

in (4.12a) and reversing the limits of integration, we obtain

$$\Psi(x, t + \epsilon) = \frac{1}{A} \int e^{im\xi^2/2\hbar\epsilon} e^{-i\epsilon V(x)/\hbar} \Psi(x - \xi, t) \, d\xi \qquad \text{(4.13)}$$

In the limit as ϵ approaches zero, the first factor oscillates rapidly from positive to negative values as a function of ξ for all ξ much different from zero. Therefore only from ξ near zero, where the first factor in the integrand does not oscillate rapidly, do contributions to the integrand for small ϵ occur. That is, only points near x at time t contribute appreciably to the wave function at x at time $t + \epsilon$. Since we are concerned only with ξ and ϵ near zero, $\Psi = x - \xi, t)$ may be expanded in a Taylor's series:

$$\Psi(x, t + \epsilon) = e^{-i\epsilon V(x)/\hbar} \frac{1}{A} \int e^{im\xi^2/2\hbar\epsilon} \left[\Psi(x, t) - \xi \frac{\partial \Psi(x, t)}{\partial x} \right.$$
$$\left. + \frac{\xi^2}{2} \frac{\partial^2 \Psi(x, t)}{\partial x^2} - \dots \right] d\xi \qquad \text{(4.14)}$$

Now

$$\int_{-\infty}^{\infty} e^{im\xi^2/2\hbar\epsilon} \, d\xi = \sqrt{2\pi\hbar\epsilon i/m}$$

and

$$\int_{-\infty}^{\infty} e^{im\xi^2/2\hbar\epsilon} \xi \, d\xi = 0$$

since the integrand is odd, and

$$\int_{-\infty}^{\infty} e^{im\xi^2/2\hbar\epsilon} \xi^2 \, d\xi = (\hbar\epsilon i/m)\sqrt{2\pi\hbar\epsilon i/m}$$

Expanding the left-hand side of Eq. (4.14) also in a Taylor's series and keeping only the first two terms, we finally obtain

$$\Psi(x, t) + \epsilon \frac{\partial\Psi(x, t)}{\partial t} = \frac{\sqrt{2\pi\hbar\epsilon i/m}}{A} e^{-i\epsilon V(x)/\hbar}$$

$$\times \left[\Psi(x, t) + \frac{\hbar\epsilon i}{m} \frac{\partial^2\Psi(x, t)}{\partial x^2} + \ldots \right] \quad (4.15)$$

By expanding our wave function $\Psi(x, t + \epsilon)$ in a Taylor's series in powers of ϵ we have obtained an equation in partial differentials. We will now obtain the second-order partial differential wave equation we seek by equating terms multiplied by similar powers of ϵ. Equating the terms which are of zero order in ϵ, we find the normalization constant to be

$$A = \sqrt{2\pi\hbar\epsilon i/m} \quad (4.16)$$

In order to obtain the first-order terms in ϵ we must first expand the exponential containing $V(x)$, and we find that

$$\Psi(x, t) + \epsilon \frac{\partial\Psi(x, t)}{\partial t} = \left[1 - \frac{i\epsilon}{\hbar} V(x) \right] \left[\Psi(x, t) + \frac{\hbar\epsilon i}{2m} \frac{\partial^2\Psi(x, t)}{\partial x^2} \right] \quad (4.17)$$

The zero-order terms in ϵ now cancel, and comparing the remaining terms which are first-order in ϵ (the ϵ^2 term can be dropped since we care about $\epsilon \ll 1$) and multiplying both sides by $\hbar/i$, we obtain

$$\frac{\hbar}{i} \frac{\partial\Psi}{\partial t} = \frac{\hbar^2}{2m} \frac{\partial^2\Psi}{\partial x^2} - V(x)\Psi \quad (4.18)$$

which is the differential equation we sought. Equation (4.18) is the *time-dependent Schrödinger equation*.

If the space and time variables are separable, product wave functions may be used. Let

$$\Psi(x,t) = \psi(x)T(t) \quad (4.19)$$

If we substitute Eq. (4.19) into Eq. (4.18) and divide through by $\psi(x)T(t)$ we obtain

$$\frac{\hbar}{i} \frac{1}{T(t)} \frac{\partial T(t)}{\partial t} = \frac{\hbar^2}{2m} \frac{1}{\psi(x)} \frac{\partial^2\psi(x)}{\partial x^2} - V(x) \quad (4.20)$$

since

$$\frac{1}{\psi(x)} \frac{\partial\psi(x)T(t)}{\partial t} = \frac{\partial T(t)}{\partial t}$$

and

$$\frac{1}{T(t)} \frac{\partial^2 \psi(x) T(t)}{\partial x^2} = \frac{\partial^2 \psi(x)}{\partial x^2}$$

The left-hand side of this equation is independent of x and the right-hand side is independent of t. Since the equation is to be valid for all x and t, both sides must be equal to a constant which is independent of both x and t. Call the constant $-E$; then

$$\frac{\hbar}{i} \frac{1}{T(t)} \frac{\partial T(t)}{\partial t} = -E$$

whose solution is

$$T(t) = e^{-(i/\hbar) Et} \tag{4.21}$$

and the time-independent equation in x, *the time-independent Schrödinger equation*, is

$$-\frac{\hbar^2}{2m} \frac{\partial^2 \psi(x)}{\partial x^2} + V(x)\psi(x) = E\psi(x)$$

or

$$\frac{\partial^2 \psi}{\partial x^2} + \frac{2m}{\hbar^2} [E - V(x)]\psi(x) = 0 \tag{4.22}$$

The extension of the theory to three dimensions is trivial. The result is

$$\nabla^2 \psi + \frac{2m}{\hbar^2} [E - V(\mathbf{r})]\psi = 0 \tag{4.23}$$

for the time-independent Schrödinger equation, and

$$-\frac{\hbar}{i} \frac{\partial \Psi(x, t)}{\partial t} = -\frac{\hbar^2}{2m} \nabla^2 \Psi(x, t) + V(x)\Psi(x, t) \tag{4.24}$$

for the time-dependent Schrödinger equation.

We will always use Ψ to represent the complete wave function including the time and ψ to represent the spatial part of the wave function.

Problem 4.7: Letting

$$\int_{t_0} p\dot{x}\, dt = \int_{x_0}^{x} p\, dx$$

in Eq. (4.8), show that the one-dimensional wave function (4.8) for $\Psi(x,t)$,

$$\Psi(x, t) = \exp\left\{\frac{i}{\hbar} \int_{x_0}^{x} p\, dx - \frac{i}{\hbar} \int_{t_0}^{t} E\, dt\right\}$$

satisfies Eq. (4.24), if energy and momentum are constant.

4.5 PARTICLE WAVES

There are differences between photons and particles having mass. Just as the nonrelativistic expression for particle kinetic energy $p^2/2m$ differs from the expression for photon energy $\hbar\omega$, Eq. (1.5) differs from Eq. (4.24) in the important matter of being second-order instead of first-order in the time derivative. Equations (1.5, 4.24) are, however, both wave equations for the probability amplitudes of photons and particles and, when employed properly, they satisfactorily predict all experimentally observed phenomena to which they have been applied. Whenever the particle position must be defined more closely than permitted by the uncertainty principle, Eq. (4.24) must replace the classical description of particle behavior in which momentum and position are exact and which has proved so useful in analyzing the motion of macroscopic objects. In other words, classical mechanics must be replaced by quantum mechanics whenever critical distances (such as the distance in which particle potential energy changes appreciably, the distance of closest approach of a positively charged particle to an atomic nucleus, or any other distance which the particle must be confined in or excluded from) become less than, or of the order of, the particle wavelength h/mv. The classical equations of motion are still valid, but as we shall see in later chapters their interpretation is different, for Newton's second law, etc., and they now refer only to the *average* behavior of the particle. Quantum mechanics is the more general and completely correct theory, to which no exception has so far been found. Classical particle motion is a satisfactory approximation to quantum mechanics whenever the particle wavelength is less than any of the relevant dimensions of the problem. In the limit of very short wavelengths or large energies, the classical result must always be obtained. Similar approximations are possible in the theory of light, as is well known. Newton, for example, was able to explain many of his observations on light by means of a corpuscular or ray theory. Similarly, radio engineers are able to analyze the reflection of radio waves in the ionosphere, where the electron density changes slowly, by means of ray optics, since in this example (or for paths of light, e.g., mirages over a hot desert or sunsets) the refractive properties of the medium change little within a wavelength.

Comparison of Eq. (4.23) with Eq. (1.22) leads to the interesting observation that in a constant potential, particle probability waves follow the same path as light waves in a constant or uniform medium, the effective indices of refraction for particles and light being given respectively by

$$n^2 = \frac{E - V}{E}; \quad n^2 = \frac{k^2}{\omega^2/c^2} = \frac{k^2}{k_0^2} \tag{4.25}$$

where k_0 is the wave number of the optical radiation in a vacuum, equal to ω/c, and $2mE$ and $2m(E - V)$ are the squares of the particle momentum when the potential is zero and different from zero respectively. This comparison of light wave paths to particle paths is not a uniquely quantum mechanical result; indeed Newton's corpuscular theory of light correctly described the path of light. The difference lies in that the index of refraction for a particle theory is the ratio of group velocity in the *medium* to group velocity in the vacuum,

$$n = p_2/p_1 = \sqrt{(E - V)/E} \; ;$$

for light waves, on the other hand, n is the ratio of velocity in the vacuum to velocity in the medium.

The entire discussion of wave packets for photons in Chapter III applies with the same validity to particle wave packets. The extension of the particle wave packet in space Δx for a corresponding spread in momentum Δp is clearly given by $\Delta x = h/\Delta p$. Identically, a Gaussian wave packet in momentum space with mean square deviation $\Delta p/h$ would, under transformation, yield a Gaussian wave packet in coordinate space with mean square deviation $\Delta x = h/\Delta p$. Instead of Eq. (2.24) we would now have

$$\Psi(x, t) = \frac{1}{2\pi\hbar} \int_{-\infty}^{\infty} B e^{-(p - p_0)^2/2(\Delta p)^2} e^{(i/\hbar)(px - Et)} \, dp \qquad (4.26)$$

where the phase velocity of the wave is now given by E/p. The group velocity is given by $\partial E/\partial p$. Note that the group velocity of the wave packet is equal to the classical velocity of the particle.

Problem 4.8: If the number of remaining excited atoms at t seconds after their excitation at $t = 0$ is $N_0 e^{-10^9 t}$, what will be the energy spectrum of a measurement of the excited state? Assume the center of the line has an energy of $\hbar\omega_0$.

Problem 4.9: Find the group velocity v_g of a wave packet which describes the motion of a nonrelativistic electron in free space. How does it differ from the velocity of light in a vacuum c? (That is, is the group velocity for electrons having different momenta, and therefore wavelength, not a constant, as it is for light of different wavelengths?)

Problem 4.10: An electron with 2-ev total energy moves in a potential field as follows: $z < 0$, $V = 0$; $z > 0$, $V = 1$ volt. Find the phase and group velocities of the wave packet above and below the xy plane.

Problem 4.11: Find the two lowest energy levels for an electron in a rectangular box measuring $1 \times 2 \times 3$Å.

BIBLIOGRAPHY

Feynman, Richard P., *Revs. Modern Phys.*, **20**, 367 (1948). This contains the third development of quantum mechanics following some suggestions of Dirac (the first two being Schrödinger's and Heisenberg's original derivations in 1926), upon which this chapter is based.

Bohm, David, *Quantum Theory*. Englewood Cliffs, N. J.: Prentice-Hall, Inc., 1951. See Chapter 3.

V

USE OF THE WAVE
FUNCTION TO
OBTAIN PHYSICAL
PROPERTIES OF A
SYSTEM

5.1 OBSERVABLES

Our entire theoretical knowledge of any physical system is now contained in the wave function, and we must find operators which operate on the wave function to give us values for the different properties of the system which we can observe experimentally. Every physical observable (e.g., energy, position, or angular momentum) may be represented by an operator Q in an eigenvalue equation $Q\psi_n = q_n\psi_n$ whose eigenvalues q_n are possible results of a measurement of the physical observable. We have previously encountered an example of the important operator ∇^2 in Chapter I and have examined some of its properties. This brings up an important point about operators. They have no meaning unless they operate on something.

If we wish to calculate a particle's position we find it is impossible to obtain a unique value for it in terms of any previous value, because of the uncertainty principle. Classically a later measurement would yield a position given by $x_2 = x_1 + (p/m)t$, but the uncertainty principle requires an uncertainty in momentum. We can calculate only the *average value* or the *expected*

54

value for any measurement. If we consider only a one-dimensional space, for simplicity, the mean value of x or the *expectation of x* is given by

$$\langle x \rangle = \int \psi^*(x) x \psi(x) \, dx \tag{5.1}$$

integrated over all space. In this case the order in which we write x and ψ does not matter, although, as detailed in a later section, in general the order in which the operator representing a physical observable (position, velocity, or whatever) and the wave functions are written *does* matter. Thus, the reader may think of this integrand as $x\psi^*(x)\psi(x)$, that is, the position times the probability that the particle has that position, in concordance with the usual definition of mean value in mathematics or classical statistics.

The particle positions at two different times cannot be calculated precisely since this would require a precise knowledge of the particle's position and momentum at one time; therefore the particle velocity cannot be computed directly from values of x. We must use an operator which by operating on the particle total wave function gives us the velocity or momentum of the particle. (The particle velocity and momentum are linearly related, as we assume here, *only* in nonrelativistic motion.) If Eq. (4.8) is written (see Problem 4.7) as

$$\Psi(x, t) = \exp \left\{ \frac{i}{\hbar} \left[\int_{x_a}^{x} p \, dx - \int_{t_a}^{t} E \, dt \right] \right\} \tag{4.8a}$$

it is clear that the momentum operator is

$$p_x = \frac{\hbar}{i} \frac{\partial}{\partial x} \tag{5.2}$$

It is also true, in general, when we must sum over all possible paths between (x_a, t_a) and (x, t).

Similarly, the energy operator is given by

$$E = -\frac{\hbar}{i} \frac{\partial}{\partial t} \tag{5.3}$$

From this it follows that the Schrödinger time-dependent equation (4.24) may be written as

$$E\Psi = \left(\frac{p^2}{2m} + V(x) \right) \Psi$$

in which p^2 and E are regarded as operators. The operator $p^2 = p_x^2 + p_y^2 + p_z^2$. The sum of two operators A, B is taken: $(A + B)\Psi = A\Psi + B\Psi$. The product of two operators is taken: $AB\Psi = A(B\Psi)$.

In classical mechanics the *Hamiltonian* is the name given to the sum of the kinetic and potential energy written as a function of position and momentum $H = p^2/2m + V(x)$. The operator expression on the right-hand side of Eq. (4.24) $-(\hbar^2/2m)\nabla^2 + V$ occurs so often that we will frequently refer to this operator on the wave function by the symbol H. Hereafter we will use the letter E to represent the eigenvalue of the H operator. Thus Eqs. (4.23) and (4.24) could be written

$$H\Psi = -\frac{\hbar}{i}\frac{\partial\Psi}{\partial t} = E\Psi \tag{5.4}$$

The second equality is true only if the energy is a constant and E is an eigenvalue of the H operator when operating on Ψ.

The mean value of the momentum, or *expectation* of any measurement of the momentum, may be found by taking the average value or *expectation* of the momentum operator, Eq. (5.2), which is given by

$$\langle p_x \rangle = \int \psi^*(x)p_x\psi(x)dx = \frac{\hbar}{i}\int \psi^*(x)\frac{\partial\psi(x)}{\partial x}dx \tag{5.5}$$

In general, the expectation value of any operator Q we can find to represent a physical observable is given by

$$\langle Q \rangle = \int \psi^*Q\psi \, d\tau \tag{5.6}$$

Note that this is *not* $\int Q\psi^*\psi \, d\tau$ or $\int \psi^*\psi Q \, d\tau$.

If Ψ is an eigenfunction of the operator Q, $Q\Psi = q\Psi$ where q is an eigenvalue of the state or eigenfunction of the operator Q. q is just a number and is a possible result of the measurement of the physical property of the state represented by the operator Q. For example, in Chapter I we found that the operator ∇^2 had the eigenvalue ω^2/c^2 when operating on the wave functions of the radiation in a box. The operator ∇^2 represents the reciprocal of the square of the wavelength of the radiation ($\times(2\pi)^2$), and any given value of ω/c which satisfies Eq. (1.11) for integral k, l, and m is a possible result of the measurement of 2π times the inverse of the wavelength of the radiation in the box. A particle plane wave would yield no definite position of the particle; therefore the momentum could be uniquely specified without violating the uncertainty principle. Thus if $\psi = Ae^{(i/\hbar)p_{0x}x}$, where A is a normalization constant,

$$\langle p_x \rangle = \frac{\hbar}{i}|A|^2 \int_{-\infty}^{\infty} e^{-(i/\hbar)p_{0x}x}\frac{\partial}{\partial x}e^{(i/\hbar)p_{0x}x}dx$$

$$= \frac{\hbar}{i}|A|^2 \int_{-\infty}^{\infty} e^{-(i/\hbar)p_{0x}x}\left(\frac{i}{\hbar}p_{0x}\right)e^{-(i/\hbar)p_{0x}x}dx \tag{5.7}$$

$$= p_{0x}$$

The integral in Eq. (5.7) is linearly infinite but identical to $|A|^{-2}$ whereas $\langle x \rangle$, being a definite integral over x, would give no specific well-defined location of the particle, since $\langle x \rangle = (|A|^2/2)[\infty^2 - (-\infty)^2]$. Throughout the later sections of this book whenever we complete a discussion of a new physical observable we will have to find an operator to represent it, so that its expectation may be found by operating with it on the wave function.

A measurement of the relative number of single systems in a particular energy level gives the probability of finding each system in the corresponding group of states. For example, the intensity of light of a particular frequency emitted by a hot (optically thin) vapor, coupled with the transition probability for the atomic emission, would tell us what fraction of the vapor atoms were in the state or group of states having the energy from which the transition was made. If several states have the same energy, the energy level is said to be *degenerate*. If the level is degenerate, the relative probability of finding the system in any one particular state must be found by means of an operator which has a different eigenvalue for each state in the same energy level.

In three dimensions, the momentum operator in coordinate space operating on the wave function is

$$\mathbf{p}\psi = \frac{\hbar}{i} \nabla \psi \tag{5.8}$$

The reader is already familiar with the use of alternative representations in either one of a pair of complementary variables in Chapter II. The momentum-space wave function is defined as the Fourier transform of the coordinate-space wave function. When the wave function is a function of momentum rather than space it is said to be expressed in the momentum representation. If the momentum rather than coordinate space is the independent variable in which the probability amplitude is expressed, then we have, corresponding to Eqs. (5.2, 5.8),

$$x\Phi = -\frac{\hbar}{i} \frac{\partial}{\partial p_x} \Phi$$

or

$$\mathbf{r}\Phi = -\frac{\hbar}{i} \nabla_\mathbf{P} \Phi \tag{5.9}$$

where Φ is the wave function expressed in momentum space (the momentum transform of $\Psi(x, t)$ and $\nabla_\mathbf{P}$ is the gradient in momentum space, the derivatives being taken with respect to the degrees of freedom in momentum space (e.g., p_x, p_y, p_z). In this connection it might be noted that

$$p_x\varphi(p_x) \neq \frac{\hbar}{i} \frac{\partial \varphi(p_x)}{\partial x}$$

because $\varphi(p_x)$ is a function of p_x only and the right-hand side is meaningless. ($\Phi(\mathbf{p}, t), \varphi(\mathbf{p})$ are analogous to $\Psi(\mathbf{r}, t), \psi(\mathbf{r})$.)

The probability of finding a particle in a given volume of momentum space is given by

$$\int \varphi^*(\mathbf{p})\varphi(\mathbf{p}) \, d\mathbf{p} \tag{5.10}$$

where the region of integration is the given volume.

5.2 COMMUTATION RELATIONS

Commutation consists in reversing the order of two quantities in an algebraic operation. For example, the scalar product of two vectors is the same when the order of the two vectors is reversed. That is, since $\mathbf{A} \cdot \mathbf{B} = \mathbf{B} \cdot \mathbf{A}$, $\mathbf{A}$ is said to commute with $\mathbf{B}$ in this operation. In the vector product of two vectors, on the other hand, the two quantities do not commute:

$$\mathbf{A} \times \mathbf{B} \neq \mathbf{B} \times \mathbf{A} \tag{5.11}$$

Application of commutation relations between appropriate operators reveals a great deal about the relationships between various physical observables. Commutation relations are of considerable utility and furnish a powerful tool for the further elucidation of our subject.

The average value or expectation value of any operator f is calculated in the coordinate representation from

$$\langle f \rangle = \psi^*(x) f \psi(x)\, d\tau \tag{5.12}$$

As we have already seen

$$\langle x \rangle = \int \psi^*(x) x \psi(x) d\tau$$

$$\langle p_x \rangle = \left\langle \frac{\hbar}{i} \frac{\partial}{\partial x} \right\rangle = \int \psi^*(x) \frac{\hbar}{i} \frac{\partial \psi(x)}{\partial x} d\tau \tag{5.13}$$

where the operator must be expressed in coordinate space since the state is represented in coordinate space. We can now easily demonstrate that operators may not commute, so that the order in which they are written is important.

$$p_x x \psi = \frac{\hbar}{i} \frac{\partial(x\psi)}{\partial x} = \frac{\hbar}{i}\left(x \frac{\partial \psi}{\partial x} + \psi\right)$$

$$= \left(\frac{\hbar}{i} + x p_x\right)\psi$$

$$\neq x p_x \psi$$

Therefore, a coordinate and its complementary momentum do not commute.
The *commutator* of x and p_x is defined as

$$[x, p_x] = x p_x - p_x x \tag{5.14}$$

and is a nonzero operator on the wave function, with the expectation

$$\langle [x, p_x] \rangle = \frac{\hbar}{i}\left\{\int \psi^* x \frac{\partial \psi}{\partial x} d\tau - \int \psi^* \psi d\tau - \int \psi^* x \frac{\partial \psi}{\partial x} d\tau\right\}$$

$$= -\frac{\hbar}{i} \tag{5.15}$$

This is not too surprising, however, since $\langle x\, p_x \rangle$ is a meaningless symbol anyway, because x and p_x cannot be measured simultaneously with infinite precision. (On the other hand, a quantity such as $(\hbar/x) + p_x$ can be measured in principle even if $\hbar/x$ and p_x cannot be measured separately.) *In general, only operators of simultaneously measurable physical observables commute.* x and p_x are not simultaneously measurable, for example, and as we have just seen, they do not commute.

It is possible to know the value of a sum $A + B$ without knowing the separate terms. In fact, it may not be possible to measure both terms simultaneously. For example, the energy may have a precise value even though $p^2/2m$ and $V(x)$ are not separately simultaneously measurable and cannot both have a precise value since they do not commute with one another. The expectation value of the energy is found, from the relation

$$H = \frac{p^2}{2m} + V(\mathbf{r}) = E$$

to be

$$\langle H \rangle = \int \psi^* \left\{ -\frac{\hbar^2}{2m}\nabla^2\psi + V(\mathbf{r})\psi \right\} d\tau \tag{5.16}$$

We now show how to compute the time derivative of an expectation value.

$$\frac{d\langle A \rangle}{dt} = \frac{d}{dt}\int \Psi^* A \Psi\, d\tau$$

$$= \int \frac{\partial\Psi^*}{\partial t} A\Psi\, d\tau + \int \Psi^* \frac{\partial A}{\partial t}\Psi\, d\tau + \int \Psi^* A \frac{\partial\Psi}{\partial t}\, d\tau$$

and

$$H\Psi = -\frac{\hbar}{i}\frac{\partial\Psi}{\partial t}\,; \qquad \Psi^*H = \frac{\hbar}{i}\frac{\partial\Psi^*}{\partial t}$$

where the operator on the complex conjugate wave function is written on the right of the wave function. Therefore, after substituting for partial derivatives,

$$\frac{d\langle A \rangle}{dt} = \frac{i}{\hbar}\int \left\{ \Psi^* HA\Psi - \Psi^* AH\Psi + \frac{\hbar}{i}\Psi^*\frac{\partial A}{\partial t}\Psi \right\} d\tau$$

$$= \frac{i}{\hbar}\langle [H, A] \rangle + \langle \partial A/\partial t \rangle \tag{5.17}$$

If in particular the operator A does not have an explicit time dependence we have for example,

$$\frac{d\langle A \rangle}{dt} = \frac{i}{\hbar}\langle [H, A] \rangle$$

$$\frac{d}{dt}\langle x \rangle = \frac{i}{\hbar}\langle [H, x] \rangle$$

$$= -\frac{i}{\hbar}\frac{\hbar^2}{2m}\int \Psi^*\left[\frac{\partial^2}{\partial x^2}(x\Psi) - x\frac{\partial^2\Psi}{\partial x^2}\right]d\tau$$

$$-\frac{i}{\hbar}\frac{\hbar^2}{2m}\int \Psi^*\left[\frac{\partial^2}{\partial y^2}(x\Psi) + \frac{\partial^2}{\partial z^2}(x\Psi) - x\frac{\partial^2\Psi}{\partial y^2} - x\frac{\partial^2\Psi}{\partial z^2}\right]d\tau$$

$$+ \frac{i}{\hbar} \int \Psi^*[V(x, y, z)x - xV(x, y, z)\Psi d\tau \tag{5.18}$$

Since x commutes with

$$\frac{\partial^2}{\partial y^2}, \ \frac{\partial^2}{\partial z^2}, \ \text{and} \ V(x, y, z)$$

only the first integral yields a nonzero result.
Since

$$\frac{d^2(x\Psi)}{dx^2} = 2\frac{d\Psi}{dx} + x\frac{d^2\Psi}{dx^2} \tag{5.19}$$

Eq. (5.18) becomes

$$\frac{d}{dt}\langle x \rangle = \frac{\hbar}{2im} \int \left\{ 2\Psi^*\frac{\partial\Psi}{\partial x} + \Psi^*x\frac{\partial^2\Psi}{\partial x^2} - \Psi^*x\frac{\partial^2\Psi}{\partial x^2} \right\} d\tau$$

$$= \frac{\hbar}{im} \int \Psi^*\frac{\partial\Psi}{\partial x} d\tau$$

$$= \frac{1}{m}\langle p_x \rangle \tag{5.20}$$

a result which is not surprising.

As a second example let us find the time derivative of the expectation of p_x:

$$\frac{d}{dt}\langle p_x \rangle = \frac{i}{\hbar}\langle Hp_x - p_xH \rangle$$

$$= \frac{i}{\hbar}\left\langle \left(\frac{p^2}{2m} + V\right)p_x - p_x\left(\frac{p^2}{2m} + V\right)\right\rangle$$

$$= \frac{i}{\hbar}\langle Vp_x - p_xV \rangle$$

since p_x commutes with p^2. The rest of the commutation operation is given by

$$\frac{d}{dt}\langle p_x \rangle = \int \left(\Psi^*V\frac{\partial\Psi}{\partial x} - \Psi^*\frac{\partial V}{\partial x}\Psi - \Psi^*V\frac{\partial\Psi}{\partial x} \right) d\tau$$

$$= -\left\langle \frac{\partial V}{\partial x} \right\rangle = -\langle (\text{grad } V)_x \rangle = \langle F_x \rangle \tag{5.21}$$

where F_x is the x component of the applied force, the force of gravity, for example, or an electrostatic force. Equation (5.21) is the quantum mechanical version of Newton's second law.

The equivalence of operating on the term to the left of the operator and operating on the term to the right [for example, in Eq. (5.18) it could be shown that the operator $\partial^2/\partial x^2$ has the same expectation when operating on ψ^* as when it operated on ψ] is a generally valid result for all operators representing physical observables, provided the complex conjugate of the operator is taken when the operator operates on the term to the left. Such operators as can be used in this way are called *Hermitian operators* and they are defined as fulfilling the equality

$$\int \psi_i^*(\mathbf{r})[Q\psi_j(\mathbf{r})]\, d\tau = \int \psi_j(\mathbf{r})[Q\psi_i(\mathbf{r})]^*\, d\tau \tag{5.22}$$

where ψ_i, ψ_j are eigenfunctions of the operator denoted Q. *All operators Q which represent physical observables are Hermitian.* We will prove this equality valid for expectation values of all operators representing physical observables. The eigenvalue of the operator Q or Q^* must be a real number since the eigenvalue has a direct physical interpretation. The proof of their being Hermitian consists in first noting that

$$(Q\psi)^* = (q\psi)^* = q\psi^*$$

and

$$Q\psi = q\psi$$

Multiplying on the left both sides of the first of these equalities by ψ and the third equality by ψ^*, subtracting one equation from the other, rearranging and integrating over all space, we find that

$$\int \psi(Q\psi)^*\, d\tau - \int \psi^*Q\psi\, d\tau = q\int \psi\psi^*\, d\tau - q\int \psi^*\psi\, d\tau = 0$$

Usually we will use the notation $\int (\psi_j^* Q^*)\psi_i\, d\tau$ for the quantity we have written in (5.22) as $\int \psi_i(\mathbf{r})[Q\psi_j(\mathbf{r})]^*\, d\tau$.

Problem 5.1: Show that $\langle p_x \rangle$ may be found in momentum space from the equation

$$\langle p_x \rangle = \int \varphi^*(\mathbf{p})p_x\varphi(\mathbf{p})\, dp_x dp_y dp_z$$

where the φ's are the Fourier transforms of the ψ's into momentum space.

Problem 5.2: Show that p_x and p_y commute. Show that x and p_y commute. Show that x and y commute. Show that x and $f(x, y, z)$ commute.

Problem 5.3: Show that

$$[f(p_x, p_y, p_z), x] = \frac{\hbar}{i}\frac{\partial f}{\partial p_x}$$

and

$$[p_x, f(x, y, z)] = \frac{\hbar}{i}\frac{\partial f}{\partial x}$$

(Hint: Use the momentum representation to prove the first relation.)

Problem 5.4: Determine which of the following operators are Hermitian:

$$x,\ x^2,\ x^3,\ \frac{d}{dx},\ \frac{d^2}{dx^2},\ \frac{d^3}{dx^3}$$

$$i\frac{d}{dx},\ x\frac{d}{dx},\ \left(x\frac{d}{dx} - \frac{d}{dx}x\right),\ i\left(x\frac{d}{dx} + \frac{d}{dx}x\right)$$

5.3 PARTICLE CURRENTS

In discussing scattering problems or particle motion it is necessary to find a suitable operator to represent the particle current or flow density. We will next discuss exactly what the particle current density is in terms of the particle wave functions. The complete Schrödinger wave equation is

$$i\hbar \frac{\partial \Psi}{\partial t} = -\frac{\hbar^2}{2m} \nabla^2 \Psi + V\Psi \tag{5.23}$$

The equation of the complex conjugate wave function is

$$-i\hbar \frac{\partial \Psi^*}{\partial t} = -\frac{\hbar^2}{2m} \nabla^2 \Psi^* + V\Psi^* \tag{5.24}$$

Multiplying Eq. (5.23) by Ψ^* and Eq. (5.24) by Ψ and subtracting (5.24) from (5.23), we find

$$i\hbar \left(\Psi^* \frac{\partial \Psi}{\partial t} + \Psi \frac{\partial \Psi^*}{\partial t} \right) + \frac{\hbar^2}{2m} (\Psi^* \nabla^2 \Psi - \Psi \nabla^2 \Psi^*) = 0$$

or

$$\frac{\partial}{\partial t} (\Psi^* \Psi) + \frac{\hbar}{2im} \nabla \cdot (\Psi^* \nabla \Psi - \Psi \nabla \Psi^*) = 0 \tag{5.25}$$

$\Psi^* \Psi$ is the expected particle density. The conservation of particle number is expressed by saying that the time derivative of the particle density ρ plus the divergence of the particle current density $\mathbf{j}$ must add up to zero; that is,

$$\frac{\partial \rho}{\partial t} + \nabla \cdot \mathbf{j} = 0 \tag{5.26}$$

Equation (5.26) is known as the equation of continuity. From Eqs. (5.25) and (5.26) we are led to postulate that the average value of particle current density $\mathbf{j}$ is given by

$$\mathbf{j} = \frac{\hbar}{2im} (\Psi^* \nabla \Psi - \Psi \nabla \Psi^*) \tag{5.27}$$

Free particles are represented by the plane wave solutions

$$\Psi = A \, \exp\left\{ \pm \frac{i}{\hbar} \mathbf{p} \cdot \mathbf{r} - \frac{i}{\hbar} Et \right\}$$

of the Schrödinger equation when the potential energy $V = 0$. By substituting the plane wave solutions into Eq. (5.27), we find that

$$\mathbf{j} = \frac{\hbar}{2im} \left[\left(\frac{i}{\hbar} \mathbf{p} \right) - \left(\frac{-i}{\hbar} \mathbf{p} \right) \right] \Psi^* \Psi = \frac{\mathbf{p}}{m} \Psi^* \Psi = \mathbf{v} \Psi^* \Psi \tag{5.28}$$

Thus the particle current density is given by the particle velocity $\mathbf{v}$ times the particle density $\Psi^* \Psi$. If the z direction is taken parallel to the incident beam of particles, the beam is represented by a plane wave e^{ikz}. In case the beam has a finite width it might be represented by a Gaussian wave packet

$$\exp\left(ikz - \frac{x^2 + y^2}{a^2}\right)$$

where a^2 is the mean square deviation of $(x^2 + y^2)^{1/2}$.

Problem 5.5: Find the flux of particles through a sphere of radius a concentric with the origin if the wave function is given by

$$\Psi = \frac{A}{r}\exp\left\{\frac{i(\mathbf{p}\cdot\mathbf{r} - Et)}{\hbar}\right\}$$

5.4 CENTER-OF-MASS COORDINATES

In atomic physics, the nucleus has essentially infinite mass compared with the electrons and one may ignore the negligible part of an electron's energy which is required to conserve the translational momentum of the complete interacting system. That is, the translational energy and momentum of the atom is negligible compared with the potential and kinetic energy of the electron in the atom. In nuclear physics, where the interacting particles may have equal masses, the translational energy and momentum of the complete system in the laboratory frame of coordinates may not be neglected, and the problem is handled best in a center-of-mass coordinate frame, in which the translational momentum of the complete system is zero. For two particles of mass m_1 and m_2 having laboratory momenta $\mathbf{p}_1$ and $\mathbf{p}_2$, the total kinetic energy is

$$T = \frac{\mathbf{p}_1^2}{2m_1} + \frac{\mathbf{p}_2^2}{2m_2} \tag{5.29a}$$

while the potential energy of their interaction is a function of $(\mathbf{r}_1 - \mathbf{r}_2)$; that is,

$$V = f(\mathbf{r}_1 - \mathbf{r}_2) \tag{5.29b}$$

Let

$$\mathbf{P} = \mathbf{p}_1 + \mathbf{p}_2, \qquad \mathbf{p} = \mathbf{p}_1 - \mathbf{p}_2$$

$$m = \frac{m_1 m_2}{m_1 + m_2}, \qquad \mathbf{v} = \mathbf{v}_1 - \mathbf{v}_2 \tag{5.30a}$$

where $\mathbf{v}$ is the relative velocity of one particle with respect to the other and m is the *reduced mass* of the particles. The position of the origin of the center-of-mass coordinate frame is given by

$$\mathbf{R} = \frac{m_1\mathbf{r}_1 + m_2\mathbf{r}_2}{m_1 + m_2} \tag{5.30b}$$

while the separation of the two particles is given by

$$\mathbf{r} = \mathbf{r}_1 - \mathbf{r}_2 \tag{5.30c}$$

By substituting Eqs. (5.30) into Eqs. (5.29), we may represent the Hamiltonian in the new coordinate frame:

$$H = \frac{P^2}{2(m_1 + m_2)} + \frac{p^2}{2m} + V(\mathbf{r})$$

$$= -\frac{\hbar^2}{2(m_1 + m_2)}\nabla_R^2 - \frac{\hbar^2}{2m}\nabla_r^2 - V(\mathbf{r}) \tag{5.31}$$

where ∇_R^2 means that the derivatives in the Laplacian ∇^2 are taken with respect to the coordinates of the point $\mathbf{R}$, and ∇_r^2 means derivatives with respect to $\mathbf{r}$.

Since the variables are separable we may write the wave function as the product of the two wave functions in $\mathbf{R}$ and $\mathbf{r}$ coordinates. That is, we let

$$\Psi(\mathbf{R}, \mathbf{r}) = \chi(\mathbf{R})\psi(\mathbf{r}) \tag{5.32}$$

By inserting Eq. (5.32) into the wave equation and then dividing by $\chi(\mathbf{R})\psi(\mathbf{r})$, we find that

$$\frac{1}{\chi(\mathbf{R})\psi(\mathbf{r})}\left(\frac{-\hbar^2}{2(m_1 + m_2)}\right)\nabla_R^2\chi(\mathbf{R})\psi(\mathbf{r})$$

$$-\frac{1}{\chi(\mathbf{R})\psi(\mathbf{r})}\frac{\hbar^2}{2m}\nabla_r^2\chi(\mathbf{R})\psi(\mathbf{r}) + \frac{1}{\chi(\mathbf{R})\psi(\mathbf{r})}V(\mathbf{r})\chi(\mathbf{R})\psi(\mathbf{r})$$

$$= \frac{1}{\chi(\mathbf{R})\psi(\mathbf{r})}E\chi(\mathbf{R})\psi(\mathbf{r})$$

Noting that ∇_R^2 operates only on $\chi(\mathbf{R})$ and ∇_r^2 only on $\psi(\mathbf{r})$, rearranging terms, and making cancellations wherever possible, we obtain

$$-\frac{\hbar^2}{2(m_1 + m_2)}\frac{1}{\chi(\mathbf{R})}\nabla_R^2\chi(\mathbf{R}) = \frac{\hbar^2}{2m}\frac{1}{\psi(\mathbf{r})}\nabla_r^2\psi(\mathbf{r}) - V(\mathbf{r}) + E \tag{5.33}$$

The left-hand side of this equation is a function of $\mathbf{R}$ but not $\mathbf{r}$ while the right-hand side is not a function of $\mathbf{R}$. If this equation is to be valid for any value of $\mathbf{R}$ or $\mathbf{r}$, either side must be equal to a constant. Call it E_{cm}, and let $E_r = E - E_{cm}$. Then

$$\frac{\hbar^2}{2(m_1 + m_2)}\nabla_R^2\chi + E_{cm}\chi = 0 \tag{5.34}$$

is the equation for the center-of-mass motion, which is the same as that for a free particle having a mass $m_1 + m_2$, and

$$\frac{\hbar^2}{2m}\nabla_r^2\psi + [E_r - V(\mathbf{r})]\psi = 0 \tag{5.35}$$

is the equation dealing with the interaction of the particles and their relative motion. Equation (5.34) is of no further interest and Eq. (5.35) will be our only concern. In all of the following, the center-of-mass coordinate system is assumed and only the relative motion of the particles will be treated.

Problem 5.6:   Prove Eq. (5.31).

BIBLIOGRAPHY

Merzbacher, E. *Quantum Mechanics.* New York: John Wiley & Sons, Inc., 1961. Chapters 4 and 8 especially.

Sherwin, C. W., *Introduction to Quantum Mechanics.* New York: Henry Holt and Co., 1959.

VI

ONE-DIMENSIONAL PROBLEMS

To emphasize the differences between our new description of particle behavior and the older, more familiar classical description, we take up in this chapter the extreme wave-mechanical particle behavior which occurs, just as it does for light, where significant changes (in potential energy) occur within a distance comparable to a wavelength. The simplest potential of this type which we could examine is a so-called square well, that is, a potential which is constant everywhere except at only two points where it changes suddenly from one constant value to another constant value (see Figure 6.6). Such a potential is extremely useful for many problems in physics, particularly in low-energy nuclear physics, where the wavelengths of the particles are much longer than the distance in which large changes occur in the nuclear force field. In such cases the details of the shape of the force field cannot be resolved by means of particles whose wavelength is too long to distinguish any short-range potential, and the square well will frequently yield answers just as correct as, and in fact indistinguishable from, answers obtained with more complicated shapes.

Geometrical optics, that is, corpuscular or ray optics, yields excellent approximations for many studies of electromagnetic radiation. For example, light waves passing through the atmosphere, or radio waves traveling through slowly varying changes in index of refraction caused by the slowly varying electron concentration in the ionosphere, follow raylike paths. Geometrical optics is incomplete, however, when applied to such sharp changes in the index of refraction as occur when light is transmitted through a glass lens or a drop of water. It is well known that the path of the ray may be correctly described by ray optics, as Newton first showed, but the reflection of a fraction of the light by a glass-air surface can be predicted only by wave theory. Coated lenses, for example, are a practical application of wave or physical optics.

Force is deduced from the gradient of a potential, and we note that a square well will lead to an infinite force at one point and no force at all in the rest of space. (Such abrupt changes in potential energy do not occur in macroscopic force fields.) When the total energy is conserved, an infinite force resulting from an abrupt change in potential energy causes an abrupt change in kinetic energy and thus an abrupt change in momentum for particles or in wave number for waves. A similar abrupt change in kinetic energy caused by an infinite force acting for an infinitesimal distance would result if a baseball were thrown through a window. (The analogy to a particle's experiencing a sudden change in potential energy is incomplete in that the potential energy of the ball is unchanged by penetrating the window, but the kinetic energy in both instances is abruptly lessened as a result of an infinite opposing force acting for an infinitesimal distance.) A similar sudden change in wave number occurs when light penetrates into glass, water, or some other solid medium having an index of refraction different from that of air. In the absence of a gravitational or other force field the path of the baseball thrown through the window could be shown classically to be the same as the path of the light ray in the parallel circumstance of a sudden change in refractive index. The wave-theoretical difference (from ray optics) is that even though the "breaking strength" of the glass may be exceeded, the baseball will have·a finite (although minute) probability of being reflected.

6.1 PARTICLES INCIDENT ON A SQUARE BARRIER

Example 1 Reflection of a particle at a square barrier infinite in extent (see Figure 6.1). The barrier may be represented by

$$x < 0: \quad V = 0$$
$$x > 0: \quad V = V_0 \tag{6.1}$$

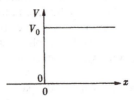

FIGURE 6.1 Illustration of a square one-dimensional barrier of infinite extent.

We further assume the particle energy E to be $> V_0$ and that the particle is incident on the barrier from the left. The potential energies are constant in the two regions on either side of $x = 0$ so that Eq. (4.8) immediately leads to simple plane wave solutions. Since E is a constant we may neglect the time-dependent part of the wave $e^{-(i/\hbar)Et}$. Letting the particle momenta on the two sides of the barrier be

$$p_1 = \sqrt{2mE}$$

and

$$p_2 = \sqrt{2m(E - V_0)}$$

we find the solutions

$$x < 0: \quad \psi = A\, e^{(i/\hbar)p_1 x} + B\, e^{-(i/\hbar)p_1 x}$$
$$x > 0: \quad \psi = C\, e^{(i/\hbar)p_2 x} \tag{6.2}$$

where $e^{(i/\hbar)p_1 x}$ represents the incident wave, $e^{-(i/\hbar)p_1 x}$ the reflected wave, and $e^{(i/\hbar)p_2 x}$ the transmitted wave. Equations (6.2) are presented, in customary treatments of this problem, as solutions of the time-independent Schrödinger wave equation, Eq. (4.22), which is

$$\frac{\partial^2 \psi}{\partial x^2} + \frac{2m}{\hbar^2}(E - V)\psi = 0 \tag{6.3}$$

and since the total probability is never greater than one and a particle's energy must be finite, Eq. (6.3) may be satisfied only if the second derivative of the wave function is finite everywhere. Hence a further boundary condition is obtained by requiring that both the wave function and its first derivative be everywhere continuous. Thus at $x = 0$, this boundary condition by use of Eq. (6.2) yields

$$A + B = C \tag{6.4a}$$

and

$$(i/\hbar)p_2 C = (i/\hbar)p_1 A - (i/\hbar)p_1 B \tag{6.4b}$$

whence

$$B = \frac{p_1 - p_2}{p_1 + p_2}A \quad \text{and} \quad C = \frac{2p_1}{p_1 + p_2}A \tag{6.5}$$

The reflection and transmission are obtained from the currents to the left and right of $x = 0$. These are given by the particle velocity in the medium p/m times the relative density of particles or the probability per unit volume of finding the particle to the left or right of the potential change. Since the density of particles moving to the left per unit volume is proportional to $|Be^{-(i/\hbar)p_1 x}|^2 = B^2$,

$$\frac{\text{Reflected current}}{\text{Incident current}} = \left|\frac{B}{A}\right|^2 \frac{p_1}{p_1} = \left(\frac{p_1 - p_2}{p_1 + p_2}\right)^2$$

$$\frac{\text{Transmitted current}}{\text{Incident current}} = \left|\frac{C}{A}\right|^2 \frac{p_2}{p_1} = \frac{4p_1 p_2}{(p_1 + p_2)^2}$$

(6.6)

Note that if there is no potential change, $p_2 = p_1$, and no reflection takes place; that in the case $E = V_0$, $p_2 = 0$, and the particle is 100% reflected; that particle probability current (cf. Section 5.3) is conserved, for the sum of Eqs. (6.6) is unity; and that Eqs. (6.6) exactly correspond to formulas for the same quantities when electromagnetic radiation is perpendicularly incident on a sharp boundary between media of different indices of refraction. In this case the reflection, for example, is given by

$$R = (n_2 - n_1)^2/(n_1 + n_2)^2$$

where n_2/n_1 is the ratio of the velocity of light in medium one to medium two just as p_1/p_2 is the same ratio for particles.

If the particle energy E is less than V_0 we obtain different expressions on the right of the barrier. The particle momentum $\sqrt{2m(E - V_0)}$ on the right of the barrier is imaginary. Letting $q_2 = \sqrt{2m(V_0 - E)}$, we find the solution of Eq. (4.8) to the right of the barrier to be

$$x > 0: \qquad \psi = Ce^{-(q_2/\hbar)x} \qquad (6.7)$$

This is, the exponent is real and no longer represents an oscillating wave function. [Note that $e^{+q_2\hbar/x}$ is also a solution of Eq. (6.3) in customary solutions of this problem which use Eq. (6.3), but since ψ must be finite everywhere and Eq. (6.7) must hold at $x = +\infty$, this added solution must have a zero coefficient.] Fitting solutions (6.2) for negative x and (6.7) for positive x as we did in obtaining Eq. (6.5), we find in place of (6.5) that for $E < V_0$

$$B = \frac{q_2 + ip_1}{-q_2 + ip_1}A$$

$$C = \frac{2ip_1}{-q_2 + ip_1}A$$

(6.5a)

and

$$\text{Reflection} = \left|\frac{B}{A}\right|^2 = 1$$

$$\text{Transmission} = \left|\frac{C}{A}\right|^2 \frac{q_2}{p_1} = 4\frac{p_1 q_2}{(p_1^2 + q_2^2)}e^{-(2q_2/\hbar)x}$$

(6.6a)

Since the transmission is zero at $x = \infty$ and no particle absorption has been included in the problem, 100% reflection is also required by particle conservation. Although there is no transmitted current, note that there is a finite probability of finding the particle on the right of the barrier, where classically it cannot be.

Example 2 Reflection and transmission through a finite square barrier depicted in Figure 6.2.

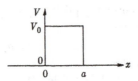

FIGURE 6.2   Illustration of a square one-dimensional barrier of limited extent.

In this case the potential is given by

$$\begin{aligned} x < 0: \quad & V = 0 \\ 0 < x < a: \quad & V = V_0 \\ a < x: \quad & V = 0 \end{aligned} \tag{6.8}$$

Again we first assume $E > V_0$ and that the wave is incident from the left, and find the following wave solutions in the three regions representing incident and reflected waves in the left and center regions and only a transmitted wave in the right region:

$$\begin{aligned} x < 0: \quad & \psi = Ae^{(i/\hbar)p_1 x} + Be^{-(i/\hbar)p_1 x} \\ 0 < x < a: \quad & \psi = Ce^{(i/\hbar)p_2 x} + De^{-(i/\hbar)p_2 x} \\ a < x: \quad & \psi = Ee^{(i/\hbar)p_1 x} \end{aligned} \tag{6.9}$$

The reflected amplitude B (made up of a wave reflected at both $x = 0$ and a) and the transmitted amplitude E may be found by requiring again that the wave functions and their first derivatives be continuous at $x = 0$ and $x = a$. The solution requires considerable tedious algebraic manipulations, with the result that

$$\begin{aligned} \text{Reflection} &= \left|\frac{B}{A}\right|^2 = \left[1 + \frac{4p_1^2 p_2^2}{(p_1^2 - p_2^2)^2}\csc^2\frac{p_2 a}{\hbar}\right]^{-1} \\ \text{Transmission} &= \left|\frac{E}{A}\right|^2 = \left[1 + \frac{(p_1^2 - p_2^2)^2}{4p_1^2 p_2^2}\sin^2\frac{p_2 a}{\hbar}\right]^{-1} \end{aligned} \tag{6.10}$$

Note that if there is no barrier and $p_2 = p_1$, there is 100% transmission. There is the further very interesting circumstance that if a, the barrier width, is equal to an integral number of half wavelengths, 100% transmission occurs, analogous to the behavior of coated lenses in optics. Otherwise we have the nonclassical result of partial reflection at the barrier.

If $E < V_0$, then classically the particle does not have sufficient energy to jump the barrier, and is 100% reflected. In this case the second of Eqs. (6.9), for $0 < x < a$, no longer represents the correct solution for that region. The correct solution when $E < V_0$ is

$$0 < x < a: \quad \psi = Ce^{(q_2/\hbar)x} + De^{-(q_2/\hbar)x} \qquad (6.9a)$$

where $q_2 = \sqrt{2m(V_0 - E)}$. Equation (6.9a) represents exponentially increasing and decreasing solutions within the barrier. Solving the simultaneous algebraic equations resulting from requiring that the wave function and its first derivatives be continuous at both boundaries, we find that

$$\text{Reflection} = \left|\frac{B}{A}\right|^2 = \left[1 + \frac{4p_1^2 q_2^2}{(p_1^2 + q_2^2)^2} \operatorname{csch}^2 \frac{q_2 a}{\hbar}\right]^{-1} \qquad (6.11)$$

$$\text{Transmission} = \left|\frac{E}{A}\right|^2 = \left[1 + \frac{(p_1^2 + q_2^2)}{4p_1^2 q_2^2} \sinh^2 \frac{q_2 a}{\hbar}\right]^{-1} \qquad (6.12)$$

If $a \gg \hbar/q_2$, then the transmission coefficient is asymptotically given by

$$\left[\frac{q_2}{p_1} + \frac{p_1}{q_2}\right]^{-2} 16 \exp\left(-2\frac{q_2 a}{\hbar}\right)$$

Note that the transmission is nonzero if a is not very many wavelengths thick. Gamow has made an amusing comparison of this situation to a billiard ball which will not remain on the top of the billiard table but will eventually penetrate the sides of the table and drop onto the floor. Figure 6.3 illustrates the wave behavior at the barrier for a roughly comparable to λ and E less than but comparable to V_0.

The particle problem in which $V_0 > E$ has an optical analog in an imaginary index of refraction such as can occur for radio waves in the ionosphere or for light waves incident at greater than the critical angle in an optically dense medium. Prism surfaces in binoculars, for example, are arranged so that light is incident on the reflecting surface at greater than the

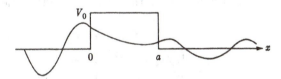

FIGURE 6.3 Illustration of the wave behavior at a finite barrier for barrier width roughly comparable to particle wavelength and particle kinetic energy less than but comparable to the barrier height. The particle has insufficient energy, classically, to surmount the barrier. The sum of the incident and reflected waves is shown to the left of $x = 0$; the transmitted wave is illustrated to the right of $x = a$.

critical angle for glass-to-air transmission, so that 100% internal reflection will occur for light incident on the air back of the prism.

If an optically flat surface were placed close to the optically flat reflecting surface of the prism, however, some of the light would be transmitted, as could be readily calculated by a formula analogous to Eq. (6.12). This was the principle of an ingenious device composed of two prisms by which, during the Second World War, German scientists attempted to audiomodulate a light beam. One of the prisms in Figure 6.4 was attached to a solenoid activated by an audiofrequency current signal so that the space between the two prisms changed at audiofrequency as the prism was made to vibrate. When the prisms were close together and the separation distance a was small, the system transmitted light to the right across the barrier presented by the air space of thickness a; conversely, when a was large light was reflected from the barrier presented by the air space. Unfortunately the prism could not be made to vibrate well at audiofrequencies.

Equations (6.11) and (6.12) have a useful application in the estimation of radioactive nuclei lifetimes from alpha particle decay. Alpha particles within radioactive nuclei have a negative binding energy, so that when the alpha particle is separated from the nucleus several Mev of negative binding energy are released. Outside the notably short range of attractive nuclear forces, however, the alpha particle's electrostatic potential energy, due to repulsion by the Coulomb field of the nucleus, far exceeds its total energy. This situation corresponds to a ball on a billiard table which is able to release energy in falling to the floor but does not do so, classically, since it has insufficient energy to surmount the barrier surrounding the table. Quantum mechanically, particles have a finite probability of penetrating finite barriers according to Eq. (6.12) and we should expect some alpha particles to penetrate the barrier illustrated in Figure 6.5.

An alpha particle with a kinetic energy of a few Mev within the nucleus has a velocity of roughly 10^9 cm/sec and strikes the nuclear wall on the order of 10^{21} times per second, since heavy nuclei have a radius of about 10^{-12} cm. Equation (6.12) gives the probability of transmission at each encounter of a particle with the nuclear surface. The thickness of the barrier is energy-

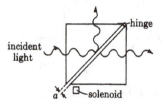

FIGURE 6.4   Illustration of the utilization of the optical analog of penetration through a square barrier in quantum mechanics. The transmitted light is modulated in intensity by varying the width (barrier thickness) of the air space between the prisms.

dependent and is very roughly 2×10^{-12} cm; the height of the barrier is of the order of 20 Mev, so that for an alpha particle having a negative binding energy of 5 Mev (i.e., an energy of 5 Mev greater than the potential energy at infinity) by use of Eq. (6.12) the probability of transmission is

$$T \sim 16 \exp \left(-\frac{4 \cdot 10^{-12}}{10^{-27}} \sqrt{2 \cdot 6.7 \cdot 10^{-24} (20 - 5) \cdot 1.6 \cdot 10^{-6}} \right)$$

$$\sim 16 \exp (-72) \qquad\qquad (6.13)$$

$$\sim 10^{-30}$$

The probability of decay per second would be

$$10^{21} \times 10^{-30} = 10^{-9}/\text{sec}$$

which corresponds roughly to a decay time of 10^9 seconds or 30 years. That the answer at all resembles the lifetimes of actual nuclei is fortuitous

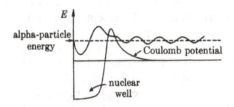

FIGURE 6.5 Illustration of total energy (— —) and potential energy (———) in alpha particle radioactive decay. The wave function is also plotted as a solid line.

in view of the enormous crudities in estimating the exponent from Eq. (6.12), but the procedure at least serves to illustrate the essential features of alpha decay. A more precise calculation (first made by Gamow and Condon) requires the use of the Coulomb barrier rather than a square barrier, and we will make a better estimate in the next chapter.

Problem 6.1 Compute the barrier penetration for electrons having a kinetic energy of 4 ev for a rectangular (one-dimensional) barrier 10 Å wide and 5 ev high. (This problem is analogous to electron tunneling in the reverse direction in a crystal rectifier.)

6.2 UNBOUND AND BOUND PARTICLES IN A SQUARE WELL

For this case, as illustrated in Figure 6.6,

$$
\begin{aligned}
x < 0 \qquad & V = 0 \\
0 < x < a \qquad & V = -V_0 \\
a < x \qquad & V = 0
\end{aligned}
$$

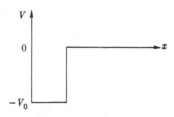

FIGURE 6.6 Illustration of square well potential.

We might examine first the case of $E > 0$. This case is similar to the case of a square barrier with $E > V_0$ except that here $p_2 = \sqrt{2m(E + V_0)}$. Again we should and do obtain transmission resonances at wavelengths equal to twice the well thickness divided by an integer, or, expressed alternatively in terms of the wavelength (h/mv), we obtain transmission resonances for wells which are an integral number of half wavelengths thick. In this case we do not have to go to optical applications such as coated lenses or a Fabry-Perot interferometer for illustration, for particle physics furnishes us with an example which is another signal triumph of the wave theory for particles. Before the development of quantum mechanics, Ramsauer observed experimentally that in the scattering of hot electrons from gas atoms, transmissions or zero-scattering resonances occurred for some atoms at particular electron energies. Instead of increasing with decreasing electron energy as expected, the total collision cross sections of the gas atoms decreased remarkably for the atoms of some elements, as illustrated in Figure 6.7. Quantum mechanics cleared up the mystery of this phenomenon. Basically, a careful, three-dimensional calculation using a screened Coulomb field can be interpreted as predicting transmission resonances whenever the effective diameter of atoms is an integral number of electron wavelengths (the wavelengths to be taken while the electron is in the screened Coulomb potential).

For the case $E < 0$, $q_1 = q_3 = \sqrt{-2mE} = \sqrt{2mW}$ where $E = -W$, and the wave functions to the left and right of the well vary exponentially. Only inside the well is a sinusoidal wave possible. Therefore, for

$$0 < x < a: \quad \psi = Be^{(i/\hbar)p_2 x} + Ce^{-(i/\hbar)p_2 x} \qquad (6.14)$$

where $p_2 = \sqrt{2m(V_0 + E)} = \sqrt{2m(V_0 - W)}$. Since ψ is finite at $\pm\infty$, for

$$x < 0: \quad \psi = Ae^{(q_1/\hbar)x}$$
$$a < x: \quad \psi = De^{-(q_1/\hbar)x} \qquad (6.15)$$

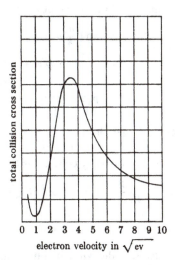

total collision cross section

0 1 2 3 4 5 6 7 8 9 10

electron velocity in $\sqrt{\text{ev}}$

FIGURE 6.7 Graph of total collision cross section of gas atoms to incident electrons having energies up to 100 ev.

representing exponentially decaying functions to the left and right. In this case there is no transmission out of the well, and we are not interested in the values of the coefficients which represent a standing wave within the well. We are interested, however, in finding any further conditions on our solutions necessary to make them acceptable solutions satisfying the usual boundary conditions. Requiring continuity of functions and their first derivatives at the boundaries, we find

$$A = B + C; \quad A = i\frac{p_2}{q_1}(B - C) \tag{6.16}$$

$$D\,e^{-(q_1 a/\hbar)} = B\,e^{(i/\hbar)p_2 a} + C\,e^{-(i/\hbar)p_2 a} \tag{6.17}$$

$$D\,e^{-(q_1 a/\hbar)} = -\,i\frac{p_2}{q_1}(B\,e^{(i/\hbar)p_2 a} - C\,e^{-(i/\hbar)p_2 a}) \tag{6.18}$$

$$B = +\frac{p_2 - iq_1}{2p_2}A; \quad C = \frac{p_2 + iq_1}{2p_2}A \tag{6.19}$$

Dividing (6.18) by (6.17) and substituting (6.19) in the result, we find that

$$1 = -i\frac{p_2}{q_1}\frac{(p_2 - iq_1)\,e^{(i/\hbar)p_2 a} - (p_2 + iq_1)\,e^{-(i/\hbar)p_2 a}}{(p_2 - iq_1)\,e^{(i/\hbar)p_2 a} + (p_2 + iq_1)\,e^{-(i/\hbar)p_2 a}} \tag{6.20}$$

which with the indicated algebraic manipulation becomes

$$\frac{q_1}{p_2} = \frac{p_2 \sin(p_2/\hbar)a - q_1 \cos(p_2/\hbar)a}{q_1 \sin(p_2/\hbar)a + p_2 \cos(p_2/\hbar)a} = \frac{\tan(p_2/\hbar)a - (q_1/p_2)}{1 + (q_1/p_2)\tan(p_2/\hbar)a} \tag{6.21}$$

Let $\varphi = \tan^{-1}(q_1/p_2)$, then Eq. (6.21) becomes

$$\tan \varphi = \tan \left(\frac{p_2}{\hbar} a - \varphi \right) \qquad (6.22)$$

This equation can be satisfied only when $\varphi = (p_2/\hbar)a - \varphi + n\pi$ where n is an integer. That is,

$$\varphi = \frac{p_2 a}{2\hbar} + \frac{n\pi}{2} \qquad (6.23)$$

Substituting Eq. (6.23) into definition of φ and using the definitions of q_1 and p_2 in terms of W and V_0, we obtain the necessary relation between E, V_0, and a in order for the boundary conditions to be satisfied:

for even n: $\qquad \sqrt{\dfrac{W}{V_0 - W}} = \tan \sqrt{2m(V_0 - W)}\,\dfrac{a}{2\hbar} \qquad (6.24a)$

for odd n: $\qquad \sqrt{\dfrac{W}{V_0 - W}} = -\cot \sqrt{2m(V_0 - W)}\,\dfrac{a}{2\hbar} \qquad (6.24b)$

Equation (6.24) illustrates the vital quantum mechanical result that *the energy of a bound particle can have only certain discrete values.* The result is similar to the familiar condition in wave motion for standing waves. For example, if a wave in a string is constrained between two fixed points, standing waves of discrete frequencies occur in the string. The reader may readily surmise that this result contains the essence of the explanation for the discrete energies possessed by electrons in atoms, as revealed by the bright-line spectra they radiate. Every bound particle can exist only in discrete energy states, similar to the constraints Planck discovered for the radiation waves in a black body. In the case of the massive planets bound to the sun, these different energy levels are so close together in value as to be totally undiscernible to any experimenter, and classical theory, which would predict a continuum for planetary motion, can thus certainly be thought of as exact. If h were much larger than it is, however, and classical physics thus did not apply, collision with another astronomical body would result in no disturbance of the earth's motion unless enough energy were exchanged for the earth to make a quantum jump to another energy state. For sub-microscopic systems, the prediction that no collisional excitation takes place unless the energy transferred is greater than some minimum value is well brought out by the Frank and Hertz experiments on electron beams in a gaseous discharge tube, discussed in Section 3.1.

Equation (6.24) results because the derivative as well as the wave function must be continuous at the boundaries of the well. Only for unique values of the energy (or the wavelength) can the wave functions inside be smoothly joined to the decaying exponential functions outside the well. Figure 6.8 illustrates the three lowest energy wave functions inside the well which satisfy the boundary conditions.

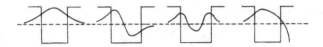

FIGURE 6.8 Illustration of the three lowest energy wave functions possible in a square well potential and a fourth impossible wave function requiring a physically unrealizable exponentially increasing wave function outside the well on the right for continuity at the well boundary.

We can apply Eq. (6.24) to find the energy binding the neutron and the proton in a deuteron. As mentioned earlier, at low energies as in this case the shape of the potential describing nuclear interactions is indistinguishable from a square well. The deuteron interaction occurs in three dimensions and thus will be treated in spherical coordinates.

The deuteron ground state is a spherically symmetric state. A spherically symmetric standing wave depends only on the radial coordinate and is given by the spherical form of Eq. (4.8). This is analogous to a spherically expanding wave in physical optics emanating from a pinhole.

$$r < a: \quad \psi = A_1[e^{(i/\hbar)p_2 r}/r] + A_2[e^{-(i/\hbar)p_2 r}/r]$$
$$r > a: \quad \psi = Be^{-q_1/\hbar r}/r \tag{6.25}$$

where a is the range of nuclear forces between neutron and proton in the deuteron, $p_2 = \sqrt{M(V_0 - W)}$, and $q_1 = \sqrt{MW}$. M is the mass of a nucleon, being nearly identical for neutron and proton, and W is the binding energy equal to $-E$, where E is negative for a bound state. Since ψ must be finite at $r = 0$, $A_1 = -A_2$, and Eq. (6.25) may be written

$$r < a: \quad u = A \sin(p_2 r/\hbar)$$
$$r > a: \quad u = Be^{-q_1 r/\hbar} \tag{6.25a}$$

where $u(r) = r\psi(r)$. Equation (6.25) may also be obtained from the Schrödinger wave equation in spherical coordinates.

In Chapter VIII it will be shown how a spherically symmetric problem may be rigorously reduced to a one-dimensional problem by making the substitution $u(r) = r(\psi)r$. Borrowing this result, we obtain the following wave equations:

$$r < a: \quad \frac{d^2u}{dr^2} + \frac{M}{\hbar^2}(V_0 - W)u = 0$$
$$r > a: \quad \frac{d^2u}{dr^2} - \frac{M}{\hbar^2}Wu = 0 \tag{6.26}$$

The solutions are as given in Eqs. (6.25a) for boundary conditions requiring ψ to be finite at the origin and at infinity.

In order for u and its derivative to be continuous at $r = a$, the derivatives with respect to r of the logarithm of u on both sides of the boundary must be equal. Therefore

$$p_2 \cot \frac{p_2 a}{\hbar} = -q_1 \qquad (6.27a)$$

that is,

$$-\sqrt{W/(V_0 - W)} = \cot(\sqrt{M(V_0 - W)}a/\hbar) \qquad (6.27b)$$

If $W \ll V_0$, as in the lowest energy state, then

$$\sqrt{MV_0}a/\hbar \sim (\pi/2) + \sqrt{W/V_0}$$

Inserting typical theoretical values into the preceding equation, $a = 2.15 \times 10^{-13}$ cm and $V_0 = 30$ Mev, we find that $-W = E = -2.3$ Mev. Actually the binding energy of the deuteron has been known for some time to be 2.22 Mev. Application of Eq. (6.27) to obtain the best V_0 and a compatible with $W = 2.22$ Mev gave some of the best early estimates of the size and range of the potential of the neutron-proton interaction. In Section 15.10 it is shown that additional, but still incomplete, information about the deuteron can be obtained by considering the neutron-proton system in an unbound state, i.e., neutron-proton scattering.

Problem 6.2 Determine the reflectivity and transmission for a particle with $E > 0$ incident from the left on a square well as illustrated in Figure 6.6.

Problem 6.3 Using Feynman probability amplitude wave functions as in Problem 4.1, determine the transmission for the example of Problem 6.1 by first summing up all the probability amplitudes for the waves (1) transmitted at both surfaces ($x = 0$ and a), (2) transmitted at the front surface ($x = 0$), reflected from the back ($x = a$), and then from the front, and finally transmitted at $x = a$, (3) transmitted at $x = a$ after four internal reflections, (4) transmitted after six internal reflections, etc., for an infinite summation. Assume that the transmission and reflection coefficients for each encounter with a surface are given correctly by Eqs. (6.5).

Problem 6.4 Find the energies available to an electron in a cube of metal 100 μ on a side if the work function for the metal is 5 ev. Idealize the problem by working in only one dimension.

Problem 6.5 For $V_0 \gg W$, Eqs. (6.24) give the approximate result that

$$W = V_0 - \frac{n^2 \pi^2 \hbar^2}{2ma^2}$$

where n is an integer. If the force exerted by a particle on the sides of a one-dimensional square well is given by $F = \partial W/\partial a$, find the force that the particle exerts on a side of the well.

Problem 6.6 Assuming the values for V_0 and a given in the text, show whether a second bound state for the deuteron, i.e., the first excited state, is possible. How can an unbound transient state arise if $E > 0$? (See Problem 6.2.)

6.3 THE LINEAR HARMONIC OSCILLATOR

The harmonic oscillator is of interest in many different fields of physics since it can often be used as a good approximation for physical systems. At this time we shall consider the one-dimensional or linear harmonic oscillator both because of its intrinsic interest and because the techniques used in the solution will be useful later on in Chapter VIII when we consider the hydrogen atom and square well problem in three dimensions.

The harmonic oscillator potential in one dimension may be written:

$$V = \frac{1}{2}kx^2$$

where k is a spring constant $= m\omega^2$ for angular frequency of oscillation ω. The Schrödinger wave equation is therefore

$$\frac{d^2\psi}{dx^2} + \frac{2m}{\hbar^2}\left(E - \frac{1}{2}kx^2\right)\psi = 0 \qquad (6.28)$$

For large enough values of $x(x^2 \gg 2E/k)$ it is readily seen that the solution for ψ will be approximately of the form

$$\psi \propto e^{\pm(\sqrt{mk}/2\hbar)x^2}$$

However, since ψ must be finite everywhere, only $e^{-(\sqrt{mk}/2\hbar)x^2}$ is acceptable.

The form of the solution of the wave equation for large x suggests that we try a solution for all x of the form

$$\psi = e^{-\xi^2/2}F(\xi) \quad \text{where} \quad \xi = \left(\frac{\sqrt{mk}}{\hbar}\right)^{1/2}x$$

In transforming variables from x to ξ we make use of the following relationships:

$$\frac{d}{dx} = \frac{d\xi}{dx}\frac{d}{d\xi} = \left(\frac{\sqrt{mk}}{\hbar}\right)^{1/2}\frac{d}{d\xi}$$

$$\frac{d^2}{dx^2} = \frac{d^2\xi}{dx^2}\frac{d}{d\xi} + \frac{d\xi}{dx}\frac{d}{dx}\left(\frac{d}{d\xi}\right) = \left(\frac{d\xi}{dx}\right)^2\frac{d^2}{d\xi^2} = \frac{\sqrt{mk}}{\hbar}\frac{d^2}{d\xi^2}$$

Equation (6.28) then becomes, after multiplication by $(\hbar/\sqrt{mk})\,e^{\xi^2/2}$,

$$\frac{d^2F}{d\xi^2} - 2\xi\frac{dF}{d\xi} + \left(\frac{2E}{\hbar}\sqrt{\frac{m}{k}} - 1\right)F = 0 \qquad (6.29)$$

The form of this equation further suggests that it may be solved by the power series method in which we set

$$F(\xi) = \sum_{i=0}^{\infty} a_i \xi^i$$

With this substitution

$$\frac{d^2F}{d\xi^2} = 2a_2 + 3\cdot2a_3\xi + \cdots + (i+2)(i+1)a_{i+2}\xi^i + \cdots$$

$$= \sum_{j=0}^{\infty} (j+2)(j+1)a_{j+2}\xi^j$$

where $j = i + 2$

$$-2\xi\frac{dF}{d\xi} = -2a_1\xi - 4a_2\xi^2 \cdots -2ia_i\xi^i \cdots = \sum_{j=0}^{\infty} -2ja_j\xi^j$$

where $j = i + 1$

$$\left(\sqrt{\frac{m}{k}}\frac{2E}{\hbar} - 1\right)F = \left(\sqrt{\frac{m}{k}}\frac{2E}{\hbar} - 1\right)a_0 + \left(\sqrt{\frac{m}{k}}\frac{2E}{\hbar} - 1\right)a_1\xi$$

$$+ \cdots \left(\sqrt{\frac{m}{k}}\frac{2E}{\hbar} - 1\right)a_i\xi^i + \cdots = \sum_{j=0}^{\infty} \left(\sqrt{\frac{m}{k}}\frac{2E}{\hbar} - 1\right)a_j\xi^j$$

where $j = i$
or substituting back in Eq. (6.29) we obtain

$$\sum_{j=0}^{\infty} (j+2)(j+1)a_{j+2}\xi^j - 2ja_j\xi^j + \left(\sqrt{\frac{m}{k}}\frac{2E}{\hbar} - 1\right)a_j\xi^j = 0 \qquad (6.30)$$

If Eq. (6.30) is to be true for all $\xi(=x(\sqrt{mk}/\hbar)^{1/2})$, it must hold independently for all coefficients of the same power of ξ. Therefore, in order for the coefficients of the various powers of ξ to vanish, we must have

$$2a_2 + \left(\sqrt{\frac{m}{k}}\frac{2E}{\hbar} - 1\right)a_0 = 0$$

$$3\cdot2a_3 - 2a_1 + \left(\sqrt{\frac{m}{k}}\frac{2E}{\hbar} - 1\right)a_1 = 0$$

or in general

$$(j+2)(j+1)a_{j+2} + \left(\sqrt{\frac{m}{k}}\frac{2E}{\hbar} - 1 - 2j\right)a_j = 0$$

so

$$a_{j+2} = \frac{2j+1 - \sqrt{\frac{m}{k}}\frac{2E}{\hbar}}{(j+2)(j+1)}a_j \qquad (6.31)$$

Equation (6.31) is called a *recursion* formula because it allows a_{j+2} to be calculated, once a_j is known. To completely specify all coefficients a_j we need to know only a_0 and a_1 and we can have a series in either even or odd powers of ξ depending on initial choices for a_0 or a_1.

The series diverges for infinite j, behaving like e^{ξ^2} at large ξ, so that ψ will be of the form $e^{\xi^2/2}$ which is not finite and hence not a physically acceptable solution. However, we can avoid this difficulty by making the series finite, i.e., we must terminate the series at some particular term. The product of the finite series and $e^{-\xi^2/2}$ then will vanish at $\xi \to \infty$, thereby becoming an acceptable solution for ψ.

The series terminates when any particular $a_j = 0$. (The recursion formula then shows that all succeeding terms must also vanish.) From the recursion formula we see that the condition for termination of the series is

$$2j + 1 = \sqrt{\frac{m}{k}}\,\frac{2E}{\hbar}$$

$$\text{or} \quad E = \sqrt{\frac{k}{m}}\,\hbar\left(j + \frac{1}{2}\right)$$

or in customary notation $E_n = \hbar\omega(n + \frac{1}{2})$, $n = 0, 1, 2, 3, \ldots$

From this result we see that by causing the series to terminate in order to make the resulting ψ a physically acceptable solution, we have automatically imposed certain restrictions on the allowed energy values for the system. This result is characteristic of all solutions for problems in quantum mechanics in which a particle is confined by a potential to a limited region of space.

When E equals one of the foregoing values for E_n the series for a_j becomes a polynomial for one of the two possible coefficients a_0 or a_1. Since the series can be terminated for one series starting with a_0 or a_1, but not both, by setting the other coefficient (a_1 or a_0) equal to zero we then can always obtain a particular physically acceptable solution appropriate for the E_n in question. The polynomials F_n which result are known as *Hermite polynomials*. A few of these Hermite polynomials are listed in Table 6.1.

The solution for a particular acceptable value of the energy E_n is:

$$\psi_n = F_n(\xi)e^{-\xi^2/2}$$

or changing back from ξ to x

$$\psi(x) = F_n\left(\left(\frac{\sqrt{mk}}{\hbar}\right)^{1/2} x\right)e^{-(\sqrt{mk}/2\hbar)x^2}$$

Using the properties of the Hermite polynomials, the normalization integral is found to be $= \sqrt{\pi\hbar/m\omega}\,2^n n!$ so that the normalized wave functions become

$$\psi_n(x) = \left(\frac{m\omega}{\hbar\pi}\right)^{\frac{1}{4}} \frac{1}{\sqrt{2^n\,n!}}\, e^{-m\omega x^2/2\hbar}\, F_n\left(\sqrt{\frac{m\omega}{\hbar}}x\right)$$

where in the foregoing expressions k has been replaced by $m\omega^2$.

TABLE 6.1

$$F_0(\xi) = 1$$
$$F_1(\xi) = 2\xi$$
$$F_2(\xi) = 4\xi^2 - 2$$
$$F_3(\xi) = 8\xi^3 - 12\xi$$
$$F_4(\xi) = 16\xi^4 - 48\xi^2 + 12$$
$$F_5(\xi) = 32\xi^5 - 160\xi^3 + 120\xi$$

Problem 6.7 Express $f(x) = e^{-2x^2}$ as a sum of harmonic oscillator wave functions in one dimension using the first five terms.

Problem 6.8 A linear oscillator of mass m and angular frequency ω has its state represented by

$$\psi = e^{-m\omega x^2/2\hbar}$$

What are the relative probabilities that its energy will be measured as $\hbar\omega/2$, $3\hbar\omega/2$, and $5\hbar\omega/2$?

Problem 6.9 A linear oscillator of mass m and angular frequency ω has its state represented by

$$\psi = e^{-m\omega x^2/2\hbar}$$

What is the probability that its momentum is larger than $\sqrt{\hbar m\omega}$?

BIBLIOGRAPHY

Rojansky, Vladimir, *Introductory Quantum Mechanics*. Englewood Cliffs, N.J.: Prentice-Hall, Inc., 1946.

Bohm, David, *Quantum Theory*. Englewood Cliffs, N.J.: Prentice-Hall, Inc., 1951.

Fong, Peter, *Elementary Quantum Mechanics*. Reading, Mass.: Addison-Wesley Publishing Company, Inc., 1962.

Bethe, Hans A. and Phillip Morrison, *Elementary Nuclear Theory*. New York: John Wiley & Sons, Inc., 1956.

VII

WKB APPROXIMATION AND THE BOHR-SOMMERFELD QUANTUM CONDITIONS

In the last chapter, to emphasize the nonclassical aspects of particle behavior revealed by quantum mechanics, we treated the extreme case of potentials which change discontinuously along a particle path, i.e., square potentials. Since classical physics is known to be an excellent approximation in many instances, it is instructive to see how classical mechanics may arise as an approximation to our more general theory. The wave behavior of particles has been shown to arise from the Heisenberg uncertainty principle, $\Delta p \Delta x \geqslant \hbar$, which would agree with classical physics if $\hbar = 0$. This suggests that we should look for a series expansion in powers of $\hbar$ in our probability amplitude solutions to the wave equation. We should further expect this method of approximating wave equation solutions to be very useful when the potential energy of the particle changes slowly within a particle wavelength, in contrast to the discontinuous changes treated in the previous chapter.

Up to this point we have examined solutions of the wave equation in regions where V (and therefore p) was a constant. In these circumstances the particle wavelength could be and was unique. This was possible only because no questions were asked about the particle position. In regions where the potential energy is a function of position, however, the wave function clearly could not be written as a simple plane wave with unique momentum. The wave function $\exp\{(i/\hbar)\int p\,dx\}$ corresponds to a sort of frequency-modulated wave; and in a slowly varying potential the sum of this wave function over all possible paths might be approximated by the constant potential wave form times a more slowly varying fudge factor. We will now show that just such an approximate wave function will result from an attempt to expand our wave function in increasing powers of $\hbar$.

7.1 THE WENTZEL-KRAMERS-BRILLOUIN METHOD

Letting $p = \sqrt{2m(E - V)}$, we can rewrite the one-dimensional time-independent wave equation as

$$\frac{d^2\psi}{dx^2} + \frac{p^2}{\hbar^2}\,\psi = 0 \tag{7.1}$$

Since p is now a function of position, the solution is no longer of the form $e^{(i/\hbar)px}$, but this form will be shown to be a reasonable first approximation when certain conditions are met. We first choose a new dependent variable u:

$$u = \frac{\hbar}{i}\frac{1}{\psi}\frac{d\psi}{dx}, \qquad \psi = \exp\left(\frac{i}{\hbar}\int^x u\,dx\right) \tag{7.2}$$

so that Eq. (7.1) becomes

$$\frac{\hbar}{i}\frac{du}{dx} = p^2 - u^2 \tag{7.3}$$

We seek solutions whose first approximations give the classical results, by expanding u in powers of $\hbar/i$,

$$u = u_0 + \frac{\hbar}{i}u_1 + \left(\frac{\hbar}{i}\right)^2 u_2 + \dots \tag{7.4}$$

Substituting (7.4) into (7.3) and keeping terms of order no higher than $\hbar/i$ we obtain

$$\frac{\hbar}{i}\frac{du_0}{dx} = p^2 - u_0^2 - 2\frac{\hbar}{i}u_1 u_0 \tag{7.5}$$

The zero-order approximation amounts to neglecting $\hbar\,du/i\,dx$ in Eq. (7.3). Equating coefficients of equal powers of $\hbar/i$ we obtain

$$u_0 = \pm p, \qquad u_1 = -\frac{1}{2}\frac{1}{u_0}\frac{du_0}{dx}$$

leading to the two alternative solutions

$$u_+ = p - \frac{\hbar}{i}\frac{1}{2p}\frac{dp.}{dx} = p - \frac{\hbar}{i}\frac{d}{dx}\log_e\sqrt{p}$$

$$u_- = -p - \frac{\hbar}{i}\frac{1}{2p}\frac{dp}{dx} = -p - \frac{\hbar}{i}\frac{d}{dx}\log_e\sqrt{p}$$

(7.6)

Substituting Eq. (7.6) into Eq. (7.2) we find the approximate wave function

$$\psi = \frac{A}{\sqrt{p}}\exp\left(\frac{i}{\hbar}\int_{x_1}^{x} p\,dx\right) + \frac{B}{\sqrt{p}}\exp\left(-\frac{i}{\hbar}\int_{x_1}^{x} p\,dx\right)$$

(7.7)

In the limit of short wavelength, i.e., large momenta, Eq. (7.7) resembles a plane wave with phase given by the action $\int p\dot{q}\,dt = \int p\,dq$ in units of $\hbar$. As $\hbar/p$ approaches zero, the wave function given in Eqs. (7.2, 7.7) becomes a rapidly oscillating function of x so that noncancelling contributions to the total wave function from neighboring paths occur only in that region of x where the action varies least rapidly with change in path. This region, where the action is an extremum, is also the region for which Hamilton's principal function, $\int L\,dt$ (where L is the Lagrangian, Chapter IV), is an extremum. The fact that a light ray follows that path for which the action is also either a minimum or a maximum is derived from *Fermat's principle* in optics. It is the classical path taken by a particle or a light ray. The Wentzel–Kramers–Brillouin or WKB method has given us this solution plus additional terms beyond the classical expression, of which we have taken only the first. This additional term in the exponent results in a factor $p^{-1/2}$ in the wave function. With this change, the probability of finding a particle per unit volume $\psi^*\psi$ becomes $1/p$ times the constant-potential probability. The time a particle spends in any one spot is inversely proportional to its velocity, and the first-order correction to the plane wave case has given us just this result, which is also entirely in accord with classical mechanics. For example, the bob in a clock pendulum is most likely to be found at any given instant at the end of its swing where its velocity is zero.

The series expansion expressed in Eq. (7.4) is an asymptotic series which diverges after the first few terms. The first few terms give a good approximation, however, as long as

$$\frac{\hbar}{p^2}\frac{dp}{dx} \ll 1$$

This condition, which is seen to follow from requiring that the second term in Eq. (7.6) be much smaller than the first term, is the usual condition for validity in ray optics or classical physics: namely, that the wave number of the wave function ($\hbar/p$ for particles, ω/nc for light where n is the index of refraction) experience little change over one wavelength. Put more precisely, the condition is that the fractional change in a wavelength within a distance equal to the wavelength be much less than unity. As the wavelength grows larger, that is, as p approaches zero, u_1, u_2, and higher terms in the expansion all diverge, as is indicated in Eq. (7.6), and the approximation is no longer valid; in this case, in general, one does better to go to the other extreme and seek a long-wavelength or physical-optics type of solution to the Schrödinger equation. That is, it is preferable to solve a square well or barrier problem as a first approximation when the potential change is large within a wavelength.

We now apply our method to the potential illustrated in Figure 7.1, for which $p = \sqrt{2m(E - V)}$ is imaginary in regions I and III and real in region II. Since ψ must be finite everywhere and p_1 and p_{III} are imaginary, u_+ is an unacceptable solution for $x < x_1$ and u_- is unacceptable for $x > x_2$. Only exponentially decreasing solutions are possible in regions I and III, for which, respectively, $A_I = 0$ and $B_{III} = 0$. In region II both positive and negative exponential solutions are acceptable and may be combined and rewritten as

$$\psi_{II} = \frac{2C_{II}}{\sqrt{p}} \cos\left(\frac{1}{\hbar} \int_{x_1}^{x} p\, dx + \delta\right) \tag{7.8}$$

with

$$A_{II} = C_{II} e^{i\delta} \quad \text{and} \quad B_{II} = C_{II} e^{-i\delta}$$

whereas

$$\psi_I = (B_I/\sqrt{ip}) \exp\left(-\frac{i}{\hbar} \int_x^{x_1} p\, dx\right), \quad \psi_{III} = (A_I/\sqrt{ip}) \exp\left(\frac{i}{\hbar} \int_{x_2}^{x} p\, dx\right)$$

Conditions will be imposed on these solutions in the next section.

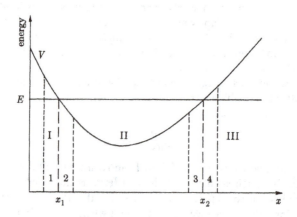

FIGURE 7.1   Energy graph for a particle with energy E bound in a potential well for which $E < V$ in $x < x_1$ and $x > x_2$.

7.2 CONNECTION FORMULAS

The solutions in the three regions must now be matched at x_1 and x_2 by the usual conditions of continuity. The difficulty is that at the points x_1 and x_2, $p = 0$, and Eq. (7.7) becomes an unsatisfactory approximation to the wave equation. At these points V may be approximated as a linear function of x by the first term in a Taylor's series expansion of V about the points $x = x_1$ and $x = x_2$. Hence the wave equation near x_1 may be written

$$\frac{d^2\psi}{dx^2} + n(x - x_1)\psi = 0 \qquad (7.9)$$

where n is $2m/\hbar^2$ times the negative slope of the potential at x_1, that is, $(2m/\hbar^2)$ $(-\partial V/\partial x)$, where $-\partial V/\partial x$ is evaluated at $x = x_1$. The solution of Eq. (7.9) is exact, and to the right of x_1 it is

$$\psi = D\sqrt{x - x_1}J_{1/3}(\tfrac{2}{3}\sqrt{n}[x - x_1]^{3/2}) + E\sqrt{x - x_1}J_{-1/3}(\tfrac{2}{3}\sqrt{n}[x - x_1]^{3/2})$$

$$(7.10)$$

where $J_{1/3}$ is the Bessel function of order $\tfrac{1}{3}$. A graph of $J_{1/3}(x)$ and $J_{-1/3}(x)$ is shown in Figure 7.2.*

As x increases without limit, Eq. (7.10) asymptotically approaches a cosine function:

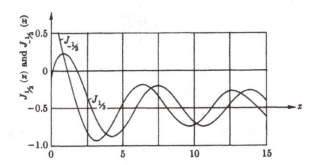

FIGURE 7.2 Graph of $J_{1/3}(x)$ and $J_{-1/3}(x)$. Note that for large x both functions become sinusoidal with slowly decreasing amplitude and with period equal to 2π. (From Jahnke, Emde, and Lösch, *Tables of Higher Functions*, Stuttgart: B. G. Teubner Verlagsgesellschaft, 1960).

---

* For a description of Bessel functions see a calculus text such as *Methods of Advanced Calculus*, P. Franklin, McGraw-Hill, N.Y. (1944) or *Applied Mathematics for Engineers and Physicists*, L. A. Pipes, McGraw-Hill, N.Y. (1946).

$$\psi \simeq \sqrt{\frac{3}{\pi n^{1/2}}} \frac{D}{(x - x_1)^{1/4}} \cos\left[\frac{2}{3}\sqrt{n}\,(x - x_1)^{3/2} - \frac{5\pi}{12}\right]$$
$$+ \sqrt{\frac{3}{\pi n^{1/2}}} \frac{E}{(x - x_1)^{1/4}} \cos\left[\frac{2}{3}\sqrt{n}\,(x - x_1)^{3/2} - \frac{\pi}{12}\right] \qquad \textbf{(7.11a)}$$

To the left of x_1 the same solution (7.10) results except that the argument of the Bessel function is imaginary; the asymptotic expansion for large absolute value of $(x - x_1)$ includes positive and negative exponentials:

$$\psi \simeq \sqrt{\frac{3}{4\pi n^{1/2}}} \frac{D}{(x - x_1)^{1/4}} [e^{(2/3)\sqrt{n}(x_1-x)^{3/2}} + e^{-i\pi/6}e^{-(2/3)\sqrt{n}(x_1-x)^{3/2}}]$$
$$- \sqrt{\frac{3}{4\pi n^{1/2}}} \frac{E}{(x - x_1)^{1/4}} [e^{(2/3)\sqrt{n}(x_1-x)^{3/2}} + e^{-5i\pi/6}e^{-(2/3)\sqrt{n}(x_1-x)^{3/2}}]$$

$$\textbf{(7.11b)}$$

These asymptotic expressions (for the linear potential approximation at the turning points) must fit the solutions we have already found for regions I, II, and III. In order that ψ be zero at $-\infty$, the WKB solution in region I must be

$$\psi_I = (B_I/\sqrt{-ip}) \exp\left(-\frac{i}{\hbar}\int_{x_1}^{x} p\,dx\right) \qquad \textbf{(7.12)}$$

where

$$p = \sqrt{2m(E - V)}$$

This solution matches Eq. (7.11b) if

$$D = E = \frac{\sqrt{4\pi/3\hbar}}{2\sin(\pi/3)} B_I$$

since

$$e^{-i\pi/6} - e^{-5i\pi/6} = e^{-i\pi/2}\, 2i\sin(\pi/3)$$

For this value of D and E Eq. (7.11a) becomes

$$\psi = \frac{2}{[n\hbar^2(x - x_1)]^{1/4}} \frac{\cos(\pi/6)}{\sin(\pi/3)} B_I \cos\left[\frac{2}{3}\sqrt{n}\,(x - x_1)^{3/2} - \frac{\pi}{4}\right] \qquad \textbf{(7.13)}$$

since

$$\cos\left(a - \frac{5\pi}{12}\right) + \cos\left(a - \frac{\pi}{12}\right) = 2\cos\left(a - \frac{\pi}{4}\right)\cos\frac{\pi}{6}$$

Equation (7.8) matches Eq. (7.13) if $C_{II} = B_I$, giving

$$\psi_{II} = \frac{2B_I}{\sqrt{p}} \cos\left(\frac{1}{\hbar}\int_{x_1}^{x} p\,dx - \frac{\pi}{4}\right) \qquad \textbf{(7.14)}$$

Similarly, if

$$\psi_{III} = \frac{A_{III}}{\sqrt{-ip}} \exp\left(\frac{i}{\hbar}\int_{x_1}^{x} p\,dx\right) \qquad \textbf{(7.15)}$$

then

$$\psi_{\mathrm{II}} = \frac{2A_{\mathrm{III}}}{\sqrt{p}} \cos \left(\frac{1}{\hbar} \int_x^{x_2} p \, dx - \frac{\pi}{4} \right) \qquad (7.16)$$

If Eqs. (7.14, 7.16) are to be the same solution, then $B_{\mathrm{I}} = A_{\mathrm{III}}$ and the absolute values of the arguments of the cosine must differ only by an integral multiple of π. That is,

$$\left| \frac{1}{\hbar} \int_{x_1}^x p \, dx - \frac{\pi}{4} \right| = \left| \frac{1}{\hbar} \int_x^{x_2} p \, dx - \frac{\pi}{4} \right| + n'\pi$$

or

$$\frac{1}{\hbar} \int_{x_1}^x p \, dx - \frac{\pi}{4} = \pm \left(\frac{1}{\hbar} \int_x^{x_2} p \, dx - \frac{\pi}{4} \right) + n\pi \qquad (7.17)$$

Since this condition must be satisfied for all x the minus sign must be taken, and hence the permissible values of p and therefore E are given by

$$\frac{1}{\hbar} \int_{x_1}^{x_2} p \, dx = \left(n + \frac{1}{2} \right) \pi \qquad (7.18)$$

That is, the energy is quantized in accordance with Eq. (7.18); only certain discrete values are possible. As stated previously, *only discrete values of the energy are ever possible for particles confined to a finite volume.*

The important consequence of the WKB method of solving the Schrödinger equation for a bound state problem is that simple quantum conditions are imposed on the possible energy states of the system. A simple formula, Eq. (7.18), has been found for obtaining them. Note that this result is valid only when the potential does not change much within a wavelength and in cases where we may obtain solutions far from the points where $p = 0$. This means, as we should expect, that the approximation is valid only for slowly varying potentials and in the limit of large quantum numbers where the classical result $\Delta E/E = 0$ i.e., no uncertainty in the energy, is a reasonably good first approximation.

Equation (7.18) may be rewritten

$$\oint p \, dx = (n + \tfrac{1}{2})h \qquad (7.19)$$

since twice the path integral from x_1 to x_2 is equivalent to a complete cycle of the particle motion in its oscillation between the classical limits of its motion. This approximate quantization rule (for p and thus E) is identical, but for the extra term $h/2$, with the formula Bohr and Sommerfeld postulated a decade before the formulation of wave mechanics to predict the energy levels possible for a bound particle. For low quantum numbers the extra term $h/2$ gives a better approximation.

As originally stated, the Bohr-Sommerfeld rule was intended to apply to any periodic motion, including rotation, whereas the above derivation applies only to vibrational motion. The corresponding rule for rotation is that the integral of the angular momentum over a complete cycle must yield an integer, not half an integer, since V and hence p_θ must have an angular period of 2π for rotational motion, where p_θ is everywhere real. That is, for rotational motion where θ is the angular displacement and p_θ the angular momentum

$$\oint p_\theta \, d\theta = nh \tag{7.19a}$$

Problem 7.1:   In the next chapter it will be shown that for central (angle-independent) potentials, the wave equation for the wave function for the radial coordinate may be written

$$\frac{d^2u}{dr^2} + \frac{2m}{\hbar^2}\left[E - V(r) - \frac{\hbar^2}{2m}\frac{l(l+1)}{r^2}\right]u = 0 \tag{8.29}$$

where l is an integer and u is r times the radial wave function. ($\hbar^2 l(l+1)$ is the eigenvalue of the square of the angular momentum operator, and

$$\frac{\hbar^2}{2m}\frac{l(l+1)}{r^2}$$

represents the rotational energy of a particle about the origin.) Using the WKB method (that is, the quantization condition resulting therefrom), compute the number of energy levels per unit energy available to a nucleon for a particular value of $l = L$ in a nuclear square well of radius 10^{-12} cm and well depth 42 Mev. (Hint: The radial component of the momentum is given by $p_r^2 = 2m[E - V(r) - \hbar^2 l(l+1)/2mr^2]$.)

Problem 7.2:   Using the WKB approximation, find the permissible levels for hydrogen for a particular value of $l = L$. The central potential is $V(r) = -e^2/r$. What is the total number of bound-state levels as a result of the great extent of the Coulomb potential?

Problem 7.3:   Using the WKB approximation, find the permissible levels for a harmonic oscillator in three dimensions, with $V(r) = +\frac{1}{2}Kr^2$, for a particular value of $l = L$.

Problem 7.4:   By the WKB method, find the energy levels in a one-dimensional well with $V(x) = C|x|$.

7.3 PENETRATION THROUGH A BARRIER

We now turn our attention to another historical application of the WKB approximation. Instead of the valley shown in Figure 7.1 we now have the hill of Figure 7.3. As in the derivation of Eq. (7.18), we first find that to the right of the barrier (which we label region III) $B_{III} = 0$, with ψ_{III} representing a wave traveling to the right. The wave function inside the barrier ψ_{II} is made up of exponentials of real positive and negative argument which

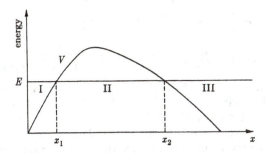

FIGURE 7.3 Energy graph for a particle with energy E incident from the left on a barrier for which $E < V$ for $x_1 < x < x_2$.

must be matched to the incident and reflected wave in region I (see Chapter 12 of Bohm, cited here in Chapter VI). Just as in Chapter VI, the transmission through the barrier is found from the ratio of the probability or intensity $\psi^*\psi$ of the transmitted wave to the intensity of the incident wave times the ratio of the velocity in region III to the velocity in region I. Since the velocities are roughly equal, the result is that

$$T = \exp - \left(\frac{2}{\hbar} \int_{x_1}^{x_1} p \, dx \right) \tag{7.20}$$

where

$$p = \sqrt{2m(V - E)}$$

Note that except for the refractive-index-like factor $(q_2/p_1 + p_1/q_2)^{-2}$ in front, which is negligible for the large values of the exponent that are required for the validity of the WKB approximation, Eq. (7.20) is similar to Eq. (6.12) used in calculating alpha particle decay. We should expect Eq. (7.20) to give a much better estimate of alpha particle half-lives than Eq. (6.12). We may approximate the barrier potential by a simple repulsive Coulomb potential beyond the attractive nuclear square well potential. That is, for an alpha particle of charge $2e$ and a residual nucleus of charge Ze,

$$V = 2Ze^2/r, \qquad r > x_1 \tag{7.21}$$

where the radius of the alpha particle-nucleus attraction is given by

$$x_1 = (1.3 + 1.35A^{1/3}) \cdot 10^{-13} \text{ cm}$$

where A is the atomic mass number of the residual nucleus.* The point at which the alpha particle energy exceeds the Coulomb potential is

$$x_2 = 2Ze^2/E$$

* J. O. Rasmussen, *Revs. Modern Phys.*, **30**, 424 (1958).

For these values of V, x_1, and x_2, Eq. (7.20) becomes

$$T = \exp\left\{ -\frac{2}{\hbar} \int_{x_1}^{x_2} \sqrt{2m[(2Ze^2/r) - E]}\, dr \right\}$$

$$= \exp\left\{ \frac{-8mZe^2}{\hbar\sqrt{2mE}} \left[\tan^{-1}\sqrt{\frac{x_2}{x_1} - 1} - \sqrt{\frac{x_1}{x_2}\left(1 - \frac{x_1}{x_2}\right)} \right] \right\} \qquad (7.22)$$

In the case of Po^{210}, $T \simeq e^{-64} \simeq 10^{-27}$. If the estimated frequency of inter-action with the walls derived in Chapter VI, 10^{21}/sec, is used, a half-life of $(1/T) \times 10^{-21} = 10^6$ sec results. This is to be compared with the measured half-life of 138 days or $\sim 10^7$ sec. Since the calculation is approximate (the estimated frequency of interaction with the wall is crude, and we have neglected the nuclear well shape and the reduction in transmission due to refractive-index-type factors), the results may be regarded as in rather good agreement with the observed half-life.

Equation (7.20) has another similar application to the theory of the Schottky effect, i.e., electron emission from metals under an applied potential. For this case $V = W - eEx$ for $x > 0$ where E is the applied electric field, x is the distance measured from the surface of the metal, and W is the work function for the metal. From (7.20), the rate of electron evaporation is thus proportional to

$$T = \exp\left[-\frac{2}{\hbar} \int_0^{x_2} \sqrt{2m(W - eEx)}\, dx \right] = \exp\left[-\frac{4}{3}\sqrt{2m}\,\frac{W^{3/2}}{\hbar eE} \right]$$

$$(7.23)$$

where x_2 is the point where $W = eEx$. Cold emission of electrons from metals exhibits just such a sensitive dependence on the work function of the metal and the applied electric field.

Problem 7.5:   What are the half-lives of U^{238}, Np^{235}, and Po^{213}? Consult the *General Electric Chart of the Nuclides* for alpha particle energies, atomic mass numbers, and observed half-lives. Note that even if the absolute value of the result is only approximate, the ratio of one half-life to another fits the observed values rather well.

Problem 7.6:   What are the barrier penetrabilities of Li, Fe, and Pb for 10-Mev protons, of Fe for 2-Mev protons, and of Li, Fe, Pb, for 10-Mev deuterons and alpha particles? [Compare the estimates of the penetrability with the curves of P. Morrison (see bibliography of this chapter).]

Problem 7.7:   If the work function in a metal is 3 ev and the applied electric fields are 10^3, 10^4, and 10^5 volts/m, what is the cold emission current for each case if 10^{28} conduction electrons arrive per cm^2-sec at the surface?

Problem 7.8:   What is the penetrability of a particle through a barrier $V = V_0 - c|x|$?

Problem 7.9: What is the penetrability of an alpha particle with energy E through a nuclear square well and Coulomb barrier if the particle has an angular momentum of $10\hbar$ (*i.e.*, $l = 10$)? (See Problem 7.1.)

Problem 7.10: Ordinarily the cold emission of metals is so affected by surface roughness that the simple WKB calculation is adequate. However, the problem can be solved exactly. On the side of the barrier where the potential changes rapidly, refractive-index-type factors result which may significantly improve the result. Obtain the exact expression for the transmission probability of an electron with wave number $k = p/\hbar$ incident from the left on a barrier changing abruptly at $x = -d$ from a constant value $-V_0$ for $-d > x$ to $V = -Ax$ for $-d < x(E = 0)$.

BIBLIOGRAPHY

Persico, E. (trans. by G. Temmer), *Fundamentals of Quantum Mechanics*. Englewood Cliffs, N.J.: Prentice-Hall, Inc., 1946.

Schiff, L., *Quantum Mechanics*. 3rd edition. New York: McGraw-Hill Book Co., 1968.

Morrison, P., in *Experimental Nuclear Physics*, Vol. 2, ed. E. Segré. New York: John Wiley & Sons, Inc., 1953. See pp. 102, 104, and 198.

VIII

ELEMENTARY THREE-DIMENSIONAL WAVE FUNCTIONS IN SPHERICAL COORDINATES

In the previous chapters we restricted our attention to problems in one dimension, because of their mathematical simplicity. Relatively few problems can be treated realistically with this simplification, however, and to understand and be able to solve most of the basic and even elementary problems in atomic and nuclear physics it is necessary to pay attention to the analysis of problems in three dimensions. In doing so we will immediately encounter differential equations different from any we have met previously, with solutions in terms of new functions which we must derive. The solutions for the fundamental bound-particle problems treated in this chapter (the rigid rotator, the spherically symmetric square well, the harmonic oscillator, and Coulomb potentials) are simple polynomials, which we will derive in detail because of their recurring importance in the study of many other more complex problems. Indeed, these polynomial functions are the basic wave functions for the analysis of the more complicated problems.

8.1 GENERAL METHOD FOR SPHERICALLY SYMMETRIC POTENTIALS

A great many problems in physics, having to do with the atom, the nucleus, or nuclear interactions, for example, may be calculated exactly by a spherically symmetric potential, that is, by a potential function $V(r)$ which depends only on r and does not involve any angle variables. For this important class of problems, the three-dimensional time-independent wave function may be separated into three product wave functions,

$$\psi(r, \theta, \varphi) = R(r)\Theta(\theta)\Phi(\varphi) \tag{8.1}$$

The time-independent Schrödinger wave equation in three dimensions is in this case

$$\nabla^2\psi + \frac{2m}{\hbar^2}[E - V(r)]\psi = 0 \tag{8.2}$$

where ∇^2, the Laplacian operator in spherical coordinates, is given by

$$\nabla^2\psi = \frac{1}{r^2}\frac{\partial}{\partial r}\left(r^2\frac{\partial\psi}{\partial r}\right) + \frac{1}{r^2\sin\theta}\frac{\partial}{\partial\theta}\left(\sin\theta\frac{\partial\psi}{\partial\theta}\right) + \frac{1}{r^2\sin^2\theta}\frac{\partial^2\psi}{\partial\varphi^2} \tag{8.3}$$

We proceed as in Chapter I, treating the radiation in a rectangular box, and substitute Eqs. (8.3, 8.1) into Eq. (8.2) and multiply (8.2) by $1/\psi$, to find

$$\frac{1}{R}\frac{1}{r^2}\frac{\partial}{\partial r}\left(r^2\frac{\partial R}{\partial r}\right) + \frac{1}{\Theta r^2\sin\theta}\frac{\partial}{\partial\theta}\left(\sin\theta\frac{\partial\Theta}{\partial\theta}\right)$$
$$+ \frac{1}{\Phi}\frac{1}{r^2\sin^2\theta}\frac{\partial^2\Phi}{\partial\varphi^2} + \frac{2m}{\hbar^2}[E - V(r)] = 0 \tag{8.4}$$

On multiplying all terms in Eq. (8.4) by $r^2\sin^2\theta$ and placing the φ-dependent term on the right, we find that while the left-hand side contains no φ dependence, the right-hand side contains no r or θ dependence, and hence

$$\frac{1}{\Phi}\frac{\partial^2\Phi}{\partial\varphi^2} = -m^2 \tag{8.5}$$

where m is an arbitrary constant. (For reasons which will appear later, m is the magnetic quantum number; it must not be confused with the mass of a particle, also written m.)

The solution of Eq. (8.5) may be recognized to be

$$\Phi = Ae^{im\varphi} + Be^{-im\varphi} \tag{8.6}$$

In order for Φ to be single-valued (i.e., to have the same value at $\varphi + 2\pi$ as at φ), m must be an integer. By substituting Eq. (8.5) into Eq. (8.4) and multiplying (8.4) by r^2 we obtain

$$\frac{1}{R}\frac{\partial}{\partial r}\left(r^2\frac{\partial R}{\partial r}\right) + \frac{2m}{\hbar^2}r^2[E - V(r)]$$

$$= -\frac{1}{\Theta \sin \theta}\frac{\partial}{\partial \theta}\left(\sin \theta \frac{\partial \Theta}{\partial \theta}\right) + \frac{m^2}{\sin^2 \theta} = K \tag{8.7}$$

where K, like m, is an arbitrary constant independent of r and θ.

Note that $\Theta(\theta)$ does *not* depend on the sign but only on the magnitude or absolute value of m. Letting $x = \cos \theta$ and noting that

$$\frac{\partial}{\partial \theta} = -\sqrt{1 - x^2}\frac{\partial}{\partial x}$$

we find

$$\frac{\partial}{\partial x}\left[(1 - x^2)\frac{\partial \Theta}{\partial x}\right] + \left(K - \frac{m^2}{1 - x^2}\right)\Theta = 0 \tag{8.8}$$

Let

$$\Theta = (1 - x^2)^{|m|/2}F(x) \tag{8.9}$$

then

$$\frac{\partial \Theta}{\partial x} = -|m|x(1 - x^2)^{(|m|/2-1)}F(x) + (1 - x^2)^{|m|/2}\frac{\partial F}{\partial x}$$

$$\frac{\partial}{\partial x}\left[(1 - x^2)\frac{\partial \Theta}{\partial x}\right] = (m^2x^2 - |m| + |m|x^2)(1 - x^2)^{(|m|/2-1)}F(x)$$

$$-2x(|m| + 1)(1 - x^2)^{|m|/2}\frac{\partial F}{\partial x} + (1 - x^2)^{(|m|/2+1)}\frac{\partial^2 F}{\partial x^2}$$

Making this substitution into Eq. (8.8) and dividing the result by

$$(1 - x^2)^{|m|/2}$$

we obtain a differential equation for F:

$$(1 - x^2)\frac{d^2F}{dx^2} - 2(|m| + 1)x\frac{dF}{dx} + [K - |m|(|m| + 1)]F = 0 \tag{8.10}$$

To find a satisfactory solution of Eq. (8.10) we try expanding $F(x)$ in a series of ascending powers of x,

$$F(x) = \sum_{i=0}^{\infty} a_i x^i \tag{8.11}$$

and substitute Eq. (8.11) into Eq. (8.10) to obtain

$$(1 - x^2)\sum_{i=2}^{\infty} i(i - 1)a_i x^{i-2} - 2(|m| + 1)\sum_{i=0}^{\infty} ia_i x^i$$

$$+ [K - |m|(|m| + 1)]\sum_{i=0}^{\infty} a_i x^i = 0 \tag{8.12}$$

In order for Eq. (8.12) to hold for any arbitrary value of x it must hold separately for each power of x in the infinite summations. That, is for the coefficients of a particular power of x, say the jth power, we find

$$(j + 2)(j + 1)a_{j+2}x^j - j(j - 1)a_jx^j$$
$$- 2(|m| + 1)ja_jx^j + [K - |m|(|m| + 1)]a_jx^j = 0$$

Thus

$$a_{j+2} = \frac{(j + |m|)(j + |m| + 1) - K}{(j + 2)(j + 1)}a_j \qquad (8.13)$$

For the start of the series where $K > (j + |m|)(j + |m| + 1)$ the coefficients of x^j in Eq. (8.11) alternate in sign. We now apply a boundary condition that the solution must satisfy, namely, that the solution must be finite over all possible values the variables may assume. The higher terms in the series all have the same sign, and for $x = \pm 1$ the series diverges unless it can be terminated. Therefore the resulting wave function can be finite only for a finite series, and this can be obtained only if for some highest value of j

$$K = (j + |m|)(j + |m| + 1) \qquad (8.14a)$$

or by letting $j + |m| = l$, whence the condition of K yielding a finite series is

$$K = l(l + 1) \qquad (8.14b)$$

where l is of course a positive integer since j and $|m|$ are positive integers. Note that the series expansion really consists of two series, an even one starting off with a_0 and an odd series starting with a_1. Both series build up in steps of two to a_l where l is an even integer for the a_0 series and odd for the a_1 series, a_{l+2} being expressible in terms of a_l. These two series may not be terminated simultaneously by the same value of K, and *therefore a solution consists of only one of the two series and contains only even powers of* $\cos \theta$ *or only odd powers but not both*. The angular wave functions. Eqs. (8.6) and (8.9), are called *spherical harmonic functions*.

Referring to Eq. (8.7), we find for the radial wave equation

$$\frac{1}{r^2}\frac{\partial}{\partial r}\left(r^2\frac{\partial R}{\partial r}\right) + \frac{2m}{\hbar^2}\left[E - V(r) - \frac{\hbar^2}{2m}\frac{l(l + 1)}{r^2}\right]R = 0 \qquad (8.15)$$

Note that $R(r)$ depends on l but not on m.

8.2 ANGULAR MOMENTUM

The angular momentum of an isolated physical system is conserved. That is, like the energy it is a constant of the motion of an isolated system. Hence the eigenvalues of the angular momentum operators are very important quantities which may be used to identify a state. This is especially true for those cases when there is more than one state having the same energy, and the Hamiltonian or energy operator alone is not sufficient to specify completely a state in terms of its eigenvalue, the energy. Moreover, the angular momentum operators can be used to separate that part of the particle motion which is rotational (and as we shall see, is the same for all spherically symmetric potentials) from the rest of the motion which differs for each of the many dissimilar central force problems occurring in nature. Thus, theorems on angular momentum have general validity and can be applied to all spherically symmetric systems independent of what the particular radial dependence of the potential is.

In this section we shall show that the operator for the total angular momentum is given by the angular derivative terms in the Laplacian operator Eq. (8.3) (for which we found the eigenvalue was some multiple of $l(l + 1)$). The letter l is, therefore, the quantum number for the total angular momentum. As we have seen, however, a state with given l can have different values of m. This corresponds to the fact that the eigenvalue l specifies the magnitude of the angular momentum but not its direction. As we will soon see, m represents the projection of the total angular momentum on a given axis. The letter m is called the *magnetic* quantum number because if it represents the angular momentum of an electric charge about a given axis, the system will have a magnetic moment about that axis. The states with different m but the same l may then be distinguished (the degeneracy in m is "broken up") by turning on a magnetic field and observing the different energies of the different states in the magnetic field.

As a first step to understanding the crucial significance of angular momentum to a theoretical interpretation of three-dimensional systems, we will express the angular momentum operator and its axial components L_z, L_x, and L_y in spherical coordinates.

The total angular momentum is defined as

$$\mathbf{L} = \mathbf{r} \times \mathbf{p} \tag{8.16}$$

The total angular momentum *operator* expressed in coordinate space is

$$\mathbf{L} = \frac{\hbar}{i}\mathbf{r} \times \nabla \tag{8.17}$$

Its z component is

$$L_z = \frac{\hbar}{i}\left(x\frac{\partial}{\partial y} - y\frac{\partial}{\partial x}\right) \tag{8.18}$$

This expression may be transformed to spherical coordinates by noting that

$$r = \sqrt{x^2 + y^2 + z^2}, \quad \tan\theta = \sqrt{(x^2 + y^2)}/z, \quad \tan\varphi = y/x$$

and

$$\frac{\partial}{\partial x} = \frac{\partial r}{\partial x}\frac{\partial}{\partial r} + \frac{\partial\theta}{\partial x}\frac{\partial}{\partial\theta} + \frac{\partial\varphi}{\partial x}\frac{\partial}{\partial\varphi}$$

$$= \frac{x}{r}\frac{\partial}{\partial r} + \frac{xz}{r^2\sqrt{x^2+y^2}}\frac{\partial}{\partial \theta} - \frac{y}{x^2+y^2}\frac{\partial}{\partial \varphi} \qquad (8.19)$$

$$\frac{\partial}{\partial y} = \frac{y}{r}\frac{\partial}{\partial r} + \frac{yz}{r^2\sqrt{x^2+y^2}}\frac{\partial}{\partial \theta} + \frac{x}{x^2+y^2}\frac{\partial}{\partial \varphi} \qquad (8.20)$$

Hence Eq. (8.18) becomes

$$L_z = \frac{\hbar}{i}\frac{x^2}{x^2+y^2}\left(1 + \frac{y^2}{x^2}\right)\frac{\partial}{\partial \varphi} = \frac{\hbar}{i}\frac{\partial}{\partial \varphi} \qquad (8.21)$$

The other components of the total angular momentum operator may be found in spherical coordinates in a similar way. Since

$$\frac{\partial}{\partial z} = \frac{z}{r}\frac{\partial}{\partial r} - \frac{\sqrt{x^2+y^2}}{r^2}\frac{\partial}{\partial \theta} \qquad (8.22)$$

$$L_x = \frac{\hbar}{i}\left(y\frac{\partial}{\partial z} - z\frac{\partial}{\partial y}\right) = \frac{\hbar}{i}\left(\frac{-y}{\sqrt{x^2+y^2}}\frac{\partial}{\partial \theta} - \frac{xz}{x^2+y^2}\frac{\partial}{\partial \varphi}\right)$$

$$= \frac{\hbar}{i}\left(-\sin\varphi\frac{\partial}{\partial \theta} - \cot\theta\cos\varphi\frac{\partial}{\partial \varphi}\right) \qquad (8.23)$$

$$L_y = \frac{\hbar}{i}\left(z\frac{\partial}{\partial x} - x\frac{\partial}{\partial z}\right) = \frac{\hbar}{i}\left(\frac{x}{\sqrt{x^2+y^2}}\frac{\partial}{\partial \theta} - \frac{yz}{x^2+y^2}\frac{\partial}{\partial \varphi}\right)$$

$$= \frac{\hbar}{i}\left(\cos\varphi\frac{\partial}{\partial \theta} - \sin\varphi\cot\theta\frac{\partial}{\partial \varphi}\right) \qquad (8.24)$$

From the operators representing L_x, L_y, and L_z, we may now readily find the operator for the total magnitude of the angular momentum, or rather the operator representing the square of the total angular momentum:

$$L_z^2 = -\hbar^2\frac{\partial^2}{\partial \varphi^2}$$

$$L_x^2 = -\hbar^2\left(\sin^2\varphi\frac{\partial^2}{\partial \theta^2} + \cot^2\theta\cos^2\varphi\frac{\partial^2}{\partial \varphi^2} + 2\sin\varphi\cos\varphi\cot\theta\frac{\partial^2}{\partial \theta\partial \varphi}\right.$$

$$\left. - \cot^2\theta\cos\varphi\sin\varphi\frac{\partial}{\partial \varphi} - \csc^2\theta\sin\varphi\cos\varphi\frac{\partial}{\partial \varphi} + \cot\theta\cos^2\varphi\frac{\partial}{\partial \theta}\right)$$

$$L_y^2 = -\hbar^2\left(\cos^2\varphi\frac{\partial^2}{\partial \theta^2} + \sin\varphi\cot^2\theta\frac{\partial^2}{\partial \varphi^2} - 2\sin\varphi\cos\varphi\cot\theta\frac{\partial^2}{\partial \theta\partial \varphi}\right.$$

$$\left. + \cot^2\theta\cos\varphi\sin\varphi\frac{\partial}{\partial \varphi} + \csc^2\theta\sin\varphi\cos\varphi\frac{\partial}{\partial \varphi} + \sin^2\varphi\cot\theta\frac{\partial}{\partial \theta}\right)$$

$$L^2 = L_x^2 + L_y^2 + L_z^2 = -\hbar^2\left(\frac{\partial^2}{\partial \theta^2} + \cot\theta\frac{\partial}{\partial \theta} + \csc^2\theta\frac{\partial^2}{\partial \varphi^2}\right) \qquad (8.25)$$

or, by rearranging the derivatives in this expression,

$$L^2 = -\hbar^2\left[\frac{1}{\sin\theta}\frac{\partial}{\partial \theta}\left(\sin\theta\frac{\partial}{\partial \theta}\right) + \frac{1}{\sin^2\theta}\frac{\partial^2}{\partial \varphi^2}\right] \qquad (8.25a)$$

Equation (8.25a) comprises the angular derivative expression in the Laplacian operator ∇^2:

$$r^2\nabla^2 = \frac{\partial}{\partial r}\left(r^2\frac{\partial}{\partial r}\right) + \frac{1}{\sin\theta}\frac{\partial}{\partial\theta}\left(\sin\theta\frac{\partial}{\partial\theta}\right) + \frac{1}{\sin^2\theta}\frac{\partial^2}{\partial\varphi^2}$$

When spherical harmonic wave functions are used, the operator for the z component of the angular momentum, given in Eq. (8.21), has the eigenvalue $\pm m$ since the φ-dependent part of the wave function is given by one of the terms in Eq. (8.6). The square of the total angular momentum has the operator given by Eq. (8.25a), which from Eq. (8.7) and (8.14b) was found to have the eigenvalue $\hbar^2 l(l+1)$. That is, when spherical harmonic wave functions are used,

$$L^2\psi = \hbar^2 l(l+1)\psi, \quad L_z\psi = \pm m\hbar\psi \tag{8.26}$$

Thus aside from the energy eigenvalue E, the system also has angular momentum and angular momentum axial projection eigenvalues, l and m. Further, *the eigenvalue $\hbar^2 l(l+1)$ corresponds to the square of the total angular momentum of the system.* Classically, the rotational energy of a point mass at radius r is given in terms of the angular velocity ω and moment of inertia I by

$$\frac{1}{2}I\omega^2 = \frac{I^2\omega^2}{2mr^2} \longrightarrow \frac{\hbar^2 l(l+1)}{2mr^2}$$

which is the rotational energy of a rigid rotator, to be discussed in the next section.

The angular motion about the axis from which θ is measured, the rotational axis, is given by the dependence of the wave function on φ. If, for example, j were zero in Eq. (8.14a) $|m|$ would be equal to l and the magnitude of the total angular momentum would be $\hbar\sqrt{|m|(|m|+1)}$. In general, m is the projection of the total angular momentum vector on this polar axis (see Figure 8.1). The angular momentum about this axis p_φ is given by

$$p_\varphi\psi = \frac{\hbar}{i}\frac{\partial\psi}{\partial\varphi} = \hbar m\psi \tag{8.26a}$$

Note that the eigenvalue of the operator for the angular momentum projection on the polar axis is m, not $\pm\sqrt{|m|(|m|+1)}$ which, in general, is larger than m. (Note that we cannot have the total angular momentum vector absolutely parallel to the polar axis and satisfy the uncertainty principle since then p_z and z are identically zero. m is called the magnetic quantum number since the polar axis is frequently chosen parallel to the magnetic field in problems where an external magnetic field is applied, as in the Zeeman effect. l is called the *azimuthal* quantum number. Since m is integral, the projection of the angular momentum vector on the polar axis is quantized; for a given l, m may take on all $2l+1$ different values from $-l$ to $+l$. All of these states will have the same energy, and hence are called *degenerate*, unless a magnetic field is present, in which case the energies will differ. In a magnetic field the electrons or nuclei will line up their angular momenta parallel or antiparallel to the magnetic field and their energy will be larger or smaller depending on their alignment in the magnetic field.

When $m = 0$, $\Theta(\theta) = F(x)$ and the series defined by Eq. (8.11) is comprised of Legendre polynomials, the number of terms in each polynomial depending on the highest possible value of i in Eq. (8.11), which is l by the condition in Eq. (8.14) for the termination of the series. When $m \neq 0$, $\Theta(\theta)$ is called the associated Legendre function and the product $\Theta(\theta)\Phi(\varphi)$ is frequently expressed as $Y_l^m(\theta, \varphi)$ and called a *surface harmonic*. A very lucid discussion and detailed table of these functions is given in Chapter 5 of Pauling and Wilson (see bibliography of this chapter). If these functions are normalized, that is

$$\int_0^{2\pi} \int_0^{\pi} \left| Y_l^m(\theta, \varphi) \right|^2 \sin \theta \, d\theta \, d\varphi = 1$$

then they are given by

$$Y_l^m(\theta, \varphi) = \frac{1}{\sqrt{2\pi}} \Theta_l^{|m|}(\theta) e^{\pm im\varphi}$$

$$\Theta_l^{|m|}(\theta) = (-1)^l \sqrt{\frac{(2l+1)(l+|m|)!}{2(l-|m|)!}} \frac{1}{2^l l!} \frac{1}{\sin^{|m|}(\theta)} \left(\frac{d}{d(\cos \theta)} \right)^{l-|m|} \sin^{2l}(\theta)$$

Graphs of the first few associated Legendre polynomials are illustrated

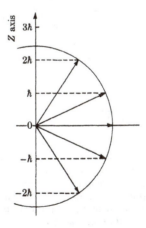

FIGURE 8.1 Illustration of the quantized projection of the angular momentum vector $1\hbar$ on the Z axis for an example for which $l = 2$. Note that the length of the vector is $\hbar\sqrt{l(l+1)} = \sqrt{6}\hbar = 2.45\hbar$ which is greater than the maximum value m may have in this case and remain integral.

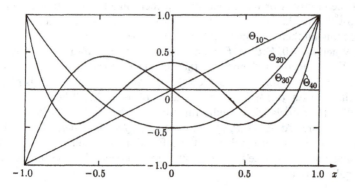

FIGURE 8.2 Zero-order ($m = 0$) associated Legendre functions $\Theta_{lm}(x)$ up to the fourth degree ($l = 4$). $x = \cos\theta$. Note that the odd-degree functions are odd functions of x while the even-degree functions are even functions of x. (From Jahnke, Emde, and Lösch, *Tables of Higher Functions*, Fig. 69. Stuttgart: B. G. Teubner Verlagsgesellschaft, 1960).

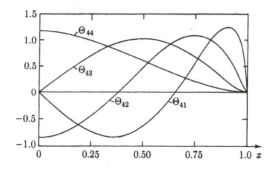

FIGURE 8.3 Associated Legendre functions, $\Theta_{4m}(x)$, of fourth degree and order m for $m = 1, 2, 3, 4$. $x = \cos\theta$. Only values for positive x are shown ($0 \leqslant \theta \leqslant \pi/2$), but as may be surmised from the figure, fourth-degree functions of odd order are odd functions of x. (Functions of odd degree and even order are also odd, while odd-degree, odd-order and even-degree, even-order functions are even.) (From Jahnke, Emde, and Lösch, *op. cit.*, Figs. 72–75.)

in Figures 8.2 and 8.3. A short table is presented in Section 8.4. Note from Table 8.1 and Figures 8.2 and 8.3 that the associated Legendre functions are even or odd functions about $\theta = \pi/2$ depending on whether $(m + l)$ is even or odd. Since also $e^{im\varphi}$ is even or odd about $\varphi = \pi$ ($e^{im(\varphi+\pi)} = (-1)^m$

$e^{im\varphi}$) depending on whether m is even or odd, the surface harmonic $Y_l^m(\theta, \varphi)$ is even or odd over the range of the θ and φ variables depending on whether $(m + l + m)$ is even or odd. Therefore, since $2m$ is always even, the surface harmonics are even or odd functions (with respect to changing θ to $\pi - \theta$ and φ to $\varphi + \pi$ which corresponds to changing $\mathbf{r}$ to $-\mathbf{r}$) depending on whether l is even or odd.

The angular wave function $\Theta(\theta)$ is spherically symmetric about the origin when $l = 0$; when $m = l$, Θ is increasingly concentrated in the xy plane as l increases. When l is constant Θ is increasingly concentrated in the xy plane as m increases as illustrated in Figure 8.4. Note that in conformity with the uncertainty principle the particle motion is not entirely confined to the xy plane even for $m = l$. For this reason the angular momentum projected on the z axis must always be less than the total angular momentum.

8.3 ANGULAR MOMENTUM COMMUTATION RELATIONS

We conclude our discussion of the angular momentum by examining a few very useful and important rules for commutation of the total angular momentum, its components, and operators representing position and linear momentum.

Problem 8.1: Prove that $[L_x, L_y] = i\hbar L_z$, where $[A, B]$ means the commutation operation of the two operators A and B; that is $[A, B] \equiv AB - BA$. Prove also that $[L_y, L_z] = i\hbar L_x$, and $[L_z, L_x] = i\hbar L_y$, that is, prove that in general $[\mathbf{L}, \mathbf{L}] = i\hbar \mathbf{L}$.

The significance of the commutation properties of two operators is that the two physical quantities they represent may be simultaneously measured and their eigenvalues used to label the same wave function, if the operators commute.

Problem 8.2: Find $[L_z, x]$, $[L_z, y]$, and $[L_z, p_x]$.

The components of $\mathbf{L}$ do not commute with one another, and thus we see that only one component of the angular momentum can be measured at a time (L_x and L_y, for example, cannot be measured simultaneously). The operator for the square of the total angular momentum does commute, however, with any component of the angular momentum. Thus, L_z, for example, and L^2 may be measured simultaneously for a given system:

$$[L^2, L_z] = L_x^2 L_z - L_z L_x^2 + [L_y^2, L_z] + [L_z^2, L_z]$$

If we use the theorem of Problem 8.1, the first three terms on the right may be shown to add up to zero

$$L_x^2 L_z = L_x(L_z L_x - i\hbar L_y) = L_z L_x^2 - i\hbar L_y L_x - i\hbar L_x L_y$$

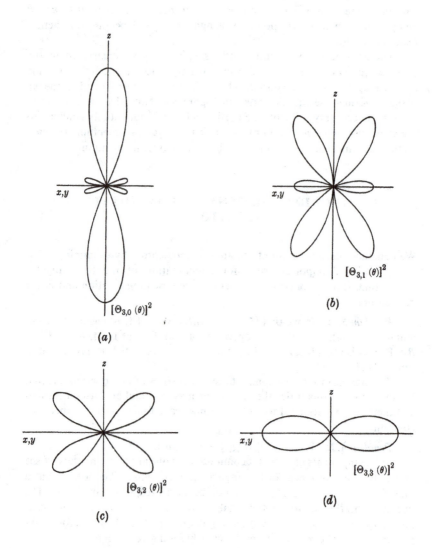

FIGURE 8.4 Polar graphs of the probability of finding a particle at θ if its wave function is $\Theta_{3m}(\theta)(l = 3)$ for $m = 0, 1, 2,$ and 3. (The projection of the total angular momentum vector on the z axis is 0, 1, 2, and 3 respectively in the four cases.)

thus

$$[L_x^2, L_z] = -i\hbar(L_xL_y + L_yL_x)$$

$$[L_y^2, L_z] = i\hbar(L_xL_y + L_yL_x)$$

Also, L_z commutes with L_z^2, and thus L^2 is proved to commute with L_z. Similarly, L_y and L_x commute with L^2, or in general, **L** commutes with L^2. Although no two components of **L** may be measured simultaneously, we know that when L_z is a maximum (that is, when $\langle L_z \rangle = m = l$), $\langle L_x \rangle = \langle L_y \rangle = 0$; in this case, however, L_x and L_y are not completely determined since

$$\langle L_x^2 \rangle = \langle L_y^2 \rangle = \tfrac{1}{2}\langle L^2 - L_z^2 \rangle = \tfrac{1}{2}[l(l+1) - m^2]\hbar^2 = \tfrac{1}{2}l\hbar^2 \qquad (8.27)$$

Problem 8.3: What is the possible result of a measurement of L_z and L^2 on the state represented by $\psi = R(r)\sin\theta\cos\varphi$?

8.4 THE RIGID ROTATOR

We shall treat rigid rotation, which means that the radial position of every point is fixed, but much of the discussion is also valid for many problems in spherical coordinates. In the case of rigid rotation all the energy is rotational and

$$E = \frac{\hbar^2}{2m}\frac{l(l+1)}{r^2} \qquad (8.28)$$

The rigid rotator wave functions are simply the angle-dependent wave functions, i.e., the spherical harmonic wave functions discussed in the previous section:

$$\psi = Y_l^m(\theta, \varphi) = \Theta_l^m(\theta)\Phi(\varphi)$$

A few of these functions are listed in Table 8.1.

TABLE 8.1

The associated Legendre functions of the first kind, $\Theta_{lm}(\theta)$, up to $l = m = 3$. The functions for $m = 0$ are proportional to the ordinary Legendre functions. Note that the evenness or oddness of the functions about $\theta = \pi/2$ depends on the evenness or oddness of $(m + 1)$.

$$\Theta_{00} = \frac{\sqrt{2l+1}}{\sqrt{2}} = \frac{1}{\sqrt{2}}$$

$\Theta_{10}(\theta) = \sqrt{\tfrac{3}{2}}\cos\theta$	$\Theta_{11}(\theta) = \frac{\sqrt{3}}{2}\sin\theta$
$\Theta_{20}(\theta) = \sqrt{\tfrac{5}{2}}(\tfrac{3}{2}\cos^2\theta - \tfrac{1}{2})$	$\Theta_{21}(\theta) = \frac{\sqrt{15}}{2}\sin\theta\cos\theta$
	$\Theta_{22}(\theta) = \frac{\sqrt{15}}{4}\sin^2\theta$
$\Theta_{30}(\theta) = \tfrac{3}{4}\sqrt{14}\cos\theta(\cos^2\theta - 1)$	$\Theta_{31}(\theta) = \frac{\sqrt{42}}{8}\sin\theta(5\cos^2\theta - 1)$
$\Theta_{32}(\theta) = \frac{\sqrt{105}}{4}\sin^2\theta\cos\theta$	$\Theta_{33}(\theta) = \frac{\sqrt{70}}{8}\sin^3\theta$

When a rotating molecular system absorbs or emits photons, the photons will have a frequency spectrum in the overwhelmingly most probable case where l changes by one unit (discussed later in Section 10.4) given by

$$\hbar\omega = \frac{\hbar^2}{2I}[(l+2)(l+1) - (l+1)l]$$

$$= \frac{\hbar^2}{2I}2(l+1)$$

where the spectrum frequencies are proportional to the integer series 1, 2, 3, 4, ... The pre-wave-theory Bohr–Sommerfeld quantum rules (which we have seen to result from the first-order WKB approximation to the wave theory) predict a frequency spectrum of

$$\hbar\omega = \frac{\hbar^2}{2I}[(l+1)^2 - l^2] = \frac{\hbar^2}{2I}(2l+1) \propto 1, 3, 5, \cdots$$

Rotational transitions are frequently coupled with vibrational transitions, so that when a vibrational transition as well as a rotational transition occur radiation may be emitted and energy conserved if l either decreases or increases. Thus the emitted or absorbed photons have a frequency spectrum twice as broad as for pure rotational transitions in which l can only decrease. Hence the spectrum is proportional to $\pm 1, \pm 2, \pm 3, \pm 4 \ldots$ on the basis of the wave theory or, $\pm 1, \pm 3, \pm 5 \ldots$ on the basis of the Bohr–Sommerfeld quantum conditions. Thus the wave theory predicts evenly spaced frequencies with a double space in the center between $+1$ and -1 whereas the pre-wave theory predicts evenly spaced lines everywhere. The observed spectrum illustrated in Figure 8.5 consists of evenly spaced lines with a gap at the center which was once a big puzzle, as the Bohr–Sommerfeld quantum conditions did not predict the experimentally observed gap.

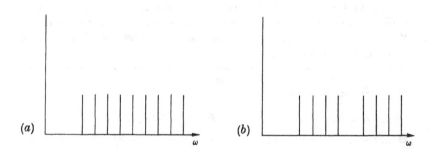

FIGURE 8.5 (a) Illustration of rotational–vibrational spectrum predicted by the Bohr–Sommerfeld quantum conditions. (b) Illustration of rotational–vibrational spectrum observed and predicted by quantum mechanics.

Problem 8.4: Represent the voltage on a sphere by a series of Legendre polynomials up to the fourth term for voltages on the two hemispheres of 100 volts for $0 < \theta < \pi/2$ and 0 volts for $\pi/2 < \theta < \pi$. (The coefficients of the spherical harmonics which are orthogonal functions are obtained as for the Fourier series coefficients in Chapter II.)

$$A_l = \int_0^\pi f(\theta)\Theta_l^0(\theta) \sin\theta \, d\theta$$

Problem 8.5: Find the energy levels of the HCl molecule at temperatures too low for vibration to be important, assuming that only rotation occurs. The moment of inertia in the appropriate coordinate system (center of mass) is given by $I = [m_H m_{Cl}/(m_H + m_{Cl})]r_0^2$ where r_0 is 1.27 Å.

Problem 8.6: The voltage on the surface of a sphere is 100 volts on one octant and 0 elsewhere. Express the voltage approximately as a sum of spherical harmonics with $l \leqslant 2$.

Problem 8.7: Express the function $x^2 - 1$ as a series of Legendre polynomials for $-1 \leqslant x \leqslant 1$.

8.5 THE SQUARE WELL

For many potentials a change is frequently made in the dependent variable in Eq. (8.15) by letting

$$R(r) = \frac{u(r)}{r}$$

Thus we obtain

$$\frac{\partial^2 u}{\partial r^2} + \frac{2m}{\hbar^2}\left[E - V(r) - \frac{\hbar^2}{2m}\frac{l(l+1)}{r^2}\right]u = 0 \qquad \textbf{(8.29)}$$

The reader will readily recognize Eq. (8.29) as the one-dimensional wave equation in r which was used to treat the deuteron in Section 6.2 with $l = 0$ for the ground state and $V(r)$ a square well. In general, if the potential is spherically symmetric and there is no rotation (that is, when l is zero), the substitution of u for R will change the radial equation into one resembling a simple one-dimensional problem in Cartesian coordinates. When l is not zero, the problem may be handled by treating the term in $l(l + 1)$ as an additional potential function. The effective potential is then given by

$$V'(r) = V(r) + \frac{\hbar^2}{2m}\frac{l(l+1)}{r^2} \qquad \textbf{(8.30)}$$

The term in l which represents the rotational energy is frequently referred to as the *centrifugal potential* or *centrifugal barrier* since it acts as the potential of a repulsive force, tending to prevent particles of high angular momentum (that is, large l) from approaching very close to the center of force ($r = 0$). In classical physics, only particles with energy greater than

$$\frac{\hbar^2}{2m} \frac{l(l+1)}{r_1^2}$$

may approach to within $r = r_1$ if energy is to be conserved. In quantum mechanics, the particle wave function falls off as r^l in regions where the centrifugal potential exceeds the particle energy, thus decreasing but not eliminating the probability of finding the particle at small r. The relevant energy diagram is shown in Figure 8.6.

For a square well potential of depth $-V_0$ and arbitrary l, Eq. (8.15) becomes for $r \leqslant a$

$$\frac{\partial^2 R}{\partial r^2} + \frac{2}{r} \frac{\partial R}{\partial r} + \left(\frac{p_1^2}{\hbar^2} - \frac{l(l+1)}{r^2}\right)R = 0 \qquad (8.31)$$

where $p_1^2 = 2m(E + V_0)$. Let $\rho = (p_1/\hbar)r$ and $R = F/\sqrt{\rho}$: then

$$\frac{\partial^2 F}{\partial \rho^2} + \frac{1}{\rho} \frac{\partial F}{\partial \rho} + \left(1 - \frac{(l+\frac{1}{2})^2}{\rho^2}\right)F = 0$$

which is Bessel's differential equation of order $l + \frac{1}{2}$. Thus for $r \leqslant a$ the radial wave function is

$$R = Ar^{-1/2}J_{l+1/2}(p_1r/\hbar) + Br^{-1/2}J_{-(l+1/2)}(p_1r/\hbar) \qquad (8.32)$$

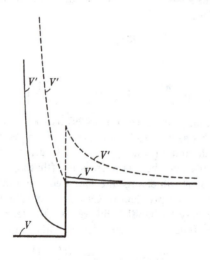

FIGURE 8.6 Energy graphs of a square well potential $V(r)$ and the effective potential $V'(r)$ which includes the centrifugal barrier in addition to the square well for two different cases, one in which bound states are possible (——) and the other for large l where V' is everywhere positive and no bound states are possible (----).

The solutions are spherical Bessel and Neumann functions, the latter being essentially Bessel functions of negative order (see Morse and Feshbach, and Schiff, cited in bibliography of this chapter). A graph of a few of these functions is presented in Figure 8.7. (The Neumann function is not, in general, a Bessel function of negative order; this is true only for fractional order.)

For $r \geqslant a$ and negative E, p^2 is negative and spherical Bessel and Neumann functions of imaginary argument result, which rearranged are called *spherical Hankel functions*, which depend exponentially on r. The permissible values for the energy of the bound particles are found by requiring the wave function and its derivative to be continuous at the boundary:

$$r \leqslant a: \quad R = \frac{A}{\sqrt{r}} J_{l+1/2}(p_1 r/\hbar) \qquad (8.32a)$$

where $p_1 = \sqrt{2m(V_0 + E)}$ and $E < 0$, and

$$r \geqslant a: \quad R = \frac{C}{\sqrt{r}} [J_{l+1/2}(p_2 r/\hbar) - J_{-(l+1/2)}(p_2 r/\hbar)]$$

$$\underset{r \to \infty}{=} (-1)^{n+1} \sqrt{\frac{2p_2}{\hbar\pi}} \frac{C}{r} e^{-p_2 r/\hbar} \qquad (8.32b)$$

where $p_2 = \sqrt{2m(-E)}$. The Neumann function which would appear in Eq. (8.32a) has been eliminated by requiring a finite wave function at the origin One of the two integration constants in the solution for $r > a$ has been chosen so that we obtain a combination of $J_{l+1/2}$ and $J_{-(l+1/2)}$ whose

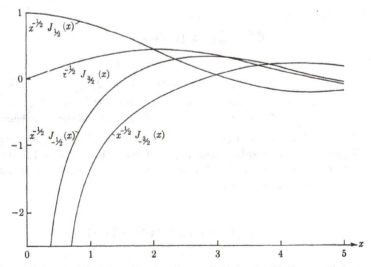

FIGURE 8.7 The first two spherical Bessel and Neumann functions as function of x. Note that the Neumann functions are infinite at the origin. The behavior of all the functions begins to approach sinusoidal, with slowly diminishing amplitude, at $x \sim \pi$.

asymptotic form does not include an increasing exponential function of r as r increases without limit. Equations (8.32a, b) and their first derivatives must be equal at $r = a$; therefore R is continuous:

$$\frac{dR}{dr} \text{ is continuous}; \quad \frac{d(\log R)}{dr} = \frac{1}{R}\frac{dR}{dr} \text{ is continuous}$$

Hence

$$[(p_1/\hbar)J'_{l+1/2}(p_1 a/\hbar) - (2a)^{-1}J_{l+1/2}(p_1 a/\hbar)]/J_{l+1/2}(p_1 a/\hbar)$$
$$= \{(p_2/\hbar)[J'_{l+1/2}(p_2 a/\hbar) - J'_{-(l+1/2)}(p_2 a/\hbar)] - (2a)^{-1}[J_{l+1/2}(p_2 a/\hbar)$$
$$- J_{-(l+1/2)}(p_2 a/\hbar)]\} \times \{J_{l+1/2}(p_2 a/\hbar) - J_{-(l+1/2)}(p_2 a/\hbar)\}^{-1} \quad \textbf{(8.33)}$$

where the prime on the Bessel function denotes the derivative with respect to the entire argument of the Bessel function. Thus a discrete energy spectrum results, for only those values of p_1 and p_2 (and therefore only certain values of E) satisfying Eq. (8.33) are possible. For many problems the energy determination may be satisfactorily simplified by making the well infinite, in which case R must vanish identically at $r = a$. The energy spectrum is then determined by examining the nodes of the spherical Bessel functions, which are

$$J_{l+1/2}(p_1 a/\hbar) = 0$$

A convenient approximation for the higher eigenvalues is given by

$$J_{l+1/2}(p_1 r/\hbar) \xrightarrow{(p_1 r/\hbar) \gg 1} \sqrt{\frac{2\hbar}{\pi p_1 r}} \cos\left(\frac{p_1 r}{\hbar} - \frac{\pi}{2}(l+1)\right)$$

so that for $p_1 a/\hbar \gg 1$

$$\frac{p_1 a}{\hbar} = \frac{\pi}{2}[(l+1) + (2n+1)]$$

where n is an integer, and

$$E \sim - V_0 + (\hbar^2/32ma^2)(2n + l + 2)^2 \quad \textbf{(8.34)}$$

Note that states of either even l or odd l are degenerate in this approximate expression in that different pairs of values of l and n yield the same energy.

Problem 8.8: An electron moves in an infinite spherical square well, that is, inside a hollow sphere with impenetrable walls of radius 1 Å. Find the lowest energy level with $l = 1$. (Note that the approximate result is not valid for the lowest energy level.)

8.6 THE HYDROGEN ATOM

If the potential energy of a particle consists of the electrostatic attraction between a positive charge Ze and a negative charge $-e$, Eq. (8.15) becomes

$$\frac{1}{r^2}\frac{d}{dr}\left(r^2\frac{dR}{dr}\right) + \frac{2m_e}{\hbar^2}\left[E + \frac{Ze^2}{r} - \frac{\hbar^2}{2m_e}\frac{l(l+1)}{r^2}\right]R = 0 \quad \textbf{(8.35)}$$

where the reduced electron mass is written as m_e. We now attempt to solve this equation by making a series expansion in positive powers of r, similar to the way we generated the associate Legendre functions. To simplify Eq. (8.35) let

$$n = [m_e(Ze^2)^2/2\hbar^2(-E)]^{1/2} = [(Ze)^2/2(-E)a_0]^{1/2} \qquad (8.36a)$$

where $a_0 = \hbar^2/m_e e^2$. Also, let

$$\rho = \sqrt{8m_e(-E)/\hbar^2}\, r \qquad (8.36b)$$

and

$$R = e^{-\rho/2}F(\rho) \qquad (8.36c)$$

We then obtain from (8.35)

$$\frac{\partial^2 F}{\partial \rho^2} + \left(\frac{2}{\rho} - 1\right)\frac{\partial F}{\partial \rho} + \left[\frac{n}{\rho} - \frac{1}{\rho} - \frac{l(l+1)}{\rho^2}\right]F = 0 \qquad (8.37)$$

where E is negative for the bound electron which this problem assumes. Equation (8.36c) is suggested by the solution of the equation to which Eq. (8.35) reduces in the limit of infinitely large r where (8.35) may be written

$$\frac{d^2 R}{dr^2} + \frac{2m_e}{\hbar^2}ER = 0$$

Let

$$F(\rho) = \sum_{s=0}^{\infty} a_s \rho^s \qquad (8.38)$$

where the sum is taken over only positive values of s if the wave function is to be well-behaved at the origin. Substitution of Eq. (8.38) into (8.37) enables us to obtain

$$\sum_{s=0}^{\infty} \{s(s-1)a_s\rho^{s-2} + 2sa_s\rho^{s-2} - sa_s\rho^{s-1}$$
$$+ (n-1)a_s\rho^{s-1} - l(l+1)a_s\rho^{s-2}\} = 0 \qquad (8.39)$$

Since this equation must hold for every value of ρ it must hold for each power of ρ taken individually. Hence for the coefficients of a particular power ρ^{j-1} we obtain

$$j(j+1)a_{j+1} + 2(j+1)a_{j+1} - ja_j + (n-1)a_j - l(l+1)a_{j+1} = 0$$
$$(8.40)$$

from which we obtain the recursion relation in the coefficients of the series,

$$a_{j+1} = \frac{j - (n-1)}{j(j+1) + 2(j+1) - l(l+1)}a_j \qquad (8.41)$$

Since j and l are both integers, the denominator will be zero for that value of s for which $(s+1)(s+2) = l(l+1)$, that is for $s = l - 1$. Unless the coefficients of ρ^s are identically zero for $s \leqslant l - 1$, the coefficients of ρ^s in Eq. (8.38) will be infinite for $s > l - 1$. Therefore the first nonvanishing term in a series representing an acceptable wave function must be $a_l\rho^l$. On the other hand the series must be terminated, since for large s, s greater than some large integer N, $a_{s+1} \sim a_s/s \sim N!a_N/s!$, and the summation becomes approximately

$$F(\rho) = \sum_{s=1}^{N} a_s\rho^s + N!a_N \sum_{s=N}^{\infty} \frac{\rho^s}{s!}$$

$$= \sum_{s=1}^{N} a_s\rho^s + N!a_N e^\rho - \sum_{s=0}^{N} \frac{\rho^s}{s!} \qquad (8.42)$$

At large ρ the first and third terms $\sum a_s\rho^s$, $\sum \rho^s/s!$ when multiplied by $e^{-\rho/2}$ lead to an acceptable wave function, but the second term blows up too rapidly. Hence only a finite series leads to an acceptable wave function for a bound electron. The series can be terminated only by requiring that

$$n = k + 1 \qquad (8.43)$$

where k is the highest term in the series and n is defined by Eq. (8.36a). Thus

$$-E = \frac{m_e(Ze^2)^2}{2\hbar^2}\frac{1}{n^2} \qquad (8.44)$$

where n must be an integer from the condition expressed by Eq. (8.43).

It is more convenient to redefine $F(\rho)$ as

$$F(\rho) = \rho^l \sum_{s=0}^{n-(l+1)} a_s\rho^s = \rho^l L(\rho) \qquad (8.45)$$

The finite summation over s, $L(\rho)$, is called an *associated Laguerre poly-·nomial*. n is called the *total quantum number* and $n - (l + 1)$ the *radial*

TABLE 8.2

The unnormalized associated Laguerre functions $L_{n,l}(\rho)$ up to $n - (l + 1) = l = 2$

$L_{10} = 1$	$L_{21} = 1$	$L_{32} = 1$
$L_{20} = 2 - \rho$	$L_{31} = 4 - \rho$	$L_{42} = 6 - \rho$
$L_{30} = 6 - 6\rho + \rho^2$	$L_{41} = 20 - 10\rho + \rho^2$	$L_{52} = 42 - 14\rho + \rho^2$

quantum number; $(2l + 1)$ is the order of the associated Laguerre polynomial, and $n - (l + 1)$, its highest power, is called its *degree*. A short table of Laguerre polynomials is presented in Table 8.2. The radial distribution of an electron in the hydrogen atom is presented in Figure (8.8) for the ground state and first few excited states. Note that the number of nodes occurring in the radial wave function is given by $n - l - 1$. (A *node* is defined as a point where the wave function is zero.) The normalization of the radial wave functions may be found from the requirement that

$$\int_0^\infty r^2 \left| R_{l,n}(r) \right|^2 dr = 1$$

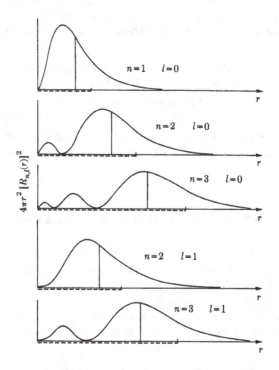

FIGURE 8.8 Graph of the probability of finding an electron at radius r in a hydrogen atom for various radial wave functions R_{nl}. The dashed lines give the spread in radial position predicted by the old Bohr theory. The vertical lines are for the average value of $r(\langle r \rangle = \int r\psi^*\psi d\tau)$.

Equation (8.44) is the well-known empirical *Balmer formula* giving the spectral lines of hydrogen. Note that for a simple Coulomb potential the energy depends only on the total quantum number n. For a given energy there are $2l + 1$ wave functions having the same n and l, and a total number of n^2 wave functions having the same n and giving the same value for the electron energy,

$$\sum_{l=0}^{n-1} (2l + 1) = (n - 1)n + n = n^2$$

since

$$1 + 2 + 3 + \cdots + l = \frac{l}{2}(l + 1)$$

The degeneracy in l is a special feature of the Coulomb field alone. This degeneracy is broken up by any small additional effects, for example a small magnetic interaction between the electron spin and its orbital motion. An externally applied magnetic field or the nuclear magnetic moment causes a further slight difference in energy between states of the same n and l but different m.

Problem 8.9: Draw the wave function of the hydrogen atom for $n = 4$, $l = 2$, $m = 0$. Draw $|\psi|^2 r^2$ as a function of r.

Problem 8.10: What is the expectation of r and r^2 in the ground state of the hydrogen atom? in the first excited state for which $l = 0$?

Problem 8.11: Find the probability that the electron of a hydrogen atom in the ground state will be found at a distance greater than $1.5 \, a_0$ from the nucleus.

8.7 THE HARMONIC OSCILLATOR

Although the one-dimensional or linear harmonic oscillator has already been dealt with in Section. 6.3, a brief treatment of the three-dimensional case is included here because the solutions of the three-dimensional harmonic oscillator problem are of great interest in connection with the individual particle or shell model of the atomic nucleus discussed in Chapter XVI.

The isotropic harmonic oscillator potential may be written

$$V = \tfrac{1}{2}kr^2 \tag{8.46}$$

where k is a spring constant $= m\omega^2$ for frequency of oscillation ω.

Thus the radial wave equation for a central harmonic oscillator potential is

$$\frac{1}{r^2}\frac{\partial}{\partial r}\left(r^2\frac{\partial R}{\partial r}\right) + \frac{2m}{\hbar^2}\left[E - \frac{1}{2}kr^2 - \frac{\hbar^2}{2m}\frac{l(l+1)}{r^2}\right]R = 0 \tag{8.47}$$

At large r only the oscillator potential contributes effectively to the energy, so that the wave equation becomes

$$\frac{1}{r^2}\frac{\partial}{\partial r}\left(r^2\frac{\partial R}{\partial r}\right) - \frac{mk}{\hbar^2}r^2R = 0 \tag{8.48}$$

whose solution at large r is approximately

$$R = Ae^{-(\sqrt{mk}/2\hbar)r^2}$$

This form of the solution of the wave equation for large r suggests that we make the substitution

$$R = e^{-\rho^2/2}F(\rho) \tag{8.49}$$

where

$$\rho = (\sqrt{mk}/\hbar)^{1/2}r$$

The solution for $F(\rho)$ is found in terms of a series expansion using exactly the same methods as were used for the linear harmonic oscillator and the other problems treated in Chapter VI. The polynomial $F_{nl}(\rho)$ is called the *spherical Hermite polynomial* and is given by

$$F_{nl} = \rho^l \sum_{s=0}^{n} a_{2s+1} \, \rho^{2s}$$

where the coefficients a_i are given by the recursion relation

$$a_{i+2} = \frac{2(2n - i)}{(i + 2)(i + 2l + 3)} a_i \qquad (8.50)$$

The energy of the three-dimensional harmonic oscillator is given by

$$E = \hbar\omega(2n + l + \tfrac{3}{2}) = \hbar\omega(n_0 + \tfrac{3}{2}) \qquad (8.51)$$

where n, l, and $n_0 = 2n + l$ are integers. *Note that the oddness or evenness of the series is entirely dependent on l.*

Table 8.3 lists the first few spherical Hermite polynomials up to $n = l = 3$. n here is the radial quantum number. (n is usually used for the radial quantum number; however, for the hydrogen atom where a special degeneracy between the radial and orbital quantum numbers occurs, it is more convenient to work with the total quantum number instead, which is also usually labelled n.) In the case of hydrogen the total quantum number giving the energy eigenvalue is given by the sum of the radial quantum number plus the orbital quantum number.

TABLE 8.3

The spherical Hermite polynomials F_{nl} (unnormalized) up to $n = l = 3$.

$F_{00} = 1$	$F_{01} = \rho$	$F_{02} = \rho^2$	$F_{03} = \rho^3$
$F_{10} = 1 - \tfrac{2}{3}\rho^2$	$F_{11} = \rho(1 - \tfrac{2}{5}\rho^2)$	$F_{12} = \rho^2(1 - \tfrac{2}{7}\rho^2)$	$F_{13} = \rho^3(1 - \tfrac{2}{9}\rho^2)$
$F_{20} = 1 - \tfrac{4}{3}\rho^2 + \tfrac{4}{15}\rho^4$		$F_{21} = \rho(1 - \tfrac{4}{5}\rho^2 + \tfrac{4}{35}\rho^4)$	
$F_{22} = \rho^2(1 - \tfrac{4}{7}\rho^2 + \tfrac{4}{63}\rho^4)$		$F_{23} = \rho^3(1 - \tfrac{4}{9}\rho^2 + \tfrac{4}{99}\rho^4)$	
$F_{30} = 1 - 2\rho^2 + \tfrac{4}{5}\rho^4 - \tfrac{8}{105}\rho^6$		$F_{31} = \rho(1 - \tfrac{6}{5}\rho^2 + \tfrac{12}{35}\rho^4 - \tfrac{8}{315}\rho^6)$	
$F_{32} = \rho^2(1 - \tfrac{6}{7}\rho^2 + \tfrac{4}{21}\rho^4 - \tfrac{8}{693}\rho^6)$		$F_{33} = \rho^3(1 - \tfrac{2}{3}\rho^2 + \tfrac{4}{33}\rho^4 - \tfrac{8}{1287}\rho^6)$	

Note that degeneracy also occurs in Eq. (8.51); that is, for a given energy, different even l states or odd l states are possible for differing n. Thus the first few energy levels of the spherical harmonic oscillator may have the following number of independent eigenfunctions: ($n = 0$, $l = 0$), only one independent wave function; ($n = 0$, $l = 1$), $2l + 1 = 3$ different values of m are possible and therefore three independent wave functions; in the third level ($n = 1$, $l = 0$) and ($n = 0$, $l = 2$), a state of $1 + 5 = 6$ or sixfold degeneracy; and for $2n + l = 3$ a tenfold degeneracy. If $2n + l = k$, then the total degeneracy is given by

$$\sum_{n=0}^{(2k+1)/4} (2k + 1 - 4n)$$

Thus, the total degeneracy is given by

$$(k + 1)\left(\frac{k}{2} + 1\right) \sim \left(k + \frac{3}{2}\right)^2 \Big/ 2$$

In Chapter XVI we make use of these results in the discussion of the existence of an energy shell structure in the nucleus.

Problem 8.12: Find the average values or expectations for x and x^2 in a one-dimensional oscillator and r in a spherically symmetric oscillator.

Problem 8.13: What is the probability of finding a particle of energy E within a distance less than a from the origin in a spherically symmetric harmonic oscillator potential if the particle's orbital momentum is zero? $E = \frac{1}{2}\hbar\omega$, $(\sqrt{mk}/\hbar)^{1/2}a \ll 1$.

BIBLIOGRAPHY

Pauling, L. and E. B. Wilson, *Introduction to Quantum Mechanics*. New York: McGraw-Hill Book Co., 1935. An exceptionally lucid and detailed treatment of the angular and radial wave functions of the hydrogen atom is given. The book also contains excellent tables of the hydrogen wave functions.

Jahnke, E., F. Emde, and F. Lösch, *Tables of Higher Functions*, 6th Ed. New York: McGraw-Hill Book Co., 1960. This is a standard reference work of formulas, tables, and graphs of many of the functions most frequently encountered in physics.

Morse, P. and H. Feshbach, *Methods of Theoretical Physics*. New York: McGraw-Hill Book Co., 1953. See page 622.

Schiff, L., *Quantum Mechanics*, p. 77 (cited in Bibliography of Chapter 7).

IX

PERTURBATION
THEORY

In the previous chapter we solved the Schrödinger equation exactly for several simple potential energy functions. In general, for any arbitrary potential, it is not possible to do this and we wish to take up in this chapter approximation methods of particular utility for problems in which the potential energy function does not differ much (i.e., is only slightly perturbed) from that in a simple, previously solved problem. There are many cases that can be handled by perturbation theory; here we will be particularly interested in establishing procedures whereby various problems in atomic spectroscopy can be handled. These will include (1) the electrons in a helium atom whose wave functions would exactly correspond to the wave function of the electron in a helium ion but for the slight disturbance or perturbation caused by the added repulsion of the two electrons; (2) the spin-orbit interaction which results from the interaction of the intrinsic magnetic moment of the electron and the effective magnetic field produced by the orbital motion of the electron about the nucleus; (3) the superposition of an externally applied magnetic field on the usual Coulomb interaction between an electron and the positively charged nucleus; and (4) the problem of emission or absorption of electromagnetic radiation by atomic electrons. Most of these applications will be discussed in succeeding chapters where we consider the question of atomic structure in some detail.

In developing the perturbation theory we will have occasion to express various wave functions in the following manner:

$$\psi = \sum_n C_n \varphi_n$$

where the C's are constants and where the φ's form a complete set of orthogonal functions like the trignometric functions discussed earlier in connection with Fourier series. The orthogonality property, it will be recalled, is

$$\int \varphi_n^* \varphi_m \, d\tau = 0, \text{ if } n \neq m,$$

where the integration is extended over all space, and the completeness property means that any arbitrary ψ can be put into this form. (If one of the φ_n's say φ_1, were missing, the others would not form a complete set, since φ_1 itself could not be expressed in terms of them; for if it were, we would have

$$\varphi_1 = \sum_{n=2}^{\infty} C_n \varphi_n$$

and consequently

$$\int |\varphi_1|^2 d\tau = \sum_{n=2}^{\infty} C_n \int \varphi_1^* \varphi_n d\tau = 0$$

because of the orthogonal property. But this is impossible, since φ_1 would then be identically zero.) Let us agree to multiply each φ by a suitable constant so as to make it satisfy the condition

$$\int |\varphi_n|^2 \, d\tau = 1,$$

which is called the *normalization* condition. Then, for any ψ, we may calculate the coefficients C_n as follows: multiply the equation for ψ given above by φ_m^* and integrate over all space.

$$\int \varphi_m^* \psi d\tau = \sum_n C_n \int \varphi_m^* \varphi_n d\tau = C_m$$

where use has been made of the orthogonality and normalization conditions. Thus, given the function ψ, we can easily calculate C_m by evaluating the integral.

Let A be any Hermitian operator, with eigenvalues a_n and eigenfunctions φ_n:

$$A\varphi_n = a_n \varphi_n$$

It can then be shown, using the Hermitian property, that any two eigenfunctions φ_n and φ_m are orthogonal, provided that the eigenvalues a_n and

a_m are distinct. If, however, φ_n and φ_m are two distinct eigenfunctions belonging to a single eigenvalue (the eigenvalue is degenerate in this case), then they need not be orthogonal, but they can be made so by a procedure to be explained presently. Thus, for any Hermitian operator, one obtains a set of orthogonal eigenfunctions, and in every case of interest it turns out to be a complete set. This applies in particular to the Hamiltonian operator, in which case the operator equation above is the time-independent Schrödinger equation. Thus the various sets of spherical wave functions derived in Chapter VIII all provide examples of complete orthogonal sets. A complete orthogonal set which has been normalized is called a *complete orthonormal set*.

Expression of the solution to a given wave equation in terms of a sum of wave functions which are solutions to a different wave equation is similar to the superposition of plane waves in simple Fourier analysis. Any function may be represented by a sum of sinusoidal functions having specified amplitudes. A single complicated electromagnetic disturbance can be represented as a sum or *superposition* of many simple plane waves of differing frequency whose amplitudes are found by the procedures given in Chapter II. Any attempt to measure the intensity of a single wave present in the summation representing the single complicated disturbance would yield a positive result with a relative intensity identical with the square of the absolute value of the computed amplitude. Similarly we may represent a given quantum mechanical state by a *superposition of states* of the complete orthonormal set of solutions to a wave equation. It is most convenient to choose solutions to a wave equation closely corresponding to the actual wave equation we wish to solve or discuss. It would involve an unnecessary amount of labor to attempt, for example, to represent solutions to the wave equation for the electron in the hydrogen atom as a superposition of solutions to the wave equation for a particle in a rectangular box.

Eigenfunctions of the wave equation having different energy eigenvalues are necessarily orthogonal, as we will now show. Let ψ_i and ψ_j be solutions of

$$\nabla^2 \psi_i + \frac{2m}{\hbar^2}(E_i - V)\psi_i = 0$$

$$\nabla^2 \psi_j^* + \frac{2m}{\hbar^2}(E_j - V)\psi_j^* = 0$$

(9.1)

Multiplying the terms in the first equation by ψ_j^* and the terms in the second by ψ_i on the right, and subtracting the second equation from the first, we obtain

$$\psi_j^* \nabla^2 \psi_i - (\nabla^2 \psi_j^*)\psi_i + \frac{2m}{\hbar^2}(E_i - E_j)\psi_j^* \psi_i = 0$$

(9.2)

Integrating over the particle coordinates after representing the system in Cartesian coordinates, we obtain

$$\int_{-\infty}^{\infty}\int_{-\infty}^{\infty}\int_{-\infty}^{\infty} \left\{ \psi_j^*\left(\frac{\partial^2 \psi_i}{\partial x^2} + \frac{\partial^2 \psi_i}{\partial y^2} + \frac{\partial^2 \psi_i}{\partial z^2}\right) \right.$$

$$- \left(\frac{\partial^2 \psi_j^*}{\partial x^2} + \frac{\partial^2 \psi_j^*}{\partial y^2} + \frac{\partial^2 \psi_j^*}{\partial z^2}\right)\psi_i \right\} dx\, dy\, dz$$

$$+ \frac{2m}{\hbar^2}(E_i - E_j)\int_{-\infty}^{\infty}\int_{-\infty}^{\infty}\int_{-\infty}^{\infty} \psi_j^* \psi_i\, dx\, dy\, dz = 0 \qquad (9.3)$$

Since

$$\int_{-\infty}^{\infty} \left\{ \psi_j^* \frac{\partial^2 \psi_i}{\partial x^2} - \frac{\partial^2 \psi_j^*}{\partial x^2}\psi_i \right\} dx = \int_{-\infty}^{\infty} \frac{\partial}{\partial x}\left\{ \psi_j^* \frac{\partial \psi_i}{\partial x} - \frac{\partial \psi_j^*}{\partial x}\psi_i \right\} dx$$

$$= \left[\psi_j^* \frac{\partial \psi_i}{\partial x} - \frac{\partial \psi_j^*}{\partial x}\psi_i \right]_{-\infty}^{\infty} = 0 \qquad (9.4)$$

because of the boundary conditions on ψ, Eq. (9.3) reduces to

$$\frac{2m}{\hbar^2}(E_i - E_j)\int_{-\infty}^{\infty}\int_{-\infty}^{\infty}\int_{-\infty}^{\infty} \psi_j^* \psi_i\, dx\, dy\, dz = 0 \qquad (9.5)$$

Therefore the normalized wave functions are orthogonal, for either $i = j$ and

$$\iiint \psi_j^* \psi_i\, d\tau = 1$$

or $i \neq j$ in which case $E_i \neq E_j$ and

$$\iiint \psi_j^* \psi_i\, d\tau = 0$$

The important case for which more than one state exists having the same energy eigenvalue will be treated later in Section 9.4.

9.1 FIRST-ORDER PERTURBATION THEORY

The Schrödinger wave equation

$$\nabla^2 \psi + \frac{2m}{\hbar^2}(E - V)\psi = 0 \qquad (9.6)$$

may contain a potential energy term V which is only slightly different from the potential energy V^0 of a problem already solved. Since we should expect the solutions of Eq. (9.6) in a perturbation problem to differ very little from the solution found with V^0, we will use the wave functions for V^0 in attempting to expand the solutions to Eq. (9.6) in terms of known functions. For a given wave function ψ_i, let

$$V = V^0 + V'$$
$$E_i = E_i^0 + E_i'$$
$$\psi_i = \psi_i^0 + \psi_i' \tag{9.7}$$
$$H^0 = -\frac{\hbar^2}{2m}\nabla^2 + V^0$$

where ψ_i^0 and E_i^0 are the solutions (eigenfunctions) and permitted energy values (eigenvalues) of the *unperturbed* Schrödinger equation

$$\nabla^2\psi_i^0 + \frac{2m}{\hbar^2}(E_i^0 - V^0)\psi_i^0 = 0 \tag{9.8}$$

which may be written more compactly as

$$H^0\psi_i^0 = E_i^0\psi_i^0$$

The subscript i, distinguishing the i independent energy eigenvalues for the system, takes on different values with any change in the separate eigenvalues (e.g., n, l, m) of the wave function. (We are assuming here that there is only one eigenfunction corresponding to each energy eigenvalue and hence we need only one subscript; we will remove this restriction in a later section on degenerate levels.) The subscript i is necessary, as we have seen in Chapter VIII, since each allowed value of E, E_i, results in a different ψ (for example, in a Hermite polynomial of different order, in the case of the harmonic oscillator). Substituting Eqs. (9.7) into Eq. (9.6) and grouping the result in ascending order of approximation, we obtain

$$[H^0 - E_i^0]\psi_i^0 + [(H^0 - E_i^0)\psi_i' + (V - E_i')\psi_i^0]$$
$$+ \text{ second-order terms such as } (V_i' - E_i')\psi_i'] = 0 \tag{9.9}$$

The first bracket of terms is zero by Eq. (9.8), leaving us with only the second bracket of terms for a first-order approximation:

$$(H^0 - E_i^0)\psi_i' + (V' - E_i')\psi_i^0 = 0 \tag{9.10}$$

We now expand ψ_i' in terms of the complete orthonormal set of solutions to Eq. (9.8), ψ_j^0. That is, we let

$$\psi_i' = \sum_{j=0}^{\infty} a_{ij}\psi_j^0 \tag{9.11}$$

which, with (9.10), gives

$$\sum_{j=0}^{\infty}(H^0 - E_i^0)a_{ij}\psi_j^0 + (V' - E_i')\psi_i^0 = 0 \tag{9.12a}$$

Note that from the second of Eqs. (9.8), Eq. (9.12a) may be rewritten

$$\sum_{j=0}^{\infty}(E_j^0 - E_i^0)a_{ij}\psi_j^0 + (V' - E_i')\psi_i^0 = 0 \tag{9.12b}$$

Multiplying Eq. (9.12b) by ψ_k^{0*} on the left and integrating over all space, we have

$$\sum_{j=0}^{\infty}(E_j^0 - E_i^0)a_{ij}\int\psi_k^{0*}\psi_j^0 d\tau + \int\psi_k^{0*}V'\psi_i^0\,d\tau - E_i'\int\psi_k^{0*}\psi_i^0 d\tau = 0$$

$$(9.13)$$

Since the ψ_i^0's are orthonormal, all but one term in the summation is zero, and Eq. (9.13) becomes

$$(E_k^0 - E_i^0)a_{ik} + \int\psi_k^{0*}V'\psi_i^0 d\tau - E_i'\int\psi_k^{0*}\psi_i^0 d\tau = 0 \qquad (9.14)$$

If $k = i$, $E_i^0 - E_k^0 = 0$ and the shift in energy, E_i', due to the perturbation is

$$E_i' = \int\psi_i^{0*}V'\psi_i^0 d\tau \qquad (9.15)$$

If $k \neq i$, the third term is zero and the a_{ik} are given by

$$a_{ik} = \frac{\int\psi_k^{0*}V'\psi_i^0 d\tau}{E_i^0 - E_k^0} \qquad (9.16)$$

a_{ii} is not given by Eq. (9.16) but we already know its value; $a_{ii} = 1$ approximately since $\psi_i \sim \psi_i^0$.

$$a_{ii}^2 = 1 - \sum_{j\neq i}a_{ij}^2 = 1$$

up to and including terms of first order in the perturbation.

Note that the first-order shift in energy from the unperturbed energy level, given in Eq. (9.15), is just the perturbed potential energy averaged over the unperturbed wave functions. We prefer to use a notation for the integrals in Eqs. (9.15, 9.16) in keeping with that of most authors, to wit,

$$V_{kl}' \equiv \int\psi_k^{0*}V'\psi_l^0 d\tau \qquad (9.17)$$

obtaining the following expressions for the energy and wave functions of the perturbed system to first order:

$$E_i = E_i^0 + V_{ii}' \qquad (9.18)$$

$$\psi_i = \psi_i^0 + \sum_{\substack{k=0 \\ k\neq i}}^{\infty}\frac{V_{kl}'}{E_i^0 - E_k^0}\psi_k^0 \qquad (9.19)$$

Note that the effect of the perturbing potential is to give the particle a small but finite probability of occupying states of the unperturbed system other than the single unperturbed state it would be in if no perturbing potential existed.

Problem 9.1: Calculate the first-order energy correction and wave functions for the ground state of a particle in a one-dimensional square well with $V = -V_0 - \Delta, -a < x < 0; V = -V_0 + \Delta, 0 < x < a; V = 0,$ $x < -a$ and $x > a.$ $(\Delta << V_0 \sim \infty.)$

Problem 9.2: Calculate the first-order energy correction and wave functions for the two lowest states of a particle in a perturbed one-dimensional square well with $V = -V_0 + cx, -a < x < a.$ $(ca << V_0 \sim \infty.)$

Problem 9.3: Calculate the first-order energy correction and wave functions for the two lowest states of a perturbed one-dimensional harmonic oscillator with $V = kx^2 + cx^3.$ $(cx_1 << k$ where x_1 is the classically permitted range of x for an oscillator with energy $E.)$

9.2 THE HELIUM ATOM

The total potential energy for the helium atom is

$$V = -\frac{Ze^2}{r_1} - \frac{Ze^2}{r_2} + \frac{e^2}{r_{12}} \tag{9.20}$$

where Z is 2 for the charge on the helium nucleus, r_1 and r_2 are the electron–nucleus separation distances for each electron, and r_{12} is the separation of the two electrons from each other. If derivatives with respect to a particular electron's position are denoted by a subscript 1 or 2, we can write the Schrödinger equation as

$$\nabla_1^2 \psi_i + \nabla_2^2 \psi_i + \frac{2m}{\hbar^2}(E_i - V_1^0 - V_2^0 - V')\psi_i = 0 \tag{9.21}$$

where V_1^0 and V_2^0 are the hydrogen-like potential functions of the two electrons in the nuclear Coulomb field and their interaction energy is treated as a perturbation. Letting a superscript zero represent the zero-order wave function, the total wave function ψ_i^0 may be written as a product $\psi_{1i}^0 \psi_{2j}^0$ of the wave functions of the two electrons taken separately, where ψ_{1i}^0 for example is a solution of

$$\nabla_1^2 \psi_{1i}^0 + \frac{2m}{\hbar^2}(E_{1i}^0 - V_1^0)\psi_{1i}^0 = 0 \tag{9.22}$$

or

$$H_1^0 \psi_{1i}^0 = E_{1i}^0 \psi_{1i}^0$$

and $E_i^0 = E_{1i}^0 + E_{2i}^0$. The ψ_i^0 are the wave functions given in Chapter VIII in terms of product functions of Laguerre functions with associated spherical harmonics. The first approximation to the energy of the helium atom is given by substituting these wave functions into Eq. (9.18), whence

$$E_i = -\frac{mZ^2e^4}{2\hbar^2}\left(\frac{1}{n_1} + \frac{1}{n_2}\right) + \int \psi_{1n_1, l_1}^{0*} \psi_{2n_2, l_2}^{0*} \frac{e^2}{r_{12}} \psi_{1n_1, l_1}^0 \psi_{2n_2, l_2}^0 \, d\tau \tag{9.23}$$

where n_1 and n_2 are the total quantum numbers of the unperturbed hydrogen-like wave functions. For the particular case of the ground state of the helium atom, $n_1 = n_2 = 1$ and $l = m = 0$. From Chapter VIII, each of the one-electron wave functions is an exponential times the zero-order Laguerre polynomial (a constant), so that the normalized wave functions are

$$\psi^0_{11,0}\psi^0_{21,0} = \sqrt{\frac{Z^3}{\pi a_0^3}}\sqrt{\frac{Z^3}{\pi a_0^3}}\, e^{-\rho_1/2}e^{-\rho_2/2}$$

where $a_0 = \hbar^2/me^2$, $\rho_1 = 2Zr_1/a_0$, and the energy E_1 is given by one-half the first term in Eq. (9.23). Since in spherical coordinates, the volume element is

$$d\tau = r_1^2 \sin\theta_1\, dr_1\, d\theta_1\, d\varphi_1\, r_2^2 \sin\theta_2\, dr_2\, d\theta_2\, d\varphi_2$$

the integral in Eq. (9.23) becomes

$$E_1' = \frac{Ze^2}{2(4\pi)^2 a_0} \int_0^\infty \int_0^\pi \int_0^{2\pi} \int_0^\infty \int_0^\pi \int_0^{2\pi} \left(\frac{e^{-\rho_1}e^{-\rho_2}}{\rho_{12}}\right)$$

$$\rho_1^2 \sin\theta_1\, d\rho_1\, d\theta_1\, d\varphi_1\, \rho_2^2 \sin\theta_2\, d\rho_2\, d\theta_2\, d\varphi_2 \tag{9.24}$$

where $\rho_{12} = 2Zr_{12}/a_0$. Note that the integration is taken over a six-dimensional space, since the overall wave function is concerned with the positions of two particles.

The integrand in Eq. (9.24) is essentially the electrostatic interaction energy between two shells of charge density $e^{-\rho_1}$ and $e^{-\rho_2}$. Its evaluation may be found in Appendix V of Pauling and Wilson (see Bibliography), but we will use the result so often that it is worthwhile to repeat the evaluation here.

To begin with let us consider just the integral over ρ_1. Let us consider further the potential at a point ρ due to a shell of thickness $d\rho_1$ at ρ_1; it is given by

$$\rho < \rho_1: \quad 4\pi\rho_1^2 \frac{e^{-\rho_1}d\rho_1}{\rho_1} = 4\pi\rho_1 e^{-\rho_1}\, d\rho_1$$

$$\rho > \rho_1: \quad 4\pi\rho_1^2 \frac{e^{-\rho_1}d\rho_1}{\rho} = \frac{4\pi}{\rho}\rho_1^2 e^{-\rho_1}d\rho_1$$

The total potential at ρ due to the infinite sphere of charge density $e^{-\rho_1}$ is given by

$$\phi(\rho) = 4\pi \int_\rho^\infty \rho_1 e^{-\rho_1}\, d\rho_1 + \frac{4\pi}{\rho}\int_0^\rho e^{-\rho_1}\rho_1^2\, d\rho_1$$

$$= \frac{4\pi}{\rho}[2 - e^{-\rho}(\rho + 2)]$$

This is the potential at a point ρ due to the entire distribution of charge density $e^{-\rho_1}$. We can now evaluate the potential energy of interaction between the distributions of charge density $e^{-\rho_1}$ and $e^{-\rho_2}$. The potential energy per unit volume of a charge density $e^{-\rho_1}$ at a point ρ_2 is given by

$$\frac{4\pi}{\rho_2}[2 - (\rho_2 + 2)e^{-\rho_2}]e^{-\rho_2}$$

Hence, the second shell of charge $4\pi\rho_2^2 e^{-\rho_2} d\rho_2$ has a potential energy of

$$[(4\pi)^2 \rho_2^2 e^{-\rho_2} d\rho_2/\rho_2] [2 - e^{-\rho_2}(\rho_2 + 2)]$$

The total electrostatic potential of the sphere of charge density $e^{-\rho_1}$ due to the sphere with charge density $e^{-\rho_1}$ is the integral over ρ_2 of this expression:

$$(4\pi)^2 \int_0^\infty \rho_2 e^{-\rho_2}[2 - e^{-\rho_2}(\rho_2 + 2)] \, d\rho_2 = \tfrac{5}{4}(4\pi)^2$$

Therefore Eq. (9.24) yields a result of

$$E_1' = \frac{5}{4}\frac{Ze^2}{2a_0} = \frac{5}{4}\frac{mZe^4}{2\hbar^2}$$

and from Eq. (9.23) with $n_1 = n_2 = 1$,

$$E_1 = -\frac{me^4}{2\hbar^2}\left(2Z^2 - \frac{5}{4}Z\right) \tag{9.25}$$

where $me^4/2\hbar^2$ is the ground state energy of the hydrogen atom and $2Z^2$ times this is the unperturbed ground state energy of the two helium electrons.

The correction to the zero-order ground state energy of the helium atom is, as we should suspect, large, being $\tfrac{5}{4}Z/2Z^2 = 5/8Z \sim 30\%$ of the zero-order energy, and is subtractive since the effect of the electron-electron interaction is to detract from or put up a shield reducing the electron-nucleus interaction. The approximation is found surprisingly good, however, in view of the size of the electron-electron interaction relative to the electron-nucleus interaction: the observed ground state energy of the helium atom is 78.6 ev, the zero-order calculated energy is 108.2 ev, and the first-order calculated energy is 74.4 ev. Thus, the zero-order calculation is found to be 38% too high while the first-order calculation is within 4.2 ev of the observed value, only 5% too low. In a later section, using the same wave functions we will take up a slightly different method which will reduce the error to only 2% of the observed value. The method yields better and better results for two-electron ions of heavier atoms as the electron-electron interaction becomes less important compared with the electron-nucleus interaction. Although the error is roughly the same (about -4 ev) for Li^+, Be^{++}, and B^{3+}, the ion binding energy increases rapidly with Z (it is proportional to Z^2) so that the first-order calculation of C^{4+} is only 0.4% too low.

Equation (9.23) is misleading in that it implies that for n_1 or $n_2 > 1$ each energy eigenvalue is associated with a unique eigenfunction specified by a pair of quantum numbers n, the total quantum number, and l, the orbital quantum number. Actually there are n^2 unique hydrogen-like wave functions associated with each energy level specified by total quantum number n. (There are $n - 1$ states of different l, and for each l, $2l + 1$ states of different m, all having the same energy.) Hence these n^2 states are mutually degenerate for hydrogen-like atoms. The energy level is a function of the sum of the radial and orbital quantum numbers and the simple theory breaks down when the energies of two different states with different quantum numbers are equal. Labeling states of different quantum numbers by different letters i and k in Eq. (9.19), we find that the energy denominator is zero for a degenerate level, giving infinite contributions of other states which have been assumed small in order for the approximation to be valid.

Problem 9.4: Calculate the first-order energy correction to the perturbed three-dimensional harmonic oscillator with potential energy

$$V = \frac{1}{2}kr^2 + cr^3 + dr^4$$

Problem 9.5: Calculate the first-order energy correction for the ground state of the valence electron in the sodium atom if the unperturbed wave function is taken to be a hydrogen wave function with $n = 3$ and $l = m = 0$, and the atomic "core" of ten electrons is treated as a perfect screen resulting from the distribution of the electrons on a spherical shell of radius 0.2Å. $V = -10e^2/r$ for $r \leqslant 0.2$Å, and beyond $r = 0.2$Å the effective nuclear charge is $+e$. (Note that what you will obtain is an incomplete solution to this problem. See Problem 9.8.) Omit states of $l \neq 0$.

9.3 SECOND-ORDER PERTURBATION THEORY

When the first-order terms in Eq. (9.4) are zero or in general if a better approximation is needed, the second-order terms must be kept. We generalize the result further by assuming an additional correction to the potential V'', so that our result may be applied also to nondegenerate cases in which a further, smaller physical effect may be estimated in addition to a larger perturbing potential. For this case we let

$$V = V^0 + V' + V''$$
$$E_i = E_i^0 + E_i' + E_i'' \tag{9.26}$$
$$\psi_i = \psi_i^0 + \psi_i' + \psi_i''$$

and substitute into Eq. (9.1) to obtain for the second-order bracket in Eq. (9.9)

$$(H^0 - E_i^0)\psi_i'' + (V' - E_i')\psi_i' + (V'' - E_i'')\psi_i^0 = 0 \tag{9.27}$$

Upon letting

$$\psi_i'' = \sum_l a_{il}'' \psi_l^0$$

and

$$\psi_i' = \sum_m a_{im}' \psi_m^0$$

multiplying Eq. (9.27) by ψ_k^{0*} on the left, and integrating over all space, we obtain

$$(E_k^0 - E_i^0)a_{ik}'' + \sum_m a_{im}'(V_{km}' - E_i'\delta_{km}) + (V_{ki}'' - E_i''\delta_{ik}) = 0 \qquad (9.28)$$

since

$$\int \psi_k^{0*} H^0 \psi_i^0 d\tau = E_i \int \psi_k^{0*} \psi_i^0 d\tau = E_i \delta_{ik}$$

where it will be recalled that $\delta_{ik} = 0$ unless $l = k$ in which case $\delta_{ik} = 1$. If $i = k$, since $V_{ii} = E_i'$ there results

$$E_i'' = V_{ii}'' + \sum_{\substack{m \neq i \\ m=0}}^{\infty} a_{im}' V_{im}'$$

$$= V_{ii}'' + \sum_{\substack{m \neq i \\ m=0}}^{\infty} \frac{V_{im}' V_{mi}'}{E_i^0 - E_m^0} \qquad (9.29)$$

When $i \neq k$, Eq. (9.28) yields the result that

$$a_{ik}'' = \frac{V_{ki}''}{E_i^0 - E_k^0} + \sum_{m \neq i} \frac{V_{km}' V_{mi}'}{(E_i^0 - E_k^0)(E_i^0 - E_m^0)} \qquad (9.30)$$

Note that the second-order perturbation to the unperturbed energy consists of the second-order potential to the first order plus the first-order potential to the second order. The second-order potential may be some additional perturbation, small compared to the first-order perturbation potential. The second term must be used whenever it happens that the first-order potential averaged over the unperturbed system gives zero. The presence of neighboring unperturbed levels, in which the particle can spend a small fraction of its time, contributes to the particle energy in this way, through the second term. The closer together the neighboring states lie in energy, the smaller the denominator becomes, and hence the greater their contribution becomes. We will return to second-order perturbation theory in Chapter XI.

Problem 9.6: A rigid body rotating in a plane has a wave function which satisfies the zero-order Schrödinger equation

$$\frac{1}{R^2}\frac{d^2\psi_0}{d\varphi^2} + \frac{2m}{\hbar^2}E\psi_0 = 0$$

If the body has an electric dipole moment μ, its potential energy in an electric field F is given by $V = -\mu F R \cos \varphi$. Calculate the ground-state energy of this rotator in the applied field F, to second order.

Problem 9.7: If an electric field F parallel to the polar axis is applied to a hydrogen atom, the perturbing potential on the electron is

$$V' = -eFr \cos \theta$$

Calculate the ground state energy to second order.

9.4 DEGENERATE LEVELS

Independent eigenfunctions (e.g. those with different l and m) having the same energy eigenvalue are said to belong to a *degenerate energy level*. Clearly the theory in Section 9.1 is incomplete for such levels because for two different eigenfunctions, ψ_i, ψ_k, the energy eigenvalues may be the same. Then Eq. (9.16), for instance, would yield an erroneous result for the relative contribution of the various states. In this section we will give a brief treatment only of the degenerate perturbation theory essential to an understanding of multi-electron atoms. (A more complete discussion will be found, for example, in Pauling and Wilson.) *We begin by considering only eigenfunctions belonging to the same energy level* and assume that the zero-order eigenfunctions are all orthogonal, as they are in actuality for the case of most interest here where the zero-order eigenfunctions are the hydrogen wave functions for electrons in the same atom.

The wave function for the perturbed system in this degenerate energy level is not as written in Eq. (9.7), but is expressed approximately in terms of the wave functions of the same energy level

$$\psi = \sum_{j=1}^{n} a_j \psi_j^0 \tag{9.7a}$$

where ψ_j^0 are the zero-order eigenfunctions for the particular energy level chosen and n is the order of the degeneracy. Note that for simplicity we have dropped the subscript i referring to the particular energy level chosen, since all the zero-order eigenfunctions of interest have the same energy eigenvalue. (Eigenfunctions belonging to a different energy level are neglected because they contribute very little to the perturbed eigenfunctions.)

Making the substitution of Eq. (9.7a) and Eqs. (9.7) into Eq. (9.6), we obtain

$$H^0 \Sigma a_j \psi_j^0 - E^0 \Sigma a_j \psi_j^0 + (V' - E') \Sigma a_j \psi_j^0$$
$$= (V' - E') \Sigma a_j \psi_j^0 = 0 \tag{9.9a}$$

If Eq. (9.9a) is multiplied on the left by $(\psi_k^0)^*$ and integrated over all space, a set of simultaneous equations for a_j can be obtained by letting k take all values from 1 to n.

$$\Sigma a_j V'_{kj} - a_k E' = 0 \tag{9.31}$$

or

$$a_1(V'_{11} - E') + a_2 V'_{12} + \ldots + a_n V'_{1n} = 0$$
$$a_1 V'_{21} + a_2(V'_{22} - E') + \ldots + a_n V'_{2n} = 0$$
$$\begin{array}{ccc} \cdot & \cdot & \cdot \\ \cdot & \cdot & \cdot \\ \cdot & \cdot & \cdot \end{array}$$
$$a_1 V'_{n1} + a_2 V'_{n2} + \ldots + \ldots + a_n(V'_{nn} - E') = 0$$

Unless the determinant of the coefficients of the a_j is zero, all the a_j must be identically zero

Therefore
$$
\begin{vmatrix}
(V'_{11} - E') & V'_{12} & \cdots & V'_{1n} \\
V'_{21} & (V'_{22} - E') & \cdots & V'_{2n} \\
\vdots & \vdots & & \vdots \\
& & \cdots & (V'_{nn} - E')
\end{vmatrix} = 0 \qquad (9.32)
$$

where the degeneracy of the energy level is n-fold. It frequently happens that the perturbation potential taken between two states is zero unless the two states are the same. That is to say, $V'_{jm} = \delta_{jm} V'_{jj}$. For example, this would occur if the perturbing potential were independent of angle, with the original wave functions being for a spherically symmetric potential. In this case the orthogonality of the Y_l^m would result in $V'_{jm} = \delta_{jm} V'_{jj}$. Equation (9.32) then becomes diagonal,

$$
\begin{vmatrix}
(V'_{11} - E') & 0 & \cdots & 0 \\
0 & (V'_{22} - E') & \cdots & 0 \\
\vdots & \vdots & & \vdots \\
& & \cdots & (V'_{nn} - E')
\end{vmatrix} = 0 \qquad (9.32a)
$$

with the particularly simple form
$$
(V'_{11} - E')(V'_{22} - E') \cdots\cdots (V_{nn} - E') = 0 \qquad (9.33)
$$
whose solutions are $E' = V'_{11}, V'_{22} \ldots, V'_{nn}$ for a level of n-fold degeneracy.

The original wave functions for this case are called the *correct zero-order wave functions*. Clearly, the problem is immensely simplified if we can find at the start a representation in which all the matrix elements not on the diagonal are zero. A spherically symmetric perturbing potential for a particle in a rectangular box, for example, would result in a nondiagonal matrix, if the original wave functions chosen were for the particle in a rectangular box. Hence, these wave functions would be an inappropriate choice for this problem.

Note that Eq. (9.33) with the V'_{jj} being given by $V'_{jj} = \int \psi_j^{0*} V' \psi_j^0 \, d\tau$ is essentially the same result as that obtained with nondegenerate perturbation theory Eq. (9.15) with the difference that in degenerate perturbation theory the perturbation potential energy V' must be evaluated for not just one eigenfunction belonging to E^0, but between all eigenfunctions in the same energy level E^0.

Problem 9.8: The $n = 3$ state for the hydrogen atom is actually a 9-fold degenerate state. Without bothering to solve the integrals obtained along the diagonal, set up the perturbation determinant for Problem 9.5. What are the off-diagonal elements?

9.5 DIRAC BRA AND KET NOTATION

To simplify calculations we will sometimes adopt a much more convenient notation for wave functions and matrix elements introduced by Dirac. Assuming ψ is a wave function for a single particle with quantum numbers n, l, m, we will now write ψ as

$$\psi_{n,l,m} = |n, l, m\rangle, \text{ a ket vector}$$

$$\psi_{n,l,m}^* = \langle n, l, m|, \text{ a bra vector}$$

and

$$\int \psi_{n,l,m}^* \psi_{n,l,m} d\tau = \langle n, l, m | n, l, m \rangle$$

where the last is a help in remembering the names of these quantities if we think of it as a bra-ket or bracket. In these expressions n is the total quantum number, l the orbital momentum, and m the magnetic quantum number. If we are not interested in one of the quantum numbers, say m, we may drop it out of our label. If the total wave functions contain additional quantum numbers of interest, we may include them also. The matrix element or expectation value of an operator is written with the operator between the vectors, e.g. $\langle n,l,m|0|n,l,m\rangle$. For example, in Section 9.1 the equation for the first-order energy shift for a given perturbation [Eq. (9.17)] could just as well be written

$$V'_{ki} = \langle k|V'|i\rangle$$

9.6 VARIATIONAL METHOD

Using the same form of the zero-order wave functions as used in perturbation theory, we may obtain a better estimate of the ground state or lowest energy of a perturbed system with little additional labor, by means of the variational method. The method rests on noting that the integral for the expectation of the energy operator in terms of an arbitrary trial wave function Φ (which is, in general, not the lowest eigenfunction of the system) is given by

$$E = \int \Phi^*\left(\frac{-\hbar^2}{2m}\nabla^2 + V\right)\Phi d\tau = \int \Phi^* H\Phi \, d\tau \qquad (9.34)$$

and overestimates the energy of the lowest state of the system. We first proceed to prove this statement. Let Φ be expanded in terms of the complete orthonormal set comprised of the true wave functions of the system:

$$\Phi = \sum_i a_i\psi_i \qquad (9.35)$$

and similarly for Φ^*. Then

$$E = \sum_i \sum_j a_i^* a_j \int \psi_i^* H \psi_j \, d\tau \qquad (9.36)$$

Since $H\psi_j = E_j\psi_j$, and the ψ's are orthogonal, Eq. (9.36) becomes

$$E = \sum_j a_j^* a_j E_j \qquad (9.37)$$

The sum over all j of $a_j^* a_j$ is 1, and therefore $E \geqslant E_0$; E is equal to E_0 only if the a_j are zero for all $j \neq 0$, in which case $\Phi = \psi_0$ which means that we have chosen the true wave function for Φ. The worse the choice made for Φ, the greater will be the admixture of states above the ground state, leading to higher estimates of E.

In general, Φ should be chosen to resemble as closely as possible the form to be expected on physical grounds. For example, for nuclear problems a rapidly attenuating wave function should be used, such as an exponentially decaying wave function. A good approximate wave function is a known wave function appropriate to a very similar potential, that is, an unperturbed ground state wave function whose potential energy is assumed only slightly different from the actual potential energy. The only alteration required in the known wave function is to make Φ a function of some unspecified parameter, say α. The integral in Eq. (9.34) is then calculated as a function of α and the result minimized with respect to α, the resultant E being the closest estimate obtainable with the chosen wave function. If we use unnormalized wave functions, Eq. (9.34) may be rewritten

$$E = \frac{\int \Phi^*(\alpha, \mathbf{r})\left(\dfrac{-\hbar^2}{2m} \nabla^2 + V\right)\Phi(\alpha, \mathbf{r})\,d\tau}{\int \Phi^*(\alpha, \mathbf{r})\Phi(\alpha, \mathbf{r})\,d\tau} \qquad (9.38)$$

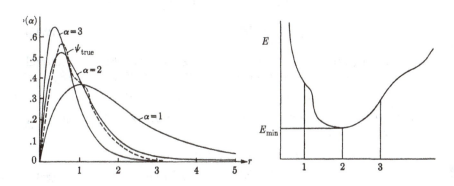

FIGURE 9.1 Illustration of how an approximate wave function $\Phi = re^{-\alpha r}$ may be adjusted to represent most closely an illustrative true wave function ψ_{true} (dashed line) by choice of the value of the parameter α that minimizes E given by Eq. (9.38) $\int \Phi H \Phi \, d\tau / \int \Phi^2 \, d\tau$.

For an illustration of what the variational method attempts to accomplish, see Figure 9.1, where trial wave functions $\Phi = re^{-\alpha r}$ are plotted for three different values of α. Φ represents an unknown wave function ψ_{true} which might have the form shown in the figure. The value of α for a best fit is found by minimizing the calculated energy given by $\int \Phi H \Phi \, d\tau / \int \Phi^2 \, d\tau$, as illustrated.

As a small example we attempt another calculation of the ground state of the helium atom, whose potential energy is given by Eq. (9.20). For Φ we choose the normalized zero-order wave functions used earlier in the chapter (e.g., in Eq. 9.24) but we let the nuclear charge be an undetermined parameter Z' representing an effective nuclear charge which is less than the actual charge, owing to the screening of the attraction of any one electron to the nucleus by the presence of the other electron. Thus

$$\Phi(Z', \mathbf{r}_1, \mathbf{r}_2) = \Phi_1 \Phi_2 = \left(\frac{Z'me^2}{\pi\hbar^2}\right)^3 e^{-(Z'me^2/\hbar^2)(r_1+r_2)} \tag{9.39}$$

where the subscripts 1, 2 refer to the two electrons. [Φ_1 and Φ_2 satisfy the hydrogen-like wave equation

$$\left(\frac{\hbar^2}{2m}\nabla_1^2 + \frac{Z'e^2}{r_1}\right)\Phi_1 = -Z'^2 W_H \tag{9.40}$$

where W_H has been obtained from Eq. (9.23) as $me^4/2\hbar^2$. Note that

$$V^0 = -\frac{Z'e^2}{r_1} - \frac{Z'e^2}{r^2} - (Z - Z')\left(\frac{e^2}{r_1} + \frac{e^2}{r_2}\right)\Big] \tag{9.41}$$

Substitution of Eqs. (9.39)–(9.41) into Eq. (9.38) yields

$$E = -2Z'^2 W_H + (Z' - Z)e^2 \int \Phi^*\left(\frac{1}{r_1} + \frac{1}{r_2}\right)\Phi d\tau$$
$$+ \int \Phi^* \frac{e^2}{r_{12}}\Phi d\tau \tag{9.42}$$

The first integral is trivial, resulting in $(Z' - Z) [4Z' W_H]$. The second integral has already been found in obtaining Eq. (9.25) to be $\frac{5}{4}W_H Z'$; hence

$$E = \{-2Z'^2 + 4Z'(Z' - Z) + \tfrac{5}{4}Z'\}W_H$$

Minimizing with respect to Z', that is, setting $\partial E/\partial Z' = 0$, we find that $Z' = Z - \frac{5}{16}$ so that

$$E = -2(Z - \tfrac{5}{16})^2 W_H$$
$$= (-2Z^2 + \tfrac{5}{4}Z - \tfrac{50}{256})W_H \tag{9.43}$$
$$= -76.9 \text{ ev}$$

Our previous result, -74.4 ev, was given by the first two terms of Eq. (9.43). The extra term $-\frac{50}{256} W_H = -2.5$ ev is an appreciable fraction of the 4.2 ev difference from the observed result of -78.6 ev .

Problem 9.9:   Find the ground state of the linear anharmonic oscillator whose potential energy is $V = kx^2 + 0.1k^2x^4$, using the trial wave function $\psi = ce^{-\alpha x^2}$.

Problem 9.10:   Assume a proton to be not a point charge but a sphere with radius a of 10^{-13} cm; then find the ground state energy of the hydrogen atom if the positive charge is distributed in such a way as to cause (use ground state hydrogenic wave functions for trial functions) $V(r) \sim -e^2/(r + a)$.

Problem 9.11:   Find the ground state energy of a particle in a spherically symmetric square well for a particle mass $M = 1.6 \cdot 10.^{-24}$ g, $V_0 = -20$ Mev, and a well radius of $1.5 \cdot 10^{-13}$ cm. Use a trial wave function of $\psi = (c/r)e^{-\alpha r}$.

Problem 9.12:   When tritium, H^3, beta-decays to He^3, the single atomic electron suddenly finds the positive nuclear charge doubled. If the initial tritium atom was in its ground state, what is the probability of finding the He^+ ion in its ground state instantaneously following the decay? (That is, take the ground state hydrogen wave function and express it as a series of helium ion wave functions (hydrogenic with $Z = 2$) and find the coefficient of the first term.)

BIBLIOGRAPHY

Pauling, L. and E. B. Wilson, cited in Chapter VIII.

X

TIME-DEPENDENT PERTURBATION THEORY

Up to this point our attention has been directed to static systems under-going no temporary interactions with other systems. Since the systems have not experienced outside forces and thus have exchanged no energy with other systems, the energy has remained constant. Under such circumstances involving time-invariant potentials, the complete wave equation (4.18),

$$\left(-\frac{\hbar^2}{2m}\nabla^2 + V\right)\Psi = H\Psi = -\frac{\hbar}{i}\frac{\partial \Psi}{\partial t} \tag{10.1}$$

is readily simplified, since the left-hand side does not involve the time, and therefore the space and time variables are separable. The time-dependent part of Ψ satisfies the differential equation

$$\frac{1}{\Psi(\mathbf{r}, t)}\frac{\hbar}{i}\frac{\partial \Psi(\mathbf{r}, t)}{\partial t} = -E \tag{10.2}$$

where E is a constant which was shown in Chapter V to be the energy of the system. The solution of Eq. (10.2) is

$$\Psi(\mathbf{r}, t) = \psi(\mathbf{r})e^{-(i/\hbar)Et} \tag{10.3}$$

After substituting (10.3) in (10.1) and then dividing out the time-dependent part of Ψ, we are left with the space-dependent part of Ψ, to which we have devoted our attention in previous chapters:

$$H\psi = E\psi \qquad (10.4)$$

There are many problems of paramount significance, however, in which an isolated static system is acted on by some external time-dependent disturbance. For example, a hydrogen atom in its ground state might experience an electromagnetic wave or photon incident upon it. The photon could cause a transition of the atom to an excited or ionized state. Calculation of the probability or cross section of transition due to the absorption of a photon is of outstanding physical importance.

For such a problem, the space and time variables in Eq. (10.1) are not separable since H or V, which includes the effect of the incident photon, is time-dependent. The resulting differential equation is of such complexity that in the past it has been solved entirely by the use of time-dependent perturbation theory. This has been a truly desperate measure in many problems in which the perturbation is not small (e.g., nuclear interactions), and that it has been virtually the only tool available is an indication of the difficulty of time-dependent problems, including static problems which have been treated as time-dependent perturbations. A major difficulty in theoretical physics, particularly with respect to the problem of nuclear forces, has been the lack of a suitable alternative approximation technique when the time-dependent potential is too large for time-dependent perturbation approximations to converge rapidly or at all. The theory will be applied here only to electromagnetic interactions for which, fortunately, the approximations converge very rapidly. Electromagnetic interactions are very weak compared with nuclear interactions.

10.1 FIRST-ORDER TIME-DEPENDENT TRANSITION PROBABILITY

Dirac derived the time-dependent perturbation approximation technique at almost the same time that Schrödinger derived the time-independent theory (Chapter IX). As will be seen shortly, the two methods are very similar and have many parallels. In the time-dependent theory the system might be assumed initially in a time-independent state whose wave function satisfies the zero-order, time-independent wave Eq. (10.4); then the system is socked momentarily by an outside disturbance, after which it is left in a possibly different state with a new solution to Eq. (10.4). To phrase the problem more exactly: A static system whose wave function satisfies Eq. (10.4) suddenly experiences a perturbing potential (possibly representing an incident photon) for a short time, during which its wave function is an inseparable function of time and space; it is then restored to its previous static condition, except for a possibly changed eigenvalue (e.g., energy or angular momentum) and eigenfunction. Since the perturbing potential is assumed small compared with the initial energy of the system, the time-dependent wave functions are expanded in terms of the complete orthonormal set of static wave functions. The coefficients of the series must, of course, be time-dependent because of the effect of the perturbing potential.

The wave functions of the initial unperturbed static system satisfy the wave equation for the static potential,

$$H_0 \Psi_n = \left(\frac{-\hbar^2}{2m} \nabla^2 + V \right) \Psi_n = -\frac{\hbar}{i} \frac{\partial \Psi_n}{\partial t} = E_n \Psi_n \qquad (10.5)$$

For the duration of the perturbation, the appropriate complete wave equation is

$$(H_0 + H')\Psi = -\frac{\hbar}{i} \frac{\partial \Psi}{\partial t} \qquad (10.6)$$

where H' represents the small, time-dependent perturbation. For example, H' could be the result of a relatively constant electric field due to the incidence of a photon having a frequency small compared with the electron's orbital frequency. We could take the case where H' does not contain the time explicitly. The perturbation begins at time zero and ends at some later time. The solution to Eq. (10.6) at any particular instant of time t may be written in terms of the complete orthonormal set of wave function solutions to Eq. (10.5):

$$\Psi(\mathbf{r}, t) = \sum_{n=0}^{\infty} a_n(t) \psi_n(\mathbf{r}) e^{-(i/\hbar) E_n t} \qquad (10.7)$$

where the $a_n(t)$ are functions only of the time and the $\psi_n(\mathbf{r})$ are functions only of the coordinates. If we were to measure some physical property of the system, e.g., an eigenvalue such as the energy, the probability of obtaining a particular value of the energy E_k of the state ψ_k would be given by $|a_k(t)|^2$. In other words the probability of finding the system in the state ψ_k at time t is given by $|a_k(t)|^2$. The solution to any problem is found, of course, by determining the time-dependent coefficients $a_n(t)$.

Just as in the time-independent perturbation theory, equations for the coefficients result from substituting Eq. (10.7) into Eq. (10.6):

$$\sum_n a_n(t) H_0 \psi_n e^{-(i/\hbar) E_n t} + \sum_n a_n(t) H' \psi_n e^{-(i/\hbar) E_n t}$$

$$= -\frac{\hbar}{i} \sum_n \dot{a}_n(t) \psi_n e^{-(i/\hbar) E_n t} + \sum_n a_n(t) \psi_n E_n e^{-(i/\hbar) E_n t} \qquad (10.8)$$

Utilizing Eq. (10.5) to eliminate the first and last terms, we obtain

$$i\hbar \sum_n \dot{a}_n(t) \psi_n e^{-(i/\hbar) E_n t} = \sum_n a_n(t) H' \psi_n e^{-(i/\hbar) E_n t} \qquad (10.9)$$

We next multiply Eq. (10.9) by Ψ_k^* (**r**, t) on the left and integrate over all configuration space. The orthogonality of the wave functions requires all resulting integrals on the left-hand side of Eq. (10.9) to be zero except for the kth term in the sum; hence

$$i\hbar\dot{a}_k(t) = \sum_n a_n(t)H'_{kn}e^{-(i/\hbar)(E_n - E_k)t} \tag{10.10}$$

where

$$H'_{kn} \equiv \int \psi_k^*(\mathbf{r})H'\psi_n(\mathbf{r})\, d\tau \tag{10.11}$$

Equation (10.10) is an infinite set of simultaneous equations for the $a_k(t)$.

If the perturbation is small and acts for times sufficiently short that the final wave function of the system has little probability of differing from that for the initial state, an approximate solution to Eq. (10.10) is readily found. Suppose the initial state of the system is ψ_{n_0}, that is, at $t = 0$ all the $a_n(0)$ are zero except a_{n_0} which is unity. In this case, Eq. (10.10) becomes

$$i\hbar\dot{a}_k(t) = H'_{kn_0}\exp\{-(i/\hbar)(E_{n_0} - E_k)t\} \tag{10.12}$$

$$a_k(t) = \frac{1}{i\hbar}\int_0^t H'_{kn_0}\exp\left\{-\frac{i}{\hbar}(E_{n_0} - E_k)t\right\} dt \tag{10.13}$$

If H'_{kn_0} is a nearly constant function of the time, that is, if it does not change appreciably over one cycle of the oscillation of the exponential function, the transition probability will be small, due to cancellations from times of opposing phase. That is, the exponential in (10.13) causes equal negative and positive contributions of the integrand to the integral if H'_{kn_0} is nearly constant, and the value of the integral will be nearly zero. If H'_{kn_0} is caused by electromagnetic radiation, $e^{i\omega t}$, or any other periodic motion, the transition probability will be large only for disturbance periods given by

$$T \text{ (max. disturbance)} = h/(E_k - E_{n_0}) \tag{10.14}$$

since destructive interference from times of opposing phase will be missing. For this reason, the electron excitation, and capture-and-loss cross sections of atoms when bombarded by other atoms or ions, show very sharp maxima when the incident particle velocity corresponds to the electron orbital velocity, as will be seen in the next section. An even more striking example is the sharp resonance in the photoelectric effect when the frequency of the incident radiation is given by the Bohr frequency condition,

$$\hbar\omega = E_k - E_{n_0} \tag{10.15}$$

If H'_{kn_0} is nearly constant over a period of the exponential, it may be taken out from under the integral sign and Eq. (10.13) can be written after a remaining trivial integration over the time

$$a_k(t) = H'_{kn_0}\frac{1 - \exp\{-(i/\hbar)(E_{n_0} - E_k)t\}}{E_k - E_{n_0}}, \quad k \neq n_0 \tag{10.16}$$

The probability of transition to the state k, that is, the probability of finding the system in the state k after a time t, is obtained from $a_k^*(t)a_k(t)$,

$$a_k^*(t)a_k(t) = |H'_{kn_0}|^2\, 2 \left[\frac{1 - \cos\,[(E_{n_0} - E_k)(t/\hbar)]}{(E_{n_0} - E_k)^2}\right] \qquad (10.17)$$

Note that if H'_{kn_0} is completely independent of the time (i.e., if the disturbance acts for an infinite time), then since the time or phase average of $1 - \cos x = (2 \sin^2 x/2)$ is $1/2$, the same result is obtained as in time-independent perturbation theory, Eq. (9.1).

Problem 10.1: What is the probability of finding the physical system (atom, molecule, or any other submicroscopic physical entity) in the state n_0 after a time t seconds? How does this compare with the wave function that would be obtained with the time-independent perturbation theory of Chapter IX? Discuss. (Hint: Integrate Eq. (10.10) over the time with $k = n_0$ and $a_{n_0}(0) = 1$.)

10.2 EXCITATION OF ATOMS BY ION BOMBARDMENT

As an illustration of the time-dependent perturbation theory we will discuss atomic excitation by alpha particles (He^+ ions), as first treated by J. A. Gaunt. In this example an atomic electron is momentarily struck by an ion whose effect can be represented by the time-dependent perturbation potential

$$V = -Ze^2/r \qquad (10.18)$$

where Z is the charge of the incident ion and r, the distance of the incident ion from an atomic electron, is given in terms of the electron coordinates in the plane of impact, x and y, as shown in Figure 10.1. Only a distant encounter of the alpha particle and the atomic electron is considered here; that is, $x^2 + y^2 \ll b^2$ where b represents the distance of closest approach (the perpendicular distance of the ion trajectory from the target atom). Let v be the ion velocity, t the time measured from the moment of closest approach, and $r_0^2 = b^2 + (vt)^2$; then if x and y are always much less than r,

$$r = [(b - x)^2 + (vt - y)^2]^{1/2}$$

$$\sim [b^2 + (vt)^2]^{1/2} \left[1 - \frac{xb + yvt}{r_0^2}\right] \qquad (10.19)$$

Substituting this potential into Eq. (10.13) and integrating over the ion trajectory, that is, over the time from minus to plus infinity, we obtain

$$a_k = \frac{-Ze^2}{i\hbar} \int_{-\infty}^{\infty} dt \int \psi_k^*(\mathbf{r}) \left\{\frac{1}{\sqrt{b^2 + (vt)^2}}\left[1 + \frac{xb + yvt}{b^2 + (vt)^2}\right]\right\} \psi_{n_0}\, d\tau$$

$$\times \exp\{-(i/\hbar)(E_{n_0} - E_k)t\} \qquad (10.20)$$

where the ψ's are, of course, the atomic electron wave functions, and the space integral is taken over the atomic electron coordinates. Since the ψ's are orthogonal, only the second term in the braces gives a finite result, and hence Eq. (10.20) becomes

$$a_k = \frac{-Ze^2}{i\hbar} \int_{-\infty}^{\infty} dt \int \frac{(\psi_k^* x \psi_{n_0})b + (\psi_k^* y \psi_{n_0})vt}{[b^2 + (vt)^2]^{3/2}} d\tau$$

$$\times \exp\{-(i/\hbar)(E_{n_0} - E_k)t\} \qquad (10.21)$$

The rest is a matter of arithmetic and the solution will not be continued. Note that the integrand in (10.21) consists of an imaginary exponential

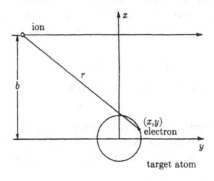

FIGURE 10.1 Illustration of collision of positive ion with atomic electron. The ion path deviates very slightly because of collision with the electron.

function of time and a cumbersome time-dependent coefficient. Note further that when the coefficient of the exponential in the integrand changes appreciably within the period of the exponential oscillation, least cancellation between contributions to the integral of opposite phase occurs. Thus, if the coefficient to the exponential in the integral changes appreciably within the time given by Eq. (10.14), a_k does indeed become a maximum. The period of maximum transition probability is given roughly by the time in which the denominator doubles in magnitude from its minimum value at $t = 0$:

$$T \sim b/v \sim h(E_{n_0} - E_k)^{-1} \sim h/E_{n_0}$$

that is, the period of the atomic electron in its original unexcited state. Since a_k is proportional to $1/b^2$, the transition probability increases as b decreases down to $b =$ orbital radius where the approximation fails. Hence, for maximum excitation or ionization probability the incident ion velocity should correspond to b/T where b is of the order of the atomic radius and T is the period of the electron ground state wave function $2\pi b/v_e$. Phrased differently, maximum atomic excitation results when the alpha particle velocity is of the order of the classical electron orbital velocity.

Problem 10.2: By integrating Eq. (10.21) over the time, estimate the conditions under which the x and the y terms are most important to a transition.

10.3 ABSORPTION OF RADIATION

The next matter to be discussed is the transitions induced by incident electromagnetic radiation, that is, the probability of absorption of a photon. For optical wavelengths (thousands of angstroms), the incident radiation consists of a time-varying electric field vector whose wavelength is so much greater than atomic dimensions that the electric field may be regarded as constant in space. The x component of the incident electric field vector may be written

$$E_x = E_{0x}(e^{i\omega t} + e^{-i\omega t}) \qquad (10.22)$$

and the perturbation potential energy is given by

$$V = E_x \sum_j e_j x_j \qquad (10.23)$$

where x_j is the x displacement of the atomic electron j in the electric field and the summation must be carried out over all the electrons in the atom. This perturbing potential is then substituted into Eq. (10.13), to yield

$$a_k(t) = \frac{E_{0x}}{i\hbar} \sum_j e_j \int_0^t \left(\int \psi_k x_j \psi_{n_0} \, d\tau \right)$$
$$\times \exp\{-(i/\hbar)(E_{n_0} - E_k)t'\}(e^{i\omega t'} + e^{-i\omega t'}) \, dt' \qquad (10.24)$$

The sum over j of e_j times the space integral is the x component of the dipole moment of the system, which we shall represent by μ, so that the x component of μ taken between the states k and n_0 will be

$$\mu_{x_{kn_0}} = \sum_j e_j \int \psi_k^*(x_j)\psi_{n_0} \, d\tau \qquad (10.25)$$

Performing the indicated integration of (10.24) over the time, we obtain

$$a_k(t) = E_{0x}\mu_{x_{kn_0}} \left\{ \frac{1 - \exp\{(i/\hbar)(E_k - E_{n_0} - \hbar\omega)t\}}{E_k - E_{n_0} - \hbar\omega} \right.$$
$$\left. + \frac{1 - \exp\{(i/\hbar)(E_k - E_{n_0} + \hbar\omega)t\}}{E_k - E_{n_0} + \hbar\omega} \right\} \qquad (10.26)$$

Only the first term in this expression yields an appreciable result, and that only for frequencies ω close to $(E_k - E_{n_0})/\hbar$, at which the denominator becomes infinitesimal. (n_0 is assumed to be the ground state.) Since

$$|\exp\{(i/2\hbar)(E_k - E_{n_0} - \hbar\omega)t\}$$
$$\times [\exp\{-(i/2\hbar)(E_k - E_{n_0} - \hbar\omega)t\} - \exp\{(i/2\hbar)(E_k - E_{n_0} - \hbar\omega)t\}]|^2$$
$$= 4\sin^2[(1/2\hbar)(E_k - E_{n_0} - \hbar\omega)t]$$

therefore

$$|a_k(t)|^2 = 4(\mu_{x_{kn_0}})^2 E_{0x}^2 \frac{\sin^2 [(1/2\hbar)(E_k - E_{n_0} - \hbar\omega)t]}{(E_k - E_{n_0} - \hbar\omega)^2} \qquad (10.27)$$

Equation (10.27) is the probability as a function of time for electromagnetic radiation (of frequency ω, intensity $2E_{0x}^2$, and polarization along the x axis) to be absorbed by an atom, which then undergoes a transition from a state n_0 to a state k as a result of the absorption.

In many cases of interest not involving resonance effects, when the incident radiation is not bright-line spectra (say from the same atoms or nuclei as are absorbing the radiation), the intensity of the incident radiation is essentially a constant function of frequency compared with the frequency dependence of the absorption function (Eq. 10.27). Under these circumstances, Eq. (10.27) must be integrated over all the frequencies contained in the incident radiation. The right side of Eq. (10.27) is negligible for all frequencies except that for which the denominator becomes zero, and therefore the integral may as well extend from zero to infinite frequency. We may note that the frequency-dependent part of (10.27) is essentially a Dirac delta function $\delta(E_k - E_{n_0} - \hbar\omega)$, for

$$\int_0^\infty \frac{\sin^2 ([\hbar\omega - (E_k - E_{n_0})]t/2\hbar)}{[\hbar\omega - (E_k - E_{n_0})]^2} \, d\omega = \frac{t}{2\hbar^2} \int_0^\infty \frac{\sin^2 (x - x_0)}{(x - x_0)^2} \, dx = \frac{\pi t}{2\hbar^2}$$

where $x_0 = (E_k - E_{n_0})t/2\hbar$. Hence integration of Eq. (10.27) over all frequencies would yield the following result for a frequency continuum of incident radiation:

$$|a_k(t)|^2 = \frac{\pi}{\hbar^2} (\mu_{x_{kn_0}})^2 E_{0x}^2 (\omega_{kn_0}) t \qquad (10.28)$$

where ω_{kn_0} is the frequency for which the denominator in Eq. (10.27) is zero and $E_{0x}(\omega_{kn_0})$ is the amplitude of the incident electric vector at that frequency. Note that because of the integration over the frequency the transition probability is linearly proportional to the time, for short times, instead of quadratically dependent on time as Eqs. (10.17, 10.27). Given the intensity of the incident radiation, it is trivial to compute the transition probability per unit time from Eq. (10.28), once the interaction integrals over all space, involving the original and final state wave functions, are known. We will now compute these integrals.

10.4 GENERAL SELECTION RULES FOR SPHERICAL HARMONIC WAVE FUNCTIONS

In Chapter VIII the general solution to a wave equation for a system with a spherically symmetric potential was shown to be expressible in product wave functions of the separated coordinate variables. For the important class of problems for which the unperturbed potential function is spherically symmetric it is well to express the dipole moment in spherical coordinates. The transformation from Cartesian to spherical coordinates is readily made. The x, y, z components of the dipole moment operator are

$$\mu_x = \mu(r) \sin \theta \cos \varphi$$
$$\mu_y = \mu(r) \sin \theta \sin \varphi \qquad (10.29)$$
$$\mu_z = \mu(r) \cos \theta$$

The dipole moment may result from radiation, an incident alpha particle, or other sources of perturbing potentials. $\mu(r)$ is a function of r alone, say er in the simple case of an atomic electron at radius r from the center of positive charge.

Hence the x component of the dipole moment between two spherical harmonic wave functions becomes

$$\mu_{x_0, f} = \iiint \{R^*_{n_f l_f}(r)\, \Theta^*_{l_f m_f}(\theta)\, \Phi^*_{m_f}(\varphi)\, \mu(r) \sin \theta \cos \varphi$$
$$\times\, R_{n_0 l_0}(r)\, \Theta_{l_0 m_0}(\theta)\, \Phi_{m_0}(\varphi)\, r^2 \sin \theta\} \, dr \, d\theta \, d\varphi \qquad (10.30)$$

where the subscript on the left-hand member has been abbreviated to 0, f for original and final states. The letter n is the total quantum number for hydrogen wave functions or the radial quantum number for most other wave functions. The orbital quantum number l and the magnetic quantum number m complete the three eigenvalues needed to specify completely a three-dimensional wave function. The three integrations may be rewritten as a product of three separate integrals. For the wave function in φ, Eq. (8.6), one obtains an integral over φ for μ_x, for example,

$$\frac{1}{\pi} \int_0^{2\pi} e^{-im_f \varphi} \left(\frac{e^{i\varphi} + e^{-i\varphi}}{2} \right) e^{im_0 \varphi} \, d\varphi \qquad (10.31)$$

since $\cos \varphi = (e^{i\varphi} + e^{-i\varphi})/2$.

The integral of $e^{-i(m_f - m_0 - 1)\varphi}$ is zero unless $m_f - 1 - m_0 = 0$, when its value is unity. Including the integral of $e^{-i\varphi}$, the entire integral over φ in Eq. (10.31) will be zero unless

$$m_f = m_0 \pm 1 \qquad (10.32)$$

in which cases it is unity. Thus we obtain our first selection rule, Eq. (10.32), giving the changes in magnetic quantum number permitted when the incident field is polarized in the x direction. For an incident field polarized in the z direction the resultant selection rule for μ_z would be $m_f = m_0$, i.e., no change in m is possible. This is understandable if one realizes that a force in the z direction can have no effect on the motion in the xy plane which is responsible for the component of the angular momentum along the z axis represented by the m, or magnetic, quantum number. A dipole transition may result from a passing ion, radiation, or any other perturbing potential.

We have obtained the universal and important result that the magnetic quantum number can change by no more than one unit in a dipole transition between unperturbed wave functions resulting from spherically symmetric potentials. That is,

$$m_f = m_0 \quad \text{or} \quad m_0 \pm 1 \tag{10.32a}$$

In a similar way selection rules for the l quantum number may be obtained from the integral over θ. For μ_z, for example, and $m = 0$, for which the Θ functions are Legendre polynomials, the θ integral in (10.30) would be

$$\int_0^1 P_{l_f}(x) x P_{l_0}(x) \, dx \tag{10.33}$$

where $x = \cos \theta$. Since the Legendre polynomials are orthogonal and x times a polynomial is a different polynomial, l_f must be different from l_0 in order that a finite result be obtained. Indeed, it is a property of the Legendre polynomials that they satisfy the relation

$$x P_l(x) = \frac{(l+1)P_{l+1}(x) + lP_{l-1}(x)}{2l+1} \tag{10.34}$$

so that Eq. (10.33) is zero unless

$$l_f = l_0 \pm 1 \tag{10.35}$$

in which cases the integral (10.33) would equal either $(l + 1)/(2l + 1)$ or $l/(2l + 1)$. When $m \neq 0$ and the associated Legendre polynomials must be used, the same selection rule for l is obtained. That is, *in a dipole transition the orbital angular momentum must change by one unit for a one-electron wave function.*

Total angular momentum is conserved as it must be always, because a photon has an intrinsic spin of unity. Photons may be represented as a superposition of circularly polarized electromagnetic waves each one of which always has one unit of angular momentum along its direction of propagation, positive or negative according to whether the light is right or left circularly polarized. It is of interest to note that such an effect can be true only for massless motion (photon or neutrino) with the speed of light. For slower particles one can always transform to a velocity frame in which the particle's motion is reversed, in which case angular momentum cannot be conserved if the particle's intrinsic spin or angular momentum is quantized along its direction of propagation.

For matrix elements different from the dipole transition expressed in Eq. (10.25), the light quantum may have additional units of angular momentum; aside from its intrinsic angular momentum, its wave function may possess orbital momentum about the origin in non-dipole matrix elements such as quadrupole or octupole transitions. In general, the electric field of an incident plane wave would be given by

$$E_x = E_{0x} e^{i(kz - \omega t)} \tag{10.22a}$$

where k is the wave number of the incident photons. A plane wave e^{ikz} may be expanded in terms of the complete orthonormal set of spherical harmonic wave functions:

$$e^{ikz} = e^{ikr\cos\alpha} = \sum_{n=0}^{\infty}(2n+1)i^n\sqrt{\frac{\pi}{2kr}}\,J_{n+1/_2}(kr)P_n(\cos\alpha) \qquad (10.36)$$

where $J_{n+1/_2}$ is a Bessel function of order $n + \frac{1}{2}$ and α is measured from the propagation vector of the plane wave, *NOT necessarily from the atomic coordinate axis* as, for example, in Eq. (10.29). If Eqs. (10.36) and (10.22a) are substituted into (10.23) and the computation repeated, we will obtain elements other than the dipole transition matrix elements for $n \neq 0$. Near the origin

$$J_{n+1/_2}(kr) \sim \frac{1}{(n+\frac{1}{2})!}\left(\frac{kr}{2}\right)^{n+1/_2}$$

Thus the expansion yields a quadrupole moment r^2, an octupole moment r^3, and so on for the higher-order terms. In virtually all electromagnetic transitions (nuclear or atomic), $kr < < 1$ for r less than the nuclear or atomic radius, which makes the expansion particularly useful since it converges rapidly.

The only remaining integral is the radial integral and that is a function of the radial dependence of the perturbing potential and the unperturbed radial wave functions. Hence no general selection rules or other results can be obtained for the radial integrals. Selection rules for the special cases of the harmonic oscillator and the hydrogen atom will be discussed in the next two sections.

10.5 SELECTION RULES FOR THE HARMONIC OSCILLATOR RADIAL QUANTUM NUMBER

The harmonic oscillator is of great physical interest since it represents electromagnetic radiation confined in a box, nucleons in a nucleus, and diatomic molecular vibrations. The dipole moment between different harmonic oscillator wave functions yields the probability of transitions between nuclear states of different energy, or of a light quantum being added to or subtracted from the total radiation in a box. The radial harmonic oscillator wave functions, Eq. (8.49), are comprised in part of an even or odd power series in r, the even or odd spherical Hermite polynomials. Similar to the case of the associated Legendre polynomials, the radial integral yields the following selection rules:

$$\begin{aligned} n_f &= n_0, n_0 - 1 \quad \text{for} \quad l_f = l_0 + 1 \\ n_f &= n_0 + 1, n_0 \quad \text{for} \quad l_f = l_0 - 1 \end{aligned} \qquad (10.37)$$

For light waves in a rectangular box, the one-dimensional harmonic oscillator integral gives

$$n_f = n_0 \pm 1$$

corresponding to the creation of one quantum in addition to those already contained in the box, or to the annihilation or loss of a quantum due to some absorption process. For a nucleon or a material oscillator representing a light wave, the wave equation in one dimension is

$$\frac{\partial^2 \psi}{\partial x^2} + \frac{2m}{\hbar^2}\left(E - \frac{m}{2}\omega_0^2 x^2\right)\psi = 0 \qquad (10.38)$$

whose solution is

$$\psi_n = A_n H_n(\sqrt{m\omega_0/\hbar}\,x)e^{-m\omega_0 x^2/2\hbar} \qquad (10.39)$$

where H_n is the Hermite polynomial of order n and A_n is a normalization constant. The energy eigenvalue of the nth wave function is

$$E_n = (n + \tfrac{1}{2})\hbar\omega_0 \qquad (10.40)$$

which may be interpreted as meaning that the material harmonic oscillator of frequency ω_0 has n quanta of energy. If a photon is emitted, one quantum is lost from the oscillator and the energy of the new state is $(n - \tfrac{1}{2})\hbar\omega_0$, the eigenvalue of the ψ_{n-1} wave function. The harmonic oscillator is a convenient way to describe the radiation field, since the energy of the radiation field is also given by $E_n = n\hbar\omega_0$. Of course the photon wave functions are quite different from the material oscillator wave functions, since they are solutions to a different wave equation.

10.6 THE HYDROGEN ATOM

Pauli evaluated some of the radial dipole moment integrals for the hydrogen atom. From the previous discussion of the spherical integrals it was learned that all spherically symmetric potentials result in the selection rules Eqs. (10.32a, 10.35) for the total orbital and magnetic quantum numbers. By use of the wave functions for the electron in the hydrogen atom, Eqs. (8.38, 8.45), the radial electric dipole moment integral was evaluated. For $l_0 = 0$, $n_0 = 1$ (the Lyman series) ($l_f = 1$ in accordance with the selection rule),

$$\mu_{n_f n_0} = \int R^*_{n_f l_f}(r)\, r\, R_{n_0 l_0}(r)\, r^2\, dr$$

$$= \frac{(n_f - 1)^{2n_f - 1}}{n_f(n_f + 1)^{2n_f + 1}} \qquad (10.41)$$

and similarly, for $l_0 = 0$ or 1 ($l_f = 1$ or $0,2$) and $n_0 = 2$ (the Balmer series),

$$\mu_{n_f n_0} = \frac{(n_f - 2)^{2n_f - 3}}{n_f(n_f + 2)^{2n_f + 3}}(3n_f^2 - 4)(5n_f^2 - 4) \qquad (10.42)$$

As can be seen from Eqs. (10.41, 10.42), no selection rule results for the radial quantum number. Thus dipole transitions are possible from any radial quantum number to any other radial quantum number, if l changes by only one unit and m changes by no more than one unit. The relative absorption probabilities for different transitions are given by the ratios of the squares of Eqs. (10.41) and (10.42). The Lyman spectra absorption probability decreases rapidly as n_f increases, behaving roughly as n_f^{-6}.

Problem 10.3: Find the probability of transition of a hydrogen atom from the ground state to the state $n,l,m = 2,1,1$ if an alpha particle passes with velocity v at a distance b from the hydrogen nucleus.

Problem 10.4: The potential for a neutron interacting with an electron might be represented by an attractive spherical square well 10^4 ev deep and 10^{-12} cm in radius. If a neutron interacts for 10^{-17} second with a hydrogen electron in its ground state, what are the probabilities of its causing transitions to the first three excited states? (Assume the neutron is located at the center of the atom during this time.)

Problem 10.5: If an electric field of 10^4 volts/m acts for 10^{-17} second on a hydrogen atom in its ground state, what is the probability of finding the atom subsequently in a 2s or 2p state?

10.7 EMISSION

So far all the theoretical development in this chapter pertains strictly to absorption. We would have to use a more difficult and sophisticated procedure for direct calculation of emission; instead, we will show by a thermodynamic argument that emission and absorption are equivalent except for a simple coefficient, and thus show how the emission may be calculated in terms of the absorption. The procedure will be to find the emission in terms of the absorption.

Let the ground state of a radiating and absorbing system be labeled i and an excited state k. The rate of absorption is linearly proportional to the density of incident radiation $u(v)$ times the number of atoms in the ground state N_i,

$$\text{Absorption rate} = Bu(v)N_i \qquad (10.43)$$

where B is a constant known as the *Einstein B* coefficient. We have calculated the absorption of radiation in Section 10.3. Since E_{0x} occurring in Section 10.3 is only one of three components of the radiation in a box (E_{0y}, E_{0z} being the others) $E_{0x}^2 = u(v)/3$. Equation (10.28) gives the transition probability for one atom in time t seconds. Hence the absorption rate per unit volume is

$$|a_k(t)|^2 N_i/t = \frac{\pi}{3\hbar^2}(\mu_{ki})^2 u(v)N_i \qquad (10.43a)$$

Comparison of (10.43a) with (10.43) yields

$$B = \frac{\pi}{3\hbar^2} |\mu_{mn}|^2 \qquad (10.44)$$

It is convenient to divide the emission process into two parts, the spontaneous emission which occurs independently of the environment and the induced emission resulting when the excited system is in a bath of radiation:

$$\text{Emission rate} = AN_k + Cu(v)N_k \qquad (10.45)$$

where A is the *Einstein A* coefficient for spontaneous emission and C is the coefficient for induced emission. The induced emission results from the second term in Eq. (10.26), and therefore it should have the same constant coefficient as the absorption resulting from the first term in Eq. (10.26). We shall soon discover by the present thermodynamic argument that this is indeed the case.

At equilibrium the absorption and emission rates must be equal, that is,

$$Bu(v)N_i = AN_k + Cu(v)N_k$$

whence

$$\frac{A}{Bu(v)} + \frac{C}{B} = \frac{N_i}{N_k} \qquad (10.46)$$

The number of systems in a state having energy E is proportional to $e^{-E/kT}$, a result borrowed from statistical mechanics which we used in Chapter I to derive Eq. (1.13). Therefore, the ratio of unexcited atoms to excited atoms, occurring on the right-hand side of (10.46), is

$$\frac{N_i}{N_k} = e^{-(E_i - E_k)/kT} = e^{hv/kT} \qquad (10.47)$$

where v is the frequency of the absorbed or emitted radiation. $u(v)$ which also occurs in (10.46) is given by Planck's formula, Eq. (1.21),

$$u(v)\, dv = \frac{8\pi hv^3/c^3}{e^{hv/kT} - 1}\, dv \qquad (1.21)$$

Thus Eq. (10.46) becomes

$$\frac{c^3}{8\pi hv^3} (e^{hv/kT} - 1) \frac{A}{B} + \frac{C}{B} = e^{hv/kT} \qquad (10.48)$$

This equation holds at any temperature or frequency. Since it must hold at $T = \infty$, $C = B$, and since it must hold at $T = 0$,

$$A = \frac{8\pi hv^3}{c^3} B \qquad (10.49)$$

Thus, we do not need to rely on quantum electrodynamics, which is beyond the scope of this text, to calculate the spontaneous emission of an excited oscillator; we may find the spontaneous emission from the simple absorption calculations given in this chapter. For example, the lifetime for an excited state pending a dipole transition is given by

$$\tau = \frac{1}{A} = \frac{3\hbar c^3}{4\omega^3} |\mu_{mn}|^{-2} \qquad (10.50)$$

Problem 10.6: What is the lifetime of a hydrogen atom in the ($n = 2$, $l = 1$, $m = 0$) state? In the ($n = 3$, $l = 2$, $m = 0$) state?

Problem 10.7: A diatomic molecule vibrates approximately like a harmonic oscillator, with a fundamental frequency of the order of 10^{10} hertz. If diatomic molecules are inside a resonant microwave cavity of the same frequency in the $(1,1,0)$ mode and $E = E_z = 10^4$ volts/m, what is the rate of transition between the ground and first excited states? What fraction of the molecules are in an excited state?

10.8 SECOND-ORDER TIME-DEPENDENT PERTURBATION THEORY

As in time-independent perturbation theory, the first-order time-dependent theory involving a direct transition from an initial state n_0 to a final state k may give a null result. That is, we may find $H'_{k n_0} = 0$. Under these circumstances it is necessary to multiply Eq. (10.9) by Ψ_j^* on the left where Ψ_j^* represents any state that may be reached by a direct transition. Thus we will have $H'_{j n_0} \neq 0$ and, if $H'_{j n_0}$ is nearly constant over a period of the exponential, we will obtain an amplitude for transition to the state Ψ_j given by

$$a_j(t) = H'_{j n_0} \frac{1 - \exp\{-(i/\hbar)(E_{n_0} - E_j)t\}}{E_j - E_{n_0}}, \quad j \neq n_0 \qquad (10.16)$$

The total wave function now becomes (as before)

$$\Psi = \sum_j a_j(t)\Psi_j$$

$$\Psi = \Psi_{n_0} + \sum_{j \neq n_0} H'_{j n_0} \frac{1 - \exp\{-(i/\hbar)(E_{n_0} - E_j)t\}}{E_j - E_{n_0}} \Psi_j \qquad (10.51)$$

where in the first term $a_{n_0} = 1$. Equation (10.51) may be used as an initial wave function in Eq. (10.11). This amounts to saying that for $j \neq n_0$ the $a_j(t)$ in Eq. (10.10) are $\neq 0$ but are given by Eq. (10.16). Hence, by repeating the calculation that gave us Eq. (10.16), we obtain

$$i\hbar \dot{a}_k(t) = \sum_j a_j(t) H'_{kj} e^{-(i/\hbar)(E_j - E_k)t} \qquad (10.10)$$

$$= H'_{k n_0} \exp\{-(i/\hbar)(E_{n_0} - E_k)t\}$$

$$+ \sum_j \frac{H'_{j n_0} H'_{kj}}{E_j - E_{n_0}} [1 - \exp\{-(i/\hbar)(E_{n_0} - E_j)t\}] \exp\{-(i/\hbar)(E_j - E_k)t\} \qquad (10.52)$$

and

$$a_k(t) = H'_{kn_0} \frac{1 - \exp\{-(i/\hbar)(E_{n_0} - E_k)t\}}{E_k - E_{n_0}}$$

$$+ \sum_j \frac{H'_{jn_0}H'_{kj}}{E_j - E_{n_0}}$$

$$\times \left[\frac{1 - \exp\{-(i/\hbar)(E_j - E_k)t\}}{E_k - E_j} - \frac{1 - \exp\{-(i/\hbar)(E_{n_0} - E_k)t\}}{E_k - E_{n_0}} \right]$$

(10.53)

The terms in $E_j - E_k$ resulting from the first term in brackets may be neglected for the usual case of $|E_j - E_k| \gg |E_{n_0} - E_k|$ (this is because if $|E_{n_0} - E_k|$ is not very small, the $a_k(t)$ are negligible); otherwise, they must be included. If the first term of (10.53) is zero due to the vanishing of H'_{kn_0}, the transition probability then becomes

$$|a_k(t)|^2 = \left| 2 \sum_j \frac{H'_{kj}H'_{jn_0}}{E_{n_0} - E_j} \right|^2 \frac{1 - \cos[(E_{n_0} - E_k)(t/\hbar)]}{(E_{n_0} - E_k)^2}$$

(10.54)

In second-order time-dependent perturbation theory, just as in time-independent perturbation theory, occupation of the final state occurs through available intermediate states whose transition probabilities must be summed over. A curious result is that the transition amplitude to a particular state depends critically on the presence of other states whose energy is nearly the same. The initial state is, as usual, a pure state ($a_i = \delta_{in_0}$), but in second-order theory, transitions to the final state occur through intermediate states which, although initially vacant, have a small but finite probability in first-order perturbation theory of becoming populated.

When the perturbation is due to continuous incident radiation (that is, when the light-intensity development leading to Eq. (10.28) is valid), we obtain for second-order perturbation theory

$$|a_k(t)|^2 = \frac{1}{\hbar^2} \left| \sum_{j \neq k, n_0} \frac{\mu_{x_{kj}}\mu_{x_{jn_0}}}{E_{n_0} - E_j} \right|^2 E_{0x}^2(\omega_{kn_0})t$$

(10.55)

which is analogous to Eq. (10.28). Equation (10.54) may be generalized to yield the third-order correction,

$$|a_k(t)|^2 = 2 \left| \sum_j \sum_i \frac{H'_{kj}H'_{ij}H'_{in_0}}{(E_{n_0} - E_j)(E_{n_0} - E_i)} \right|^2 \frac{1 - \cos[(E_{n_0} - E_k)(t/\hbar)]}{(E_{n_0} - E_k)^2}$$

which includes transitions via two intermediate states.

10.9 TRANSITIONS TO OR FROM UNBOUND SYSTEMS

Both light emission and absorption are calculated by use of matrix elements (10.11) H'_{kn_0}, which are the same for both processes since the processes are similar. It may be, though, that either the final state of the particle is unbound, as would be the case in atomic ionization due to the photoelectric effect, or the initial state is unbound as occurs in light emission caused by electron capture. *If either the initial or final particle state is free, the particle energy spectrum in this state is no longer discrete*, and given continuous radiation we no longer observe or are interested in finding the transition probability to or from a particular state, but instead concern ourselves with an energy interval between E and $E + dE$ and assume that all states within this interval have equal probability of being occupied. The probability of a transition's occurring to some state in this energy's interval will be proportional to the number of states in the interval $\rho(E)\, dE$, where $\rho(E)$ is the level density or number of states per unit energy interval. In the case of a free electron or nucleon where both spin polarizations must be included,

$$\rho(E)\, dE = p^2\, dp\, \frac{V}{(\hbar\pi)^3} \qquad (10.56)$$

from Eq. (1.12), for example, where p is the momentum of the free particle and V is the volume in which the free particle wave functions are contained. Equation (1.12) for radiation in a box applies to free particles in a box, as we know, since the particles in a box have a probability wave amplitude which satisfies a wave equation similar to the equation for the photon wave function, Eq. (1.5). The boundary conditions are also similar and even the twofold degeneracy due to particle spin values of $\pm\frac{1}{2}$ is matched by the two degrees of photon transverse polarization freedom. Thus, for a free electron or nucleon the level density obeys the proportionality

$$\rho(E)\, dE \propto p^2\, dp \propto \sqrt{E}\, dE \qquad (10.56\text{a})$$

While the level density is directly proportional to the volume in which the particle is contained, the matrix elements, on which the transition probability also depends (Eq. 10.57 below), are directly proportional to the normalization constant of the free particle wave function C_N. Since the square of the matrix elements is used to obtain the transition probability, the transition probability is proportional to the square of the normalization constant for the free particle wave function. The normalization constant is so defined that $|C_N|^2 \int \psi^* \psi\, d\tau = 1$ with the integral extending over the volume of the box in which the particle moves. Since for free particle wave functions $\psi = e^{i\mathbf{k}\cdot\mathbf{r}}$ the integral is equal to this volume, $|C_N|^2 = 1/V$ and hence the transition probability $\propto \rho(E)\,|C_N|^2$ is independent of the volume. For convenience, in the following we will assume the free particle contained in a unit volume.

The probability of transition of a particle to or from a free state by absorption or emission of a photon of frequency ω is obtained by integrating the corresponding expression for discrete initial and final states (and for continuous ω instead of continuous E), Eq. (10.27), over all the equally probable states with energies E (in the case of the *initial* state being free but unknown an *average* is taken); in the case of absorption by a system in state n_0 to any of many equally probable states in the energy interval dE,

$$|a(E)|^2 = \int 4\,|H'_{kn_0}|^2 I(\omega) \left\{ \frac{\sin^2 [(E_{n_0} - E + \hbar\omega)(t/2\hbar)]}{(E_{n_0} - E + \hbar\omega)^2} \right\} \rho(E)\,dE \qquad (10.57)$$

where the general matrix element H'_{kn_0} has been written instead of the particular matrix element for dipole transitions μ_{kn_0}. $I(\omega)$ represents the intensity of radiation of frequency ω, n_0 labels the discrete energy level (bound) state, and k any continuous (unbound) state of energy E. Note that Eq. (10.27) was derived by assuming (as in obtaining Eq. $\{0.12\}$) that the perturbation is small and does not deplete the initial state appreciably; that is, $a_{n_0} = 1$ always. As in the derivation of Eq. (10.28), the quantity in braces behaves like a delta function for $t \gg \hbar/(E_{n_0} + \hbar\omega)$, so that Eq. (10.57) becomes upon integration

$$|a(E)|^2 = (2\pi/\hbar)I(\omega)\rho(E_{n_0} + \hbar\omega)\,|H'_{kn_0}|^2 t$$

The transition probability per unit time between states of energy E_{n_0} and $E_{n_0} + \hbar\omega$ is a constant given by

$$T(E_{n_0} + \hbar\omega, E_{n_0}) = (2\pi/\hbar)I(\omega)\rho(E_{n_0} + \hbar\omega)\,|H'_{kn_0}|^2 \qquad (10.58)$$

As t becomes very large the initial state is depleted so that the initial condition $a_{n_0} \sim 1$ is no longer valid, and the theory given here based on this simple condition breaks down.

Problem 10.8: What energy dependence of the transition rate is to be expected for neutrons emitted in a photonuclear transition if H'_{kn_0} is independent of energy and the nucleus is left in its ground state?

Problem 10.9: What is the rate of ionization of hydrogen atoms in the ground state when in light of which the intensity I_0 is constant over all frequencies up to twice the minimum ionization frequency?

Problem 10.10: Fermi assumed that in some circumstances the beta-decay transition matrix H'_{kn_0} is independent of electron energy and that two particles, an electron and a neutrino, are emitted in beta decay. The sum of the electron and neutrino energy must be a constant and the sum of the momenta of the recoil nucleus, electron, and neutrino must be zero in the laboratory system. Derive the beta-particle energy spectrum resulting from these assumptions.

Problem 10.11: If in a sample of tritium H^3 there are 10^{19} tritium atoms per cubic centimeter, how many beta-decay–formed He^+ ions in the $(n = 3, l = 1)$ excited state are there per cubic centimeter? (See Problem 9.12.) (Hint: Compare the observed half-life of H^3, 12.26 yrs with the atomic deexcitation rate of He^+ in this state and the fraction of H^3 which decays initially into this state.)

BIBLIOGRAPHY

Pauling, L. and E. B. Wilson, cited in Chapter VIII.

Heitler, W. *The Quantum Theory of Radiation.* London: Oxford University Press, 1950.

XI

THE ONE-ELECTRON
ATOM

The simplest atom is the atom having only one electron, i.e., hydrogen. The results obtained for this case will also give an insight into the problem of more com: lex atoms. For a hydrogen-like atom we take the potential energy of the electron to consist simply of the electrostatic attraction between it and the positively charged proton so that

$$V(r) = \frac{-Ze^2}{r} \tag{11.1}$$

The solution of the Schrödinger equation in this case has already been discussed in detail in Chapter VIII. There we saw that the time-independent solution could be written in the form

$$\psi_{nlm}(r, \theta, \varphi) = R_{nl}(r)\, \Theta_{lm}(\theta)\Phi_m(\varphi) \tag{11.2}$$

Here

$$\Phi_m(\varphi) = Ae^{\pm im\phi}$$

$$\Theta_{lm}(\theta) = (1 - \cos^2 \theta)^{|m|/2}F(\cos \theta) \quad \text{(the associated Legendre functions)}$$

and

$$R_{nl}(\rho) = e^{-\rho/2}\rho^l L(\rho)$$

where

$L(\rho)$ are the associated Laguerre functions

and

$$\rho = \frac{\sqrt{8m_e(-E)}}{\hbar} r$$

Table 11.1 gives the ψ_{nlm} (normalized) for several values of n, l and m in terms of the parameter

$$a_0 = \frac{\hbar^2}{m_e e^2} = 0.529 \times 10^{-8}\ \text{cm}$$

a_0 is very nearly equal to the *Bohr radius* or *first Bohr orbit* which is $\hbar^2/me^2$, where m is the electron mass, not the reduced electron mass.

TABLE 11.1

Normalized Hydrogen-like Wave Functions

$$\rho = \frac{Zr}{a_0}$$

$$\psi_{100} = \frac{1}{\sqrt{\pi}}\left(\frac{Z}{a_0}\right)^{3/2} e^{-\rho}$$

$$\psi_{200} = \frac{1}{4\sqrt{2\pi}}\left(\frac{Z}{a_0}\right)^{3/2} (2 - \rho)e^{-\rho/2}$$

$$\psi_{210} = \frac{1}{4\sqrt{2\pi}}\left(\frac{Z}{a_0}\right)^{3/2} \rho e^{-\rho/2} \cos\theta$$

$$\psi_{21\pm1} = \frac{1}{8\sqrt{\pi}}\left(\frac{Z}{a_0}\right)^{3/2} \rho e^{-\rho/2} \sin\theta e^{\pm i\varphi}$$

$$\psi_{300} = \frac{1}{81\sqrt{3\pi}}\left(\frac{Z}{a_0}\right)^{3/2} (27 - 18\rho + 2\rho^2)e^{-\rho/3}$$

$$\psi_{310} = \frac{\sqrt{2}}{81\sqrt{\pi}}\left(\frac{Z}{a_0}\right)^{3/2} (6 - \rho)\rho e^{-\rho/3} \cos\theta$$

$$\psi_{31\pm1} = \frac{1}{81\sqrt{\pi}}\left(\frac{Z}{a_0}\right)^{3/2} (6 - \rho)\rho e^{-\rho/3} \sin\theta e^{\pm i\varphi}$$

$$\psi_{320} = \frac{1}{81\sqrt{6\pi}}\left(\frac{Z}{a_0}\right)^{3/2} \rho^2 e^{-\rho/3}(3\cos^2\theta - 1)$$

$$\psi_{32\pm1} = \frac{1}{162\sqrt{\pi}}\left(\frac{Z}{a_0}\right)^{3/2} \rho^2 e^{-\rho/3} \sin 2\theta e^{\pm i\varphi}$$

$$\psi_{32\pm2} = \frac{1}{162\sqrt{\pi}}\left(\frac{Z}{a_0}\right)^{3/2} \rho^2 e^{-\rho/3} \sin^2\theta e^{\pm 2i\varphi}$$

In Section 8.6 the probability of finding the electron at any radius r in the hydrogen atom for various radial wave functions R_{nl} is discussed. Similarly, the angular part of the wave function in general may be expected to give rise to an angular distribution of the probability density dependent on θ but independent of φ. This can be seen as follows:

$$\Phi_m^*(\varphi)\Phi_m(\phi) \, d\varphi \propto e^{im\phi}e^{-im\phi} \, d\varphi = d\varphi$$

whereas $\theta_{lm}^* \, \theta_{lm} \sin\theta \, d\theta$ will in general be a function of θ (see Section 8.2). However, we recall that for the simple one-electron atom all eigenstates with the same principal quantum number n are degenerate, i.e., they all have the same energy. Therefore, for any given n, eigenstates of all allowed combinations of l and m will have equal probabilities of being populated. For instance, with $n = 2$, the total normalized probability density would be $P = \frac{1}{4}[\psi_{200}^*\psi_{200} + \psi_{21-1}^*\psi_{21-1} + \psi_{210}^*\psi_{210} + \psi_{211}^*\psi_{211}]$. Substituting the expressions for the ψ's, we find that P is spherically symmetric. This is an example of the fact that without a preferred direction in space one cannot measure an asymmetry.

Problem 11.1: Prove that the probability density P is spherically symmetric for $n = 2$, as discussed in the foregoing section.

11.1 ENERGY LEVELS

From Eq. (8.44) we see that the energy levels of the hydrogen atom are

$$E_n = -\frac{m_e e^4}{2\hbar^2}\frac{1}{n^2} \qquad n = 1, 2, 3, 4, \cdots \qquad (11.3)$$

In Section 5.4 it was pointed out that we are interested only in the relative motion of a particle in the center-of-mass system, and that consequently the mass appearing in the Schrödinger equation should be the reduced mass of the particle. For an atomic electron the reduced mass is very nearly equal to its mass. Therefore, in the following often we will not bother to distinguish between them and will use the notation m_e to avoid possible confusion with the magnetic quantum number m.

If an electron changes from one state to another, there will be a corresponding change in energy of the system.

$$E_{n_2} - E_{n_1} = \frac{-m_e e^4}{2\hbar^2}\left(\frac{1}{n_2^2} - \frac{1}{n_1^2}\right)$$

These transitions will in general be accompanied by the emission or absorption of electromagnetic radiation. For instance, if $n_1 = 1$, then one gets the *Lyman series*, in which n_2 can take on values of $2, 3, 4, \ldots$. Figure 11.1 shows an energy-level diagram for the hydrogen atom with the major spectral series named after the men who first identified them.

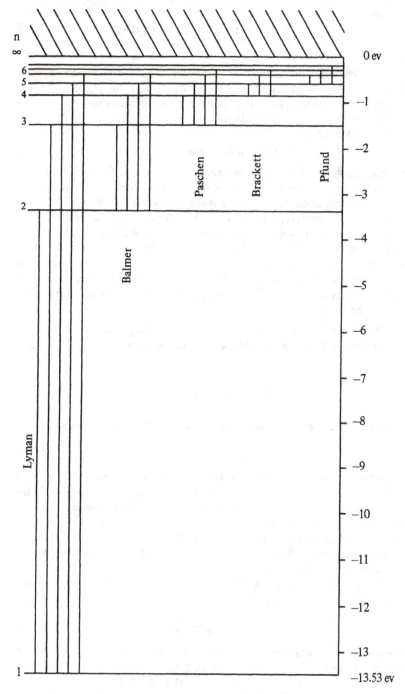

FIGURE 11.1 Energy-level diagram for hydrogen. The spectral series are named after the men who first discovered them.

Problem 11.2: A μ^- meson is a particle having a mass 206 times the rest mass of the electron and a charge equal to the electron charge. Assume a mesonic atom is formed in lead ($Z = 82$) and calculate the expected energies (in ev) of the first three levels corresponding to $n = 1, 2$, and 3. What are the corresponding radii? What do their values tell us regarding the assumption that the nucleus is a point charge? (Radius of the lead nucleus is $\approx 6 \times 10^{-13}$ cm.)

Problem 11.3: Assume that the helium nucleus is not a point charge, but is rather a sphere of uniform charge density with a radius $r_0 = 2 \times 10^{-13}$ cm. Then by using first-order perturbation theory calculate the expected shift in the energy level of singly ionized helium with the remaining electron in the $1s$ state. Note that r_0 is very small compared to the first Bohr orbit for helium. (The effect of a finite nuclear size is most pronounced for heavy atoms and has been observed for elements such as lead and mercury).

Problem 11.4: Calculate the energy shift of the first four members of the Lyman series for deuterium compared to the values for hydrogen. (It was the observed shift in these energy levels that led to the discovery of deuterium.) Note that m_e changes slightly in this case.

11.2 ORBITAL MAGNETIC MOMENT

So far we have considered the hydrogen atom to consist of a simple light negatively charged electron bound to a comparatively massive positively charged proton. In practice, various refinements must be added to this overly simplified picture. In discussing these refinements, it is desirable to use a semiclassical approach in most instances. This has the advantage of making the calculations relatively easy, as well as providing a simple physical picture to describe the system. The results of this procedure are in good agreement with those using a more formal, completely quantum mechanical approach, as can readily be demonstrated in some cases. However, we must always bear in mind that a semiclassical picture can never be completely adequate when applied to a microscopic system.

Consider an electron moving with constant speed v in a circular orbit of radius r. (According to the uncertainty principle the electron can never actually travel in a definite orbit, but this is a useful picture nevertheless.) The electron will of course have an angular momentum associated with its motion about the origin, namely,

$$\mathbf{L} = \mathbf{r} \times \mathbf{p} \text{ where } \mathbf{p} = m_e\mathbf{v} \text{ linear momentum}$$

$$L = rm_e v \text{ in the case at hand}$$

The moving electron constitutes a current:

$$j = ef = e\frac{v}{2\pi r} \text{ (cgs units)}$$

where f is the cycle frequency of the orbital motion of the electron.

Now it is well known in electromagnetism that a closed current loop acts like a magnetic dipole whose magnetic moment is given by the product of the current and the area of the loop. (Strictly speaking, this is an approximation which improves as the distance from the loop becomes larger and larger in comparison to the loop dimensions). Therefore, we may say

$$\mu_l = \frac{jA}{c} = -\frac{e}{c}\frac{v}{2\pi r}\pi r^2 = -\frac{evr}{2c} = \text{magnetic moment}$$

$$\frac{\mu_l}{L} = -\frac{evr}{2c}\frac{1}{m_e vr} = \frac{-e}{2m_e c} = -g_l\frac{\mu_B}{\hbar} \tag{11.4}$$

where we define

$$\mu_B = \frac{e\hbar}{2m_e c} = \text{Bohr magneton } (= 0.927 \times 10^{-20} \text{ erg/gauss})$$

$$g_l = \text{orbital } g \text{ factor } (= 1)$$

Consequently,

$$\boldsymbol{\mu}_l = -\frac{g_l\mu_B\mathbf{L}}{\hbar}$$

But we have already seen that the magnitude of L is $\hbar\sqrt{l(l+1)}$, thus

$$|\mu_l| = \frac{g_l\mu_B}{\hbar}\hbar\sqrt{l(l+1)} = g_l\mu_B\sqrt{l(l+1)} \tag{11.5a}$$

And also, since $L_z = \hbar m$,

$$\mu_{lz} = -\frac{g_l\mu_B}{\hbar}L_z = -g_l\mu_B m \tag{11.5b}$$

We see that, just as the projections of L on the z axis were quantized, so are the z projections of the orbital magnetic moment $\boldsymbol{\mu}_l$.

This result could of course be obtained using quantum mechanics. First, we recall the particle current density (Eq. (5.27)), $\mathbf{j}_p$:

$$\mathbf{j}_p = \frac{\hbar}{2m_e i}(\Psi^*\nabla\Psi - \Psi\nabla\Psi^*) \tag{11.6}$$

Here $\Psi = \Psi_{nlm} = R_n(r)\Theta_{lm}(\theta)\Phi(\phi)e^{-\frac{i}{\hbar}E_{nt}}$ and in spherical coordinates

$$\nabla = \hat{r}\frac{\partial}{\partial r} + \hat{\theta}\frac{1}{r}\frac{\partial}{\partial\theta} + \hat{\phi}\frac{1}{r\sin\theta}\frac{\partial}{\partial\varphi} \tag{11.7}$$

Then

$$\mathbf{j}_p = \hat{r}\frac{\hbar}{2m_e i}\Theta_{lm}^2\left[R_{nl}\frac{\partial R_{nl}}{\partial r} - R_{nl}\frac{\partial R_{nl}}{\partial r}\right]$$

$$+ \hat{\theta}\frac{\hbar}{2m_e i}R_{nl}^2\left[\Theta_{lm}\frac{\partial\Theta_{lm}}{\partial\theta} - \Theta_{lm}\frac{\partial\Theta_{lm}}{\partial\theta}\right]$$

$$+ \hat{\phi}\frac{\hbar}{2m_e i}R_{nl}^2\Theta_{lm}^2\left[e^{-im\varphi}\frac{1}{r\sin\theta}\frac{\partial e^{im\varphi}}{\partial\varphi} - \frac{e^{im\varphi}}{r\sin\theta}\frac{\partial e^{-im\varphi}}{\partial\varphi}\right]$$

$$= \hat{\phi}R_{nl}^2\Theta_{lm}^2\frac{m\hbar}{m_e r\sin\theta}\Phi_m^*\Phi_m = \varphi\frac{m\hbar\psi_{nlm}^*\psi_{nlm}}{m_e r\sin\theta} \quad \textbf{(11.8)}$$

$$= j_p\hat{\phi}$$

Notice that the expression for $j_p\hat{\phi} \longrightarrow 0$ if $m \longrightarrow 0$ so that $j_p\hat{\phi}$ vanishes unless $l > 0$.

Thus the electron motion corresponds to a net current flow circulating about the z axis when $m \neq 0$. Since $\boldsymbol{\mu}_l = (j/c)A$ is the magnetic moment associated with a current loop, we see from Figure 11.2 that the contribution to μ_l (along the z axis) from the current loop is

$$d(\mu_l)_z = \frac{\pi}{c}(r\sin\theta)^2(-ej_p)r\,dr\,d\theta$$

and

$$(\mu_l)_z = \frac{-e\pi}{c}\iint\frac{r^3\sin^2\theta\, m\hbar\psi_{nlm}^*\psi_{nlm}\,dr\,d\theta}{m_e r\sin\theta}$$

$$= \frac{-em\hbar}{2m_e c}\iint\psi_{nlm}^*\psi_{nlm}2\pi r^2\sin\theta\,dr\,d\theta$$

$$= \frac{-em\hbar}{2m_e c} = -g_l\mu_B m\left(=\frac{-g_l\mu_B}{\hbar}L_z\right) \quad \textbf{(11.5b)}$$

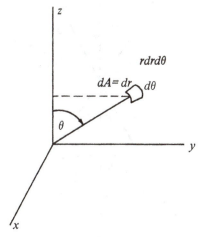

FIGURE 11.2 Coordinate system used to calculate the projection of μ_l along the z axis.

11.3 PRECESSION IN AN EXTERNAL MAGNETIC FIELD

If a dipole is placed in an appropriate external field, it will tend to line up in the direction of the field or, in other words, a torque will act on it. In this case, we are interested in the effects of a magnetic field on the orbital magnetic dipole moment μ_l. Therefore, $\tau = \mu_l \times H =$ torque acting on magnetic dipole moment in the magnetic field **H**. For simplicity, we assume **H** is uniform and in the z direction. Now if a system has a net torque acting on it, then its total angular momentum must be changing and the following relationships hold

$$\tau = \frac{d\mathbf{L}}{dt} = \mu_l \times \mathbf{H} = -\frac{g_l \mu_B}{\hbar}\mathbf{L} \times \mathbf{H} \qquad (11.9)$$

Thus, considering any arbitrary small but finite change in **L** called **ΔL**, we find

$$\Delta\mathbf{L} = -\frac{g_l \mu_B}{\hbar}\mathbf{L} \times \mathbf{H}\Delta t \qquad (11.10)$$

and it follows that the change in **L** is in a direction perpendicular both to **L** and **H**. That is, the change in **L** is in the direction of increasing or decreasing angle ϕ. This situation is illustrated in Figure 11.3. From the figure it can be

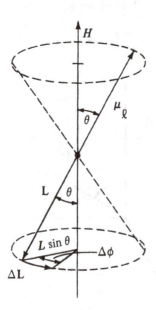

FIGURE 11.3 Illustration of the precession of a magnetic dipole in a magnetic field about the direction of the field.

seen that the change in L is represented by the angle $\Delta\phi$ given by

$$\Delta\phi = \frac{\Delta L}{L \sin\theta} = -\frac{g_l\mu_B L \times H\Delta t/\hbar}{L \sin\theta} \tag{11.11}$$

$$\Delta\phi = -\frac{\dfrac{g_l\mu_B}{\hbar}LH \sin\theta\,\Delta t}{L \sin\theta} = -\frac{g_l\mu_B}{\hbar}H\Delta t$$

Going back to the limit of an infinitesimal change $\Delta\phi$ in an infinitesimal time Δt, one can say $\omega = d\phi/dt = -g_l\mu_B H/\hbar$. ω is called the *Larmor precession frequency*.

This shows that a magnetic dipole placed in a magnetic field will precess about the direction of the field with a frequency which depends only on the strengths of the field and the dipole. Therefore, we should expect an atom to precess in an external magnetic field if its electron is in a state for which $l \neq 0$.

Problem 11.5: Calculate the Larmor precession frequency for the hydrogen atom with the electron in the $2p$ state. Compare your result with frequency of blue light ($\lambda = 4000$ Å), $H = 10^4$ Oesteds.

11.4 FORCE ON A DIPOLE IN A NON-UNIFORM MAGNETIC FIELD

In general, a dipole in an external magnetic field has potential energy due to its orientation in the magnetic field:

$$U = -\mathbf{\mu}_l \cdot \mathbf{H} \tag{11.12}$$

Moreover, in a non-uniform external field, a net force will act on the dipole if the opposite ends of the dipole are positioned in unequal field strengths because then the forces on the two poles will be different. The force on a dipole in an inhomogeneous field may be obtained from the gradient of the potential energy (11.12) since this is the sum of the positional potential energy of the two poles taken separately in the magnetic field.

$$\mathbf{F} = -\nabla U = -\nabla(-\mathbf{\mu}_l \cdot \mathbf{H}) \tag{11.13}$$

Since $\nabla \cdot H = 0$ under the conditions assumed here

$$\nabla(\mathbf{\mu}_l \cdot \mathbf{H}) = \mathbf{\mu}_l(\nabla \cdot \mathbf{H}) + (\mathbf{\mu}_l \cdot \nabla)\mathbf{H} = (\mathbf{\mu}_l \cdot \nabla)\mathbf{H}$$

and

$$\mathbf{F} = (\mathbf{\mu}_l \cdot \nabla)\mathbf{H} \tag{11.14}$$

If we take the z direction to be that of the magnetic field and if H_z is a function only of z, then the net force acting on the dipole becomes

$$F_z = (\mathbf{\mu}_l \cdot \nabla)H_z = \mu_{l_z}\frac{\partial H_z}{\partial z} = -g_l\mu_B\frac{\partial H_z}{\partial z}m \tag{11.15}$$

$$F_x = F_y = 0$$

The quantized projection of the angular momentum on a given axis can have $2l + 1$ discrete values, where l is the total orbital momentum of the atoms corresponding to the values $-l, -l + 1, \cdots, l - 1, l$ which the orbital quantum number may assume. Therefore, integrality of l requires an odd number of spots. If a beam of neutral hydrogen atoms were to move through the field H_z, discussed above and shown schematically in Figure 11.4, we should expect to see a single spot as long as the atoms are in the ground state with $n = 1, l = 0, m = 0$. However, when the experiment is carried out *two spots are observed!* For $l = 1$, four spots are observed for these atoms. (The first experiment of this type was carried out by O. Stern and W. Gerlach in 1922 using silver atoms. Subsequently, other similar Stern-Gerlach experiments were carried out; it was not until 1927 that the experiment was done using atomic hydrogen.)

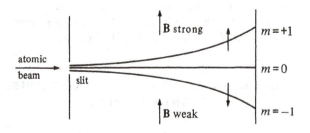

FIGURE 11.4 Illustration of the deflection of a beam of atoms having one unit of angular momentum in an inhomogeneous magnetic field.

A way of explaining the anomalous splitting is to assume there is an extra magnetic moment so far unaccounted for. G. Uhlenbeck and S. Goudsmit made the suggestion that the electron has an intrinsic magnetic moment $\mathbf{\mu}_s$ due to an inherent angular momentum or spin S of one-half unit. In analogy with the known quantum mechanical results for the orbital angular momentum they assumed

magnitude of $S = \sqrt{1/2(1/2 + 1)}\hbar$

z component of the spin $= m_s\hbar$ where $m_s = \pm 1/2$

and further that

$$\mathbf{\mu}_s = \frac{-g_s\mu_B}{\hbar}\mathbf{S}$$

assumes values of

$$\mu_{s_z} = -g_s\mu_B m_s$$

where $g_s = $ spin g factor for the electron.

With $m_s = \pm 1/2$ one should then expect the observed splitting for the hydrogen atom with $m_l = 0$ ($l = 0$) because Eq. (11.15) then becomes

$$F_z = -\frac{\partial H}{\partial z}\mu_B g_s m_s$$

Since all the other quantities in the expression for F_z can be determined, a Stern-Gerlach experiment also allows the value for g_s to be obtained. Experimentally it is found that $g_s m_s = \pm 1$ so $g_s = 2$. (A rigorous theoretical treatment gives the value of g_s to ten significant figures in agreement with the experimental value of $2.002319244 \pm 0.000000054$ obtained in recent years by H. R. Crane and co-workers.)

It should be emphasized at this point that, although according to classical electrodynamics a rotating sphere of charge should have an angular momentum and associated magnetic moment, the actual intrinsic electron angular momentum and magnetic moments are essentially nonclassical concepts. In the nonrelativistic theory using the Schrödinger equation, which we are discussing here, the intrinsic spin concept has to be added ad hoc, but in the Dirac (relativistic) theory the electron spin and magnetic moment are seen to follow as natural properties of the electron.

It now becomes apparent that for the hydrogen atom we should be using a wave function labelled with four quantum numbers instead of three, where the last quantum number m_s refers to whether the electron spin is pointing up or down with respect to some preferred direction in space. The wave function will then be written $\psi_{nlm_lm_s}$. We can construct this by simply multiplying the space part by a spin function. (See Appendix C for a more thorough discussion.)

Problem 11.6: Evaluate the magnetic moment of a hydrogen atom in its ground state. If atomic hydrogen gas at $100°K$ is subjected to a magnetic field of 10^4 gauss, what will be the percentage difference between the number of atoms aligned parallel to the field and opposite to it? (Hint: Assume all the atoms are in the ground state and make use of the Maxwell-Boltzmann distribution.)

11.5 SPIN-ORBIT INTERACTION

If an atom has both a magnetic moment due to the orbital electron motion and the intrinsic magnetic moment of the electron, then it is reasonable to expect that these two magnetic moments will interact. One can crudely represent the situation by the drawing in Figure 11.5.
The interaction energy can be written in the form

$$\Delta E_{s \cdot l} = -\mathbf{\mu}_s \cdot \mathbf{H} = \frac{g_s \mu_B}{\hbar}\mathbf{S} \cdot \mathbf{H} \tag{11.16}$$

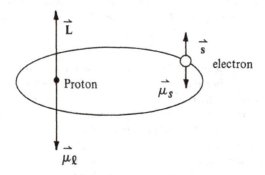

FIGURE 11.5 Schematic representation of the hydrogen atom showing the effective magnetic moment produced by the electron's orbital motion as well as the intrinsic electron magnetic moment.

where **H** is the magnetic field produced by the electron's orbital motion and is, of course, directly related to the orbital magnetic moment. The problem is to calculate this field. For $l \neq 0$ an atomic electron necessarily has a net orbital motion about the nucleus and the static electric field of the nucleus appears as a magnetic field as well when relativistically transformed to the velocity frame of the electron. If V is the nuclear electrostatic potential in which the electron moves, then

$$\mathbf{H} = \frac{1}{c} \mathbf{v} \times \mathbf{E} = \frac{1}{c} \mathbf{v} \times \frac{-\mathbf{r}}{r} \frac{\partial V}{\partial r}$$

$$= \frac{1}{cr} \frac{\partial V}{\partial r} (\mathbf{r} \times \mathbf{v}) = \frac{1}{cr} \frac{\partial V}{\partial r} \left(\frac{\mathbf{L}}{m_e} \right) \tag{11.17}$$

where **L** is used, as always, to denote the orbital angular momentum vector operator. The magnetic energy of the spinning electron in the magnetic field created by the nuclear electrostatic field and the electron's orbital motion is found by combining Eqs. (11.16, 11.17)

$$\Delta E_{s \cdot l} = -\frac{e}{2m_e^2 c^2} \frac{1}{r} \frac{\partial V}{\partial r} (\mathbf{L} \cdot \mathbf{S}) \tag{11.18}$$

where **S** denotes the angular momentum of the electron spin. (Note that owing to relativistic corrections calculated by L. H. Thomas, $\Delta E_{s \cdot l}$ in Eq. (11.18) is one-half the result of combining Eqs. (11.16) and (11.17).) $\Delta E_{s \cdot l}$ is a small quantity and can be calculated using perturbation theory. Before we undertake to evaluate this expression, it is worthwhile to discuss in more detail some of the consequences of assuming the existence of an intrinsic electron spin.

First, the atom now has a total angular momentum **J** where $\mathbf{J} = \mathbf{L} +$

S. In analogy with the previous results for **L**, we expect (and it can be proved) that

$$|J| = \sqrt{J(J+1)}\hbar$$
$$J_z = m_j\hbar \quad \text{where } m_j = -J\cdots\cdots\cdots+J$$
$$J_z = L_z + S_z \quad \text{or} \quad m_j\hbar = m_l\hbar + m_s\hbar$$

so that
$$(m_j)_{\text{max}} = l + \tfrac{1}{2}$$
$$(m_j)_{\text{min}} = -(l + \tfrac{1}{2})$$

It is instructive to illustrate the vector picture for J by showing the quantized projections on an axis representing some preferred (z) direction in space. Figure 11.6 shows the possible projections for a particular example of $l = 2$ so that $J = 5/2$ or $3/2$. (The reader will note the close similarity to Figure 8.1 illustrating the discussion in Section 8.2.)

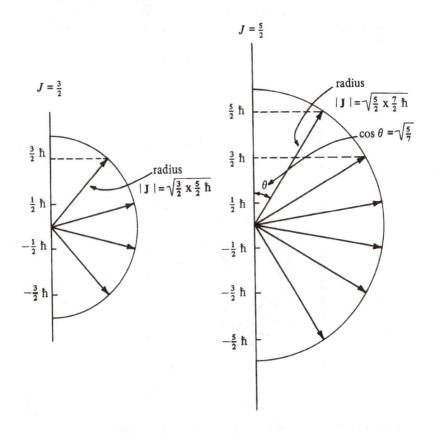

FIGURE 11.6 Illustrating the possible projections of the total angular momentum along any given axis.

If there is no spin-orbit interaction, then J, J_z, L, L_z, S, and S_z are all constants of the motion and J, m_j, l, m_l, and m_s can all be good quantum numbers. In the case of a spin-orbit interaction, we recall that the process can be thought of as an interaction between the electron magnetic moment and the magnetic field caused by the electron orbital motion. The torque produced gives a coupling between L and S. The net result is that both precess around the total angular momentum J and it is only J and J_z that are constant. If a weak external magnetic field is added to the system, J will precess slowly about the z axis (direction of the field) while L and S both continue to precess rapidly about the direction of J. This is illustrated by Figure 11.7.

All this raises an interesting point concerning the wave functions $\psi_{nlm_lm_s}$. The presence of the spin-orbit coupling means that the $\psi_{nlm_lm_s}$ are no longer eigenfunctions of the system, because the m_l and m_s are no longer good quantum numbers. Instead, we can use a new set of eigenfunctions ψ_{nljm_j} because j and m_j are still good quantum numbers. Since these $\psi_{nlm_lm_s}$ form a complete orthonormal set, ψ_{nljm_j} is expressible as a linear combination of them

$$\psi_{nljm_j} = \Sigma_{m_lm_s} A_{m_lm_s}\psi_{nlm_lm_s}$$

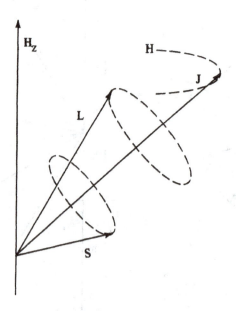

FIGURE 11.7 Precession of L and S about J and the precession of J about the direction of a weak external magnetic field in the z direction.

The actual linear combinations are somewhat complicated, but they can readily be found (see Appendix B). One can say that the spin-orbit interaction mixes states of different m_l and m_s. To calculate $\Delta E_{s \cdot l}$ using the $\psi_{nlm_lm_s}$, degenerate perturbation theory could be employed and in the process of removing the degeneracies in the $\psi_{nlm_lm_s}$ states the ψ_{nljm_j} eigenfunctions would be found. However, if we assume that the correct ψ_{nljm_j} have already been found, the perturbation calculation can be carried out much more simply using nondegenerate theory.

$$\overline{\Delta E}_{s \cdot l} = \int \psi^*_{nljm_j}(\Delta E_{s \cdot l})\psi_{nljm_j}\, dr \qquad (11.19)$$

where $(\Delta E_{s \cdot l}) \propto (1/r)(dV(r)/dr)(\mathbf{S} \cdot \mathbf{L})$ from Eq. (11.18) and $(dV(r)/dr) = (Ze^2/r^2)$. Also

$$\mathbf{S} \cdot \mathbf{L} = \tfrac{1}{2}(J^2 - L^2 - S^2)$$

Since we already have seen that $J^2 = J(J+1)\hbar^2$, $L^2 = L(L+1)\hbar^2$, and $S^2 = S(S+1)\hbar^2 = \tfrac{3}{4}\hbar^2$, and ψ_{nljm_j} is of the form ψ_{nlm}, then

$$\overline{\Delta E}_{s \cdot l} = -\frac{Ze^2\hbar^2}{4m_e^2c^2}\Big[J(J+1) - L(L+1) - \frac{3}{4}\Big] \int_0^\infty R^*_{nl}(r)\frac{1}{r^3}R_{nl}(r)r^2 dr \qquad (11.20)$$

As an illustrative example let us calculate the spin-orbit interaction of the $3d(n = 3, l = 2)$ state in the hydrogen atom. Referring to Table 11.1 for an appropriate wave function (ψ_{320}, for instance) and integrating over the angular distribution to obtain R_{32}, we find the integral in Eq. (11.20) can be expressed as

$$I = \frac{8}{(81)^2 15}\Big(\frac{m_e e^2}{\hbar^2}\Big)^7 \int_0^\infty r^3 \exp\Big\{\frac{2m_e e^2}{3\hbar^2}r\Big\}\, dr$$

Let $x = (2m_e e^2/3\hbar^2)r$ then

$$I = \frac{8}{(81)^2 15}\Big(\frac{m_e e^2}{\hbar^2}\Big)^7 \Big(\frac{3\hbar^2}{2m_e e^2}\Big)^4 \int_0^\infty x^3 e^{-x}\, dx$$

$$= \frac{6}{81 \times 30}\Big(\frac{m_e e^2}{\hbar^2}\Big)^3$$

and thus

$$\overline{\Delta E}_{s \cdot l} = -\frac{1}{810}\frac{m_e e^4}{2\hbar^2}\frac{e^4}{\hbar^2 c^2}[J(J+1) - L(L+1) - 3/4]$$

$$= -\frac{1}{810}W_H\alpha^2[J(J+1) - L(L+1) - 3/4]$$

where W_H is the binding energy of the ground state of the hydrogen atom, $\alpha = e^2/\hbar c$ and is called the *fine structure constant*, and $[J(J+1) - L(L+1) - 3/4] = +2$ or -3 depending on the orientation of the spin relative to the orbital momentum axis (i.e. $J = L + 1/2$, or $J = L - 1/2$), so that the energy would be split into two levels having an energy difference of $(5/810)W_H\alpha^2$.

It is possible to obtain general results for $\overline{\Delta E}_{s \cdot l}$ by using the expression for R_{nl} (Eq. 11.2). After evaluation of the integral one finds

$$\overline{r^{-3}} = \frac{Z^3}{a_0 n^3 l(l+\frac{1}{2})(l+1)} \qquad a_0 = \text{Bohr radius}$$

$$\overline{\Delta E}_{s \cdot l} = -\frac{Z^2 |E_n| \alpha^2}{n(2l+1)(l+1)} \qquad j = l + \frac{1}{2}, l \neq 0 \qquad \textbf{(11.21)}$$

$$= \frac{-Z^2 |E_n| \alpha^2}{nl(2l+1)} \qquad j = l - \frac{1}{2}, l \neq 0$$

$$\overline{\Delta E}_{s \cdot l} = 0 \text{ for } l = 0$$

Here $E_n = (-m_e e^4 / 2\hbar^2)(1/n^2)$.

As we have already noted, all the states in hydrogen having the same total quantum number have the same energy eigenvalue (are mutually degenerate), since the energy is a function of the sum of the radial and orbital quantum numbers. These degenerate states are split up and separated by the magnetic interaction of the electron spin with the electron orbital motion; each separate value of l gives two new slightly shifted levels. In our example where $n = 3$, there would result five levels where before there was one level: two levels for the $3d$ state $l = 2$, two less separated levels for the $3p$ state $l = 1$, and one unshifted level for the $3s$ state $l = 0$. The degeneracy in the projection of l on a given axis m remains. Note that in our example we computed only one of the three integrals for the threefold degeneracy ($3d$, $3p$, $3s$) between states of different l but same total quantum number ($n = 3$) for all the different values of E_l' in Eq. (9.33).

For atoms other than hydrogen, the central Coulombic potential is

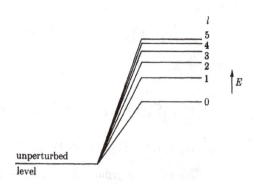

FIGURE 11.8 Illustration of the way a single valence level of total degeneracy n^2 is broken up into n levels, each having a degeneracy $(2l + 1)$, by the effect of shielding by the core electrons in an alkali atom.

modified by the shielding effect of the inner or core electrons, and this effect is very much larger than the spin-orbit interaction. In the case of hydrogen-like alkali atoms, for example, the potential is nearly Coulombic for the outer electrons and their wave functions nearly hydrogenic. A level of total quantum number n, however, is split into n different levels corresponding to different l's. The probability of finding an electron at small r in a Coulomb potential decreases as l increases even though n, the total quantum number, remains the same (as illustrated in Figure 8.8); therefore outer electrons in alkali atoms represented by hydrogen wave functions of small l penetrate further inside the electron cloud shielding the nucleus than do outer electrons represented by states of larger l. Hence states of smaller l have lower energy than states of larger l having the same n, and the degeneracy is broken up as shown in Figure 11.8.

11.6 RELATIVISTIC CORRECTIONS

So far it has been assumed that the hydrogen atom can be adequately treated using a nonrelativistic theory with the exception that we have seen that a relativistic effect (Thomas precession) was important in calculating the spin-orbit interaction. Now we will consider the effect of making a first-order correction to our basic nonrelativistic Hamiltonian to account for additional relativistic effects. If this correction causes only a small change in the energy, then this is good evidence that the nonrelativistic Schrödinger wave equation is appropriate for treating the one-electron atom. The Hamiltonian may be written relativistically as

$$H = T + V = \sqrt{p^2 c^2 + m_0^2 c^4} - m_0 c^2 + V$$

Further, we have

$$H = m_0 c^2 \left(1 + \frac{p^2}{m_0^2 c^2} \right)^{1/2} - m_0 c^2 + V$$

$$\simeq m_0 c^2 + \frac{p^2}{2m_0} - \frac{p^4}{8m_0^3 c^2} - m_0 c^2 + V$$

$$\simeq \frac{p^2}{2m_0} - \frac{p^4}{8m_0^4 c^2} + V = H_0 - \frac{p^4}{8m_0^3 c^2}$$

and thus

$$\overline{\Delta E}_{\text{rel}} = -\frac{1}{8m_0^3 c^2} \int \psi_{nljm_j}^* p^4 \psi_{nljm_j} d\tau = \frac{-\hbar^4}{8m_0^3 c^2} \int \psi_{nljm_j}^* \nabla^4 \psi_{nljm_j} d\tau \quad (11.22)$$

Another somewhat easier approach for evaluating $\overline{\Delta E}_{\text{rel}}$ is to observe that

$$\Delta E_{\text{rel}} = \frac{-p^4}{8m_0^3 c^2} \simeq \frac{-T^2}{2m_0 c^2} = \frac{-(E-V)^2}{2m_0 c^2} = \frac{-(E^2 + V^2 - 2EV)}{2m_0 c^2}$$

where

$$V = \frac{-Ze^2}{r} \text{ and } \overline{E^2} = E_n^2$$

Carrying out the necessary integration yields

$$\overline{\Delta E}_{rel} = \frac{-Z^2 |E_n| \alpha^2}{n}\left(\frac{2}{2l+1} - \frac{3}{4n}\right) \tag{11.23}$$

This energy change is small and of the same order of magnitude as the result for the spin-orbit effect which also included a relativistic correction. Because these are both small first-order perturbations, they may be combined linearly and we get a net correction factor of

$$\overline{\Delta E} = \overline{\Delta E}_{s \cdot l} + \overline{\Delta E}_{rel}$$

$$= \frac{Z^2 |E_n| \alpha^2}{4n^2}\left(3 - \frac{4n}{j+1/2}\right) \text{ for all values of } j \tag{11.24}$$

This result when combined with the zero-order energy yields a total energy surprisingly close to the result obtained using the relativistic Dirac equation.

11.7 SPECTROSCOPIC NOTATION

We have seen how a particular state for the electron is characterized by the four quantum numbers n, l, j, and m. Historically, a spectroscopic notation has been developed to describe various electronic configurations. Since it is now a universal notation we introduce it here.

First, a given value of the orbital angular momentum is specified as follows:

$$l = 0 \ 1 \ 2 \ 3 \ 4 \ 5 \ 6 \ 7 \text{ etc.}$$

$$S \ P \ D \ F \ G \ H \ I \ K$$

The first four letters came into use in describing the various spectral series that resulted from transitions between states of different l. For instance, in the alkali metals, certain series appeared to give sharp lines and others diffuse lines.

Sharp:	$nS \longrightarrow N_0P$	Where $n > N_0$
Principal:	$nP \longrightarrow N_0S$	
Diffuse:	$nD \longrightarrow N_0P$	
Fundamental:	$nF \longrightarrow N_0D$	

A subscript to the right of the letter designating the angular momentum designates the J value, while a numerical superscript to the left indicates the multiplicity due to the $2S + 1$ projections of the electron spin. For example

$$^2S_{1/2} \text{ means that } L = 0, J = \tfrac{1}{2}, S = \tfrac{1}{2}$$

$$^2P_{3/2} \text{ means that } L = 1, J = \tfrac{3}{2}, S = \tfrac{1}{2}$$

Here for a single electron the superscript can only be two; in the subsequent chapter on multi-electron atoms we will see how the electron spin states combine to yield superscripts different from two.

11.8 ENERGY-LEVEL DIAGRAMS

Using the spectroscopic notation, one can make an energy-level diagram showing the ground state and various possible excited states of the hydrogen (one-electron) atom. Figure (11.9) shows an energy-level diagram for hydrogen with just the first few levels to indicate the possible terms for various values of n. Levels having different values of L and J for a given n do not have the same energies, but the resultant level splittings are too small to be shown in the figure without greatly exaggerating the effect.

11.9 ZEEMAN AND PASCHEN-BACK EFFECTS

Let us consider again the effect of placing the one-electron atom in a uniform external magnetic field in the z direction. Then as we have already seen (Section 11.4), the electron will have an energy of orientation

$$\Delta E = -\boldsymbol{\mu} \cdot \mathbf{H} = -\mu_z H$$

where now

$$\boldsymbol{\mu} = -\frac{g_l \mu_B \mathbf{L}}{\hbar} - \frac{g_s \mu_B \mathbf{S}}{\hbar}$$

$$= \frac{-\mu_B}{\hbar}(\mathbf{L} + 2\mathbf{S}) \quad \text{since } g_l = 1 \text{ and } g_s = 2$$

$$= \frac{-\mu_B}{\hbar}(\mathbf{J} + \mathbf{S})$$

Note that the resultant magnetic moment for the electron is not proportional to the total angular momentum $\mathbf{J}$ and the magnetic moment has a contribution both from the orbital motion and the intrinsic electron spin.

Now let us further assume that the external magnetic field is weak compared to the internal field produced by the electron's orbital motion and intrinsic magnetic moment. Then we can consider that $\mathbf{L}$ and $\mathbf{S}$ will precess

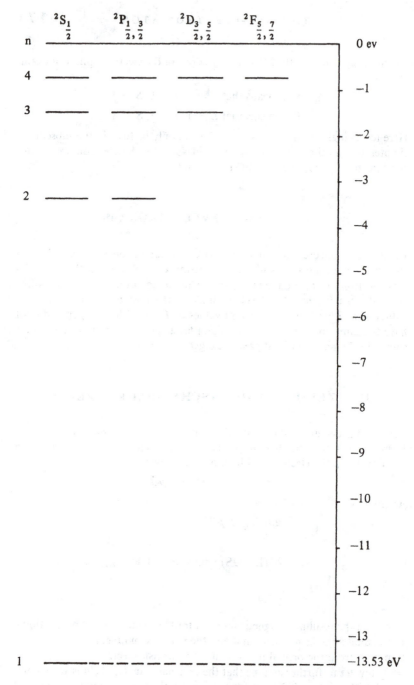

FIGURE 11.9 Energy-level diagram of the hydrogen atom illustrating the appropriate spectroscopic notation for the different levels.

rapidly about $\mathbf{J}$ and $\mathbf{J}$ will in turn precess much less rapidly about the direction of the external magnetic field H. Under these circumstances, we can evaluate μ_z by finding μ_J, the average component of $\boldsymbol{\mu}$ in the direction of $\mathbf{J}$, and then taking the component of μ_J in the z direction ($|\mathbf{H}| = H_z$). (See Figure 11.10.)

$$\mu_J = \frac{\boldsymbol{\mu} \cdot \mathbf{J}}{|\mathbf{J}|} = \frac{-\mu_B}{\hbar} \frac{(\mathbf{L} + 2\mathbf{S}) \cdot (\mathbf{L} + \mathbf{S})}{|\mathbf{J}|}$$

and

$$\mu_z = \mu_J \frac{\mathbf{J} \cdot \mathbf{H}}{|\mathbf{J}||\mathbf{H}|} = \mu_J \frac{J_z}{|\mathbf{J}|} = \frac{-\mu_B}{\hbar}(\mathbf{L} + 2\mathbf{S}) \cdot (\mathbf{L} + \mathbf{S}) \frac{J_z}{|\mathbf{J}|^2}$$

Since

$$|\mathbf{J}|^2 = |\mathbf{L}|^2 + |\mathbf{S}|^2 + 2\mathbf{L} \cdot \mathbf{S}$$

we may replace the $\mathbf{L} \cdot \mathbf{S}$ terms by the squares of $\mathbf{J}, \mathbf{L},$ and $\mathbf{S}$

$$\mu_z = \frac{-\mu_B}{\hbar}\left(|\mathbf{L}|^2 + 2|\mathbf{S}|^2 + \frac{3}{2}|\mathbf{J}|^2 - \frac{3}{2}|\mathbf{L}|^2 - \frac{3}{2}|\mathbf{S}|^2\right)\frac{J_z}{|\mathbf{J}|^2}$$

and

$$\Delta E = -\mu_z H = \frac{\mu_B}{\hbar}H(3|\mathbf{J}|^2 + |\mathbf{S}|^2 - |\mathbf{L}|^2)\frac{J_z}{2|\mathbf{J}|^2} \qquad (11.25)$$

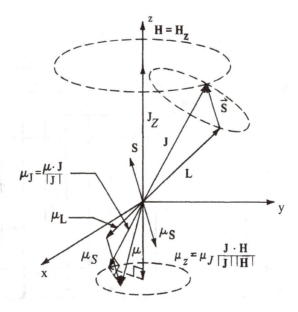

FIGURE 11.10   Vector diagram illustrating how the z component of the magnetic moment $\boldsymbol{\mu}$ can be obtained for a weak external magnetic field in the z direction.

Because we have chosen wave functions that are eigenfunctions of all the operators appearing in ΔE, using first-order perturbation theory we have

$$\overline{\Delta E} = \int \psi_T^* \Delta E \, \psi_T \, d\tau$$

$$\overline{\Delta E} = \frac{\mu_B}{\hbar} H \frac{[3J(J+1)\hbar^2 + S(S+1)\hbar^2 - L(L+1)\hbar^2]m_j\hbar}{2J(J+1)\hbar^2}$$

$$\overline{\Delta E} = \mu_B Hgm_j \text{ where } g = 1 + \frac{J(J+1) + S(S+1) - L(L+1)}{2J(J+1)}$$

$$(11.26)$$

g is called the *Landé g factor*. For $L = 0$, $g = 2$. In Chapter XII dealing with multi-electron atoms we shall see cases where $L \neq 0$, but $S = 0$, and then $g = 1$. For an illustrative example (Figure 11.11) consider the splittings of the $^2S_{1/2}$, $^2P_{3/2}$, and $^2P_{1/2}$ states in a weak external magnetic field. (The well-known sodium D lines, which are the source of bright yellow light from sodium vapor lamps, correspond to the transitions $^2P_{1/2} \rightarrow {}^2S_{1/2}$ and $^2P_{3/2} \rightarrow {}^2S_{1/2}$. Since sodium has one S electron outside a closed shell it can be considered a one-electron atom and the energy-level diagram for this one electron is very similar to that for hydrogen.)

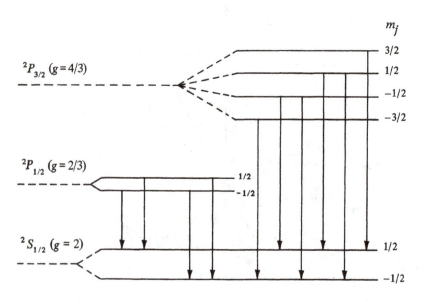

FIGURE 11.11 Illustrating the Zeeman splitting of doublet P and S states. The allowed transitions are as indicated.

By following the approach used in Chapter X one can show that the selection rule for transitions between these states is

$$\Delta m_j = 0, \pm 1 (\text{not including } m_j = 0 \longrightarrow m_j = 0 \text{ if } \Delta J = 0)$$

In general, a transition between the two states E_1 and E_2 would involve an emission or absorption of a photon of energy $\hbar\omega_{12} = E_1 - E_2$. If now an external magnetic field is applied, then each level will be perturbed.

$$\Delta E_1 = \mu_B H g_1 m_{j_1}$$

$$\Delta E_2 = \mu_B H g_2 m_{j_2}$$

and there would be a corresponding change in the photon energies, i.e.,

$$\hbar\Delta\omega_{12} = \Delta E_1 - \Delta E_2 = \mu_B H(g_1 m_{j_1} - g_2 m_{j_2}) \tag{11.27}$$

When $g_1 = g_2$, so-called *normal Zeeman* splitting occurs; when $g_1 \neq g_2$, the effect is called the *anomalous Zeeman* splitting. (This effect is named after P. Zeeman who observed as early as 1896 that placing an atom in a magnetic field caused a splitting of the emission lines. The effect could not be satisfactorily explained until after quantum mechanics was introduced.)

If we assume now that the external field is much stronger than the internal atomic magnetic field, then in first approximation the spin-orbit interaction can be neglected and both **S** and **L** will precess independently about **H** (*Paschen-Back* effect). In this case, the situation is particularly simple with

$$\boldsymbol{\mu} = \frac{-\mu_B}{\hbar}(\mathbf{L} + 2\mathbf{S})$$

$$\mu_z = \frac{-\mu_B}{\hbar}(L_z + 2S_z)$$

$$\Delta E = -\mu_z H = \frac{\mu_B H}{\hbar}(L_z + 2S_z)$$

and

$$\overline{\Delta E} = \mu_B H(m_l + 2m_s) \tag{11.28}$$

Figure (11.12) shows a schematic of the Paschen-Back effect for the ^2P and ^2S levels including the allowed transitions and the selection rules. Because there will always be some residual interaction of **L** and **S**, there should be a small correction term added to $\overline{\Delta E}$. This means that in general each line will be a closely spaced doublet, triplet, etc., depending on what the original field-free transition was. For instance, each of the lines shown in the figure will actually consist of a closely spaced doublet.

So far we have considered the problem of the hydrogen atom in a uniform external magnetic field using the same kind of semiclassical approach that was employed previously (Section 11) to explain the magnetic moment due to the orbital motion of the electron. However, these results can, of course, be obtained also directly using quantum mechanics. As an

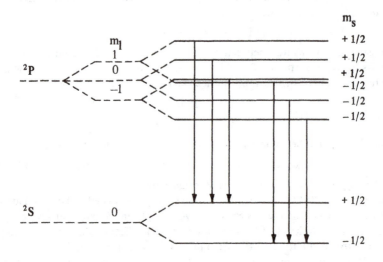

FIGURE 11.12 Illustrating the splitting of ^2P and ^2S levels in a strong external magnetic field (Paschen-Back effect). The indicated allowed transitions are governed by the selection rules, $\Delta m_l = 0, \pm 1$ and $\Delta m_s = 0$.

illustration, we consider the case of the hydrogen atom, only now for simplicity we neglect the effect of the electron spin.

In the presence of an external magnetic field the Hamiltonian for the electron (with no spin) moving in the Coulomb field of the proton can be written as

$$H = \frac{1}{2m_e}\left(\mathbf{p} - \frac{e\mathbf{A}}{c}\right)^2 + \frac{e^2}{r} \text{ (cgs units)} \qquad (11.29)$$

where $\mathbf{A}$ = vector potential for the magnetic field. Substituting $p \rightarrow (\hbar/i)\nabla$ and calling $H_0 = (-\hbar^2/2m_e)\nabla^2 + e^2/r$, we find the Schrödinger equation to be

$$\left\{H_0 + \frac{e\hbar i}{m_e c}\left(A_x\frac{\partial}{\partial x} + A_y\frac{\partial}{\partial y} + A_z\frac{\partial}{\partial z}\right) + \frac{e^2}{2m_e c^2}|\mathbf{A}|^2\right\}\psi = E\psi \qquad (11.30)$$

Now assume a constant field in the z direction $\mathbf{H} = H\hat{z}$. Then, since we want curl $\mathbf{A} = H\hat{z}$, it is convenient to make the following choices for $\mathbf{A}$

$$A_x = -\tfrac{1}{2}Hy$$
$$A_y = \tfrac{1}{2}Hx$$

With these choices

$$A_x\frac{\partial}{\partial x} + A_y\frac{\partial}{\partial y} + A_z\frac{\partial}{\partial z} = \frac{H}{2}\left(x\frac{\partial}{\partial y} - y\frac{\partial}{\partial x}\right) = \frac{H}{2}\frac{\partial}{\partial \phi}$$

and

$$|\mathbf{A}|^2 = H^2\left(\frac{x^2+y^2}{4}\right) = \frac{H^2 r^2}{4}\sin^2\theta$$

By making use of these results the Schrödinger equation becomes

$$\left\{H_0 + \frac{e\hbar H i}{2m_e c}\frac{\partial}{\partial\phi} + \frac{e^2 H^2 r^2}{8m_e c^2}\sin^2\theta\right\}\psi_n = E_n\psi_n \qquad (11.31)$$

If the external magnetic field is assumed to be small (Zeeman effect), then

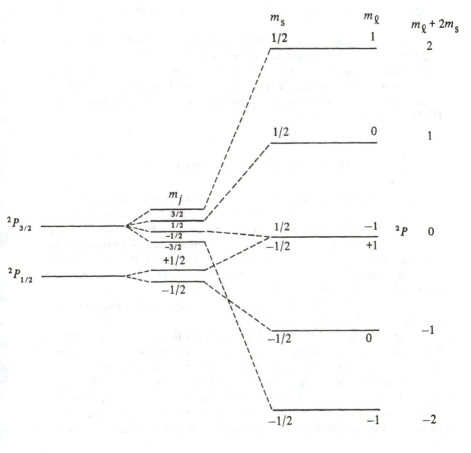

FIGURE 11.13 Illustration showing how the levels match up for Zeeman and Paschen-Back splitting of a typical ^2P level.

we need keep only the first-order term proportional to H and write

$$\left\{ H_0 - \frac{eH}{2m_ec} \frac{\hbar}{i} \frac{\partial}{\partial\phi} \right\} \psi_n = E_n\psi_n \tag{11.32}$$

However, we already know the wave functions which are the solutions of

$$H_0\psi_n = E_{n_0}\psi_n$$

and it is immediately clear that the ψ_n are also eigenfunctions of the operator $\partial/\partial\phi$ so that

$$E_n = E_{n_0} - \frac{e\hbar}{2m_ec}Hm = E_{n_0} - \mu_B Hm$$

in agreement with the result of the semiclassical approach with $g = 1$ (Eq. (11.26)).

As we have seen, it is a straightforward problem to find the level splitting for the case of either weak or strong magnetic fields. It is a much more difficult calculation for the case when the field is intermediate, and we will not attempt it here. However, it is instructive to see how the levels match up in going from a weak field to a strong field. Figure 11.13 shows a diagram for the ^2P state already discussed.

11.10 HYPERFINE STRUCTURE

As instrumentation was improved, early spectroscopists found a hyperfine structure superposed on the fine structure appearing in various spectra. In 1924 Pauli suggested that the hyperfine structure (hfs) of spectral lines of certain elements could be explained by assuming that the nucleus had an intrinsic angular momentum, analogous to that of the electron, and that it would consequently have an intrinsic magnetic moment. This being the case, one should expect interactions between the nuclear magnetic moment and the field due to the orbital motion of the electron and the intrinsic electron magnetic moment.

In general, because of its much greater mass than that of the electron, it is reasonable to assume that any intrinsic magnetic moment of the nucleus would be much smaller than the intrinsic magnetic moment of the electron since the intrinsic magnetic moment is inversely proportional to the mass. Hence, interactions involving the nuclear magnetic moment should produce much smaller perturbations than do electron spin-orbit interactions; i.e., hyperfine-structure splitting should be much smaller than fine-structure splitting.

It is now known that both the neutron and proton have an intrinsic angular momentum (*nuclear spin*) equal to the spin of the electron. Therefore, we can say that both the proton and neutron have spin I_p. Likewise both the neutron and proton have a magnetic moment. However, these magnetic moments do not equal each other or the electron magnetic moment times m_e/m_p. It is customary to express the magnetic moments as follows:

For the proton: $\mu_p = \frac{g_p \mu_N}{\hbar} |I_p|$

For the neutron: $\mu_n = \frac{g_N \mu_N}{\hbar} |I_p|$

(11.33)

where

$$\mu_N = \frac{e\hbar}{2m_p c} \approx \frac{1}{1800} \mu_B$$

In analogy with the electron case, μ_N is called the *nuclear magneton*, while the proton and neutron g factors are g_p and g_N respectively and neither is equal to unity. From the definition μ_N is found to be $= 5.05038 \times 10^{-12}$ erg gauss^{-1}; the results of various experiments show that the z component of $\mu_p = 2.79274 \, \mu_N$ and the z component of $\mu_N = -1.91303 \, \mu_N$.

Since the nucleus is known to be made up of neutrons and protons, it is not surprising that nuclei have a nuclear spin I, which can be either integer or half-integer, and a magnetic moment of the order of μ_N. In accord with previous discussions, we say then that

$$|I| = \sqrt{I(I+1)} \, \hbar$$

$$I_z = m_I \hbar \qquad m_I = -I, (-I+1), \cdots, +I$$

$$\mu_I = \frac{g_I \mu_N}{\hbar} |I|$$

$$\mu_{I_z} = g_I \mu_N m_I$$

where g_I is the nuclear g factor and m_I is the projection of I along any (z) axis.

Our description of the one-electron atom must now be enlarged to take into account the existence of the nuclear spin and magnetic moment. In particular we should expect to incorporate the following:

(a) Nuclear spin variables; operators and quantum numbers must be introduced.

(b) Interactions between the nuclear magnetic moment and various other quantities must be considered. We will need a term giving an $\mathbf{I} \cdot \mathbf{J}$ interaction in analogy to the $\mathbf{S} \cdot \mathbf{L}$ interaction between the electron magnetic moment and magnetic moment due to the orbital angular momentum. Because the nuclear magnetic moment will be of the order of $\mu_N (\ll \mu_B)$ we should expect the $\mathbf{I} \cdot \mathbf{J}$ term to be very much less than the $\mathbf{S} \cdot \mathbf{L}$ perturbation. In other words, we should expect the $\mathbf{I} \cdot \mathbf{J}$ interaction to produce a hyperfine splitting of the levels which will be small compared to the fine-structure splitting produced by the $\mathbf{S} \cdot \mathbf{L}$ interaction.

A calculation of the hyperfine splitting using perturbation theory shows it to have the following form:

$$\overline{\Delta E}_{I.J} = K[F(F+1) - J(J+1) - I(I+1)] \qquad (11.34)$$

where K is a constant and $\mathbf{F}$ is the new total angular momentum of the system.

$$\mathbf{F} = \mathbf{I} + \mathbf{J} = \mathbf{I} + \mathbf{L} + \mathbf{S}$$

$$|\mathbf{F}| = \hbar\sqrt{F(F+1)} \text{ and } F_z = m_F\hbar, \; m_F = -F \cdots + F \qquad (11.35)$$

In general, there will be fine-structure splitting into levels of different J and there will then be a further hyperfine splitting of these levels. As an example, consider the case where $J = 2$ and $I = 3/2$ shown in Figure 11.14. Of course, as was the case with the fine structure, there will be selection rules governing possible dipole radiative transitions between levels of different F. They are $\Delta F = \pm 1, 0$, but not $F = 0 \rightarrow F = 0$.

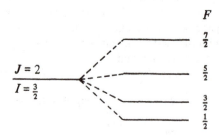

FIGURE 11.14 Illustrating the nuclear hyperfine splitting of a level with $J = 2$ and $I = \frac{3}{2}$.

If one now introduces an external magnetic field, there will be a further splitting of the levels in analogy to the results of the discussion of Section 11.9. As an example, we illustrate in Figure 11.15 the case where $J = \frac{1}{2}$, $I = \frac{3}{2}$; then $F = 2$ or 1.

Notice the great similarity between Figure 11.15 and Figures 11.11 and 11.13. The splitting of the hyperfine structure in the presence of the external magnetic field can be used to determine the nuclear spin in favorable cases. For instance, if the same number of hyperfine lines is obtained for two terms having different J values (assumed known) then the number of hyperfine lines $2I + 1$ tells us the nuclear spin I.

Problem 11.7: What are the values of L, S, and J and the multiplicities of levels having the following term designations?

$$^{1}S_{0}, \, ^{3}D_{2}, \, ^{4}P_{5/2}, \, ^{1}D_{2}, \, ^{2}F_{7/2}, \, ^{6}I_{13/2}$$

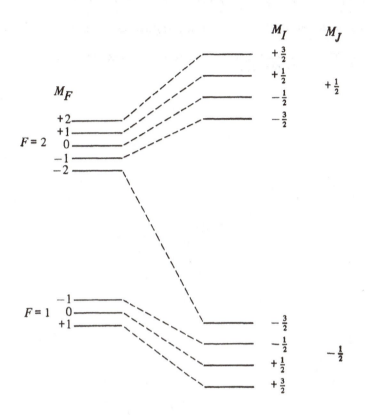

FIGURE 11.15 Illustration of Zeeman splitting of hyperfine levels formed from a $J = \frac{1}{2}$ and $I = \frac{3}{2}$ state. The indicated allowed transitions are governed by the selection rule, $\Delta m_F = 0, \pm 1$ (not including $m_F = 0 \longrightarrow m_F = 0$ if $\Delta F = 0$).

Problem 11.8: Find the Zeeman splittings for the $^2D_{5/2}$ and $^2F_{7/2}$ levels of hydrogen in a magnetic field of 2×10^4 gauss. Show the expected structure of the spectral line resulting from the transition between the two levels in the presence of the magnetic field.

BIBLIOGAPHY

Herzberg, Gerhard, *Atomic Spectra and Atomic Structure*. Englewood Cliffs, N.J.: Prentice-Hall, Inc., 1944.

Kuhn, H. G., *Atomic Spectra*. New York: Academic Press Inc., 1963.

Eisberg, R. M., *Fundamentals of Modern Physics*. New York: John Wiley & Sons, Inc., 1961

Leighton, Robert B., *Principles of Modern Physics*. New York: McGraw-Hill Book Co., 1959.

Fong, Peter, *Elementary Quantum Mechanics*. Reading, Mass.: Addison-Wesley Publishing Co. Inc., 1962.

XII

MULTI-ELECTRON ATOMS

In the foregoing discussions, we have considered the particularly simple case of the one-electron atom and we have seen how to account for the various interactions, including the effect of an external magnetic field. Now we are ready to consider the situation where there is more than one electron in the atomic system. To do this, we first take up the problem of a system containing two identical particles and then apply the results to the two-electron helium atom.

12.1 IDENTICAL PARTICLES—SYMMETRY

For a system containing two identical particles (such as the two electrons in the helium atom), one cannot distinguish between two eigenfunctions of the system which differ only in that the particles are interchanged. They are both eigenfunctions of the same eigenvalue. For example, if x denotes spin and space coordinates, then in

$$H \psi(x_1, x_2) = E \psi(x_1, x_2)$$
$$H \psi(x_2, x_1) = E \psi(x_2, x_1)$$

$\psi(x_1, x_2)$ and $\psi(x_2, x_1)$ are degenerate eigenfunctions of the H operator. The exchange operation consists in exchanging the two particles, and if ψ is an eigenfunction of the exchange operator then

$$P \psi(x_1, x_2) = k \psi(x_2, x_1)$$

where k is the eigenvalue of the exchange operator. When this exchange operation is taken twice, one must end up with the original eigenfunction:

$$P^2\psi(x_1, x_2) = k^2\psi(x_1, x_2) = \psi(x_1, x_2)$$

Therefore, $k = \pm 1$, whence

$$P \psi(x_1, x_2) = +\psi(x_2, x_1)$$

or

$$P \psi(x_1, x_2) = -\psi(x_2, x_1)$$

The first equation defines a *symmetric* eigenfunction and the second an *antisymmetric* eigenfunction. If the state is degenerate and two or more eigenfunctions are possible for a given eigenvalue, it is possible to construct combinations of these eigenfunctions which are either entirely symmetric or entirely antisymmetric. The importance of symmetry is that the symmetry of wave functions representing actual physical systems is limited, thereby greatly restricting the number of wave functions describing such a system.

Consider a system of two identical particles such that one particle is in a state labeled α and the other state β where $\alpha(1)$ represents particle 1 in state α, etc. If particles 1 and 2 are interchanged in either the $\alpha(1) \, \alpha(2)$ or $\beta(1) \, \beta(2)$ state, the eigenfunction is unchanged and therefore these states are said to be symmetric. The other states $\alpha(1) \, \beta(2)$ and $\beta(1) \, \alpha(2)$ are neither entirely symmetric nor entirely antisymmetric, however, and it is convenient to treat them as a superposition of two other states one of which is entirely symmetric and the other antisymmetric. We therefore use linear combinations of the $\alpha(1) \, \beta(2)$ and $\beta(1) \, \alpha(2)$ states to represent the two remaining wave functions of the system:

$$\psi_S = \frac{1}{\sqrt{2}}[\alpha(1)\beta(2) + \beta(1)\alpha(2)] \qquad (12.1)$$

or

$$\psi_A = \frac{1}{\sqrt{2}}[\alpha(1)\beta(2) - \beta(1)\alpha(2)]$$

where the factor $1/\sqrt{2}$ is used for normalization. Since the $\alpha(1)$, $\beta(2)$, $\alpha(2)$, and $\beta(1)$ are all solutions of the Schrödinger equation, the ψ_S and ψ_A will also be solutions.

To illustrate we choose the case of two identical noninteracting particles in a one-dimensional box. If we let x_1 represent the position of particle 1 and x_2 that of particle 2, the Schrödinger equation is

$$\left\{-\frac{\hbar^2}{2m}\left(\frac{\partial^2}{\partial x_1} + \frac{\partial^2}{\partial x_2}\right) + V(x_1) + V(x_2)\right\}\psi(x_1, x_2)$$

$$= E\psi(x_1, x_2)$$

where $V(x) = 0$ $\qquad 0 \leqslant x \leqslant L$

$\qquad\qquad\quad = \infty$ $\qquad L \leqslant x, x \leqslant 0$

This equation can readily be solved if we set

$$\psi(x_1, x_2) = \psi_1(x_1)\,\psi_2(x_2)$$

Normalized solutions which satisfy the boundary conditions are $\psi_{nk} = (2/L) \sin n\pi(x_1/L) \sin k\pi(x_2/L)$ and the corresponding energy of the system is $(\pi^2\hbar^2/2m)((n^2 + k^2)/L^2)$.

However, we could just as well have chosen solutions

$$\psi_{kn} = \frac{2}{L}\sin\frac{k\pi x_1}{L}\sin\frac{n\pi x_2}{L}$$

so that the solutions are seen to be twofold degenerate $(n \neq k)$. In accord with the foregoing discussion we then choose symmetric and antisymmetric combinations:

$$\psi_S = \frac{1}{\sqrt{2}}(\psi_{nk} + \psi_{kn}) = \frac{\sqrt{2}}{L}\left(\sin\frac{n\pi x_1}{L}\sin\frac{k\pi x_2}{L} + \sin\frac{k\pi x_1}{L}\sin\frac{n\pi x_2}{L}\right)$$

$$\psi_A = \frac{1}{\sqrt{2}}(\psi_{nk} - \psi_{kn}) = \frac{\sqrt{2}}{L}\left(\sin\frac{n\pi x_1}{L}\sin\frac{k\pi x_2}{L} - \sin\frac{k\pi x_1}{L}\sin\frac{n\pi x_2}{L}\right)$$

It is readily seen that

$$\int_0^L \int_0^L \psi_S^* \psi_A\, dx_1 dx_2 = 0$$

Consequently, the ψ_S and ψ_A are orthogonal.

It is interesting to compare the probability densities as determined by ψ_A and ψ_S.

$$\psi_S^* \psi_S = \left(\frac{\psi_{nk}^2 + \psi_{kn}^2}{2} + \psi_{nk}\psi_{kn}\right)$$

$$\psi_A^* \psi_A = \left(\frac{\psi_{nk}^2 + \psi_{kn}^2}{2} - \psi_{nk}\psi_{kn}\right)$$

Looking at the probability density for the special case where $x_1 \approx x_2$, we find

$$\psi_S^* \psi_S(x_1 \approx x_2) \longrightarrow \text{a maximum}$$

$$\psi_A^* \psi_A(x_1 \approx x_2) \longrightarrow \text{a minimum (0)}$$

Thus it becomes apparent that where the two particles are described by a symmetric wave function they have a maximum probability of being found close together, while in the antisymmetric state they have small probability of being found close together.

Note that in the foregoing discussion we have assumed that $n \neq k$. If $n = k$, then $\psi_A = 0$ and we are left with just a single nondegenerate symmetric eigenfunction

$$\psi_{nn} = \frac{2}{L} \sin \frac{n\pi x_1}{L} \sin \frac{n\pi x_2}{L}$$

From the foregoing we see that it is possible to describe a system of the two identical particles by wave functions that are either symmetric or antisymmetric. Without any further information, we might, in general, expect that for any given system either the antisymmetric or symmetric wave functions, or possibly both, could just as well be used. However, it happens that the universe is so constituted that all systems of identical particles with integer spin must be represented by wave functions which are symmetric with respect to the exchange of any two particles. Similarly, all systems of identical particles with half-integer spin must be represented by wave functions which are antisymmetric with respect to the exchange of any two particles. For the multi-electron system, this result was first postulated by W. Pauli in 1924 even before the advent of quantum mechanics. The *Pauli exclusion principle* states that in a multi-electron atom there can never be more than one electron in the same quantum state. As will be made more clear in the following discussion, this statement can be seen to be a consequence of saying that the electron wave function must be antisymmetric.

In Section 11.4 we saw that the wave function for an electron with spin could be written as follows:

$$\bar{\psi} = \psi_{\text{space}}\psi_{\text{spin}} = \psi_{nlm_l}\sigma_{m_s}$$

so that for the helium atom with two electrons we will have to take the appropriate combination of the two wave functions $\bar{\psi}_a(1) = \psi_a(1)\sigma_a$ and $\bar{\psi}_b(2) = \psi_b(2)\sigma_b$, where $\bar{\psi}_a(1)$ represents the first electron at position 1 with a spin represented by σ_a (either $\frac{1}{2}$ or $-\frac{1}{2}$) and similarly for $\bar{\psi}_b(2)$. The total wave function for the two electrons will have to be the appropriate combination of space and spin states that makes it antisymmetric. This total wave function can of course be written as the product of a space part times a spin part; if the space part is symmetric, the spin part must be antisymmetric and vice versa. As we have seen, the space parts can be written

$$\psi_S = \frac{1}{\sqrt{2}}[\psi_a(1)\psi_b(2) + \psi_a(2)\psi_b(1)] \qquad \textbf{(12.2)}$$

$$\psi_A = \frac{1}{\sqrt{2}}[\psi_a(1)\psi_b(2) - \psi_a(2)\psi_b(1)]$$

The spin wave functions will be a bit more complicated because the separate spin angular momenta can add vectorially just as the orbital and spin angular momenta do in the case of the one-electron atom. Instead of having $\mathbf{J} = \mathbf{L} + \mathbf{S}$, here $\mathbf{S} = \mathbf{s}_1 + \mathbf{s}_2$ so that $S = 0$ or 1. The magnitude of $\mathbf{S}$ will of course be $\sqrt{S(S+1)}\, \hbar$.

For $S = 0$, $m_s = 0$ and we have just a single state while for $S = 1$, $m_s = 1, 0, -1$ and a triplet spin state results. This corresponds to the three possible orientations with respect to some preferred (z) direction in space and is equivalent to saying we can start with the following four spin combinations $\sigma_a(\tfrac{1}{2})\sigma_b(\tfrac{1}{2})$, $\sigma_a(-\tfrac{1}{2})\sigma_b(-\tfrac{1}{2})$, $\sigma_a(\tfrac{1}{2})\sigma_b(-\tfrac{1}{2})$, and $\sigma_a(-\tfrac{1}{2})\sigma_b(\tfrac{1}{2})$ to construct four symmetric and antisymmetric combinations.

S	m_s		
0	0	$\dfrac{1}{\sqrt{2}}\left[\sigma_a\left(\tfrac{1}{2}\right)\sigma_b\left(-\tfrac{1}{2}\right) - \sigma_a\left(-\tfrac{1}{2}\right)\sigma_b\left(\tfrac{1}{2}\right)\right]$	antisymmetric-singlet
1	1	$\sigma_a\left(\tfrac{1}{2}\right)\sigma_b\left(\tfrac{1}{2}\right)$	
1	0	$\dfrac{1}{\sqrt{2}}\left[\sigma_a\left(\tfrac{1}{2}\right)\sigma_b\left(-\tfrac{1}{2}\right) + \sigma_a\left(-\tfrac{1}{2}\right)\sigma_b\left(\tfrac{1}{2}\right)\right]$	symmetric-triplet
1	-1	$\sigma_a\left(-\tfrac{1}{2}\right)\sigma_b\left(-\tfrac{1}{2}\right)$	

$$\text{(12.3)}$$

The total eigenfunction must be antisymmetric and thus we can form the following combinations from Eqs. (12.2, 12.3)

$$\frac{1}{2}[\psi_a(1)\psi_b(2) + \psi_a(2)\psi_b(1)]\left[\sigma_a\left(\tfrac{1}{2}\right)\sigma_b\left(-\tfrac{1}{2}\right) - \sigma_a\left(-\tfrac{1}{2}\right)\sigma_b\left(\tfrac{1}{2}\right)\right]$$

$$\text{(12.4)}$$

and

$$\frac{1}{\sqrt{2}}[\psi_a(1)\psi_b(2) - \psi_a(2)\psi_b(1)]\begin{cases} \sigma_a\left(\tfrac{1}{2}\right)\sigma_b\left(\tfrac{1}{2}\right) \\ \left\{\dfrac{1}{\sqrt{2}}\left[\sigma_a\left(\tfrac{1}{2}\right)\sigma_b\left(-\tfrac{1}{2}\right) + \sigma_a\left(-\tfrac{1}{2}\right)\sigma_b\left(\tfrac{1}{2}\right)\right]\right\} \\ \sigma_a\left(-\tfrac{1}{2}\right)\sigma_b\left(-\tfrac{1}{2}\right) \end{cases}$$

Notice that if the spin part of the wave function is symmetric, corresponding to parallel spins, the space part must be antisymmetric. This has the interesting and important consequence that the electrons will have small probability of being found close together if they have parallel spins and a maximum probability of being found close together if they have antiparallel spins. One can say that in effect *parallel spins repel* and *antiparallel spins attract*.

Problem 12.1: The interactions of two particles in the same ground state in a one-dimensional square well of width a cm is a small square well-type potential which is V_0 ev deep and b cm wide. That is, $V' = 0$ unless $x_2 - x_1 < b$, where the x's are particle coordinates, so that the integral over x_2 extends from $x_1 - b$ to $x_1 + b$ and the subsequent integral over x_1 from 0 to a. How does their interaction affect their energy? If the particles repel one another with a potential $+V_0$, how does this affect the symmetry of the ground state?

12.2 HELIUM ATOM

We are now ready to turn to the discussion of the helium atom. Since this is a three-body problem, we can expect immediately that it cannot be solved explicitly in closed form as unfortunately no one has ever been able to solve in closed form any three-dimensional case involving more than two interacting particles. However, as we have already seen in Sections 9.2 and 9.6, approximate solutions can be obtained for the energy of the ground state. In Section 9.2 it was pointed out that the excited states of the He atom are degenerate and hence degenerate perturbation theory is necessary in order to calculate their energies and approximate wave functions. Fortunately, the simplified development of degenerate perturbation theory given in Section 9.4 is adequate to treat, as an example, the first excited state of helium. In the following discussion we will make use of the Dirac bra and ket notation introduced in Section 9.5.

For the degenerate first excited state of helium it is useful to think of a *resonance* occurring in which either electron may spend some time in the higher energy level. The *resonance*, or *exchange*, energy that will be shown to result from this situation shifts the energy of the first excited p states from the energy expected if resonance is not taken into account. The energy shift due to resonance is observed in the shift of the frequency of the bright-line emitted radiation when the helium atom is de-excited and collapses to its ground state.

The zero-order wave functions for the helium atom, in which the interaction of the two electrons is ignored, are hydrogen-like wave functions. Because of the degeneracy between states of different n, l, and m there are eight states with the same energy possible to the unperturbed first excited energy level: if, say, electron number one is in the ground state $|100\rangle$, electron number two may be in any one of the excited states $|200\rangle$, $|210\rangle$, $|211\rangle$, or $|21 - 1\rangle$, and conversely for electron number two in the ground state, the resulting four more states being physically indistinguishable from the first four. The total wave function would be written, for example,

$|100,200\rangle$ if the first electron was in the 100 state (also called $1s$) and the second in the 200 (or $2s$) state.

The interaction energy between the two electrons is treated by means of degenerate perturbation theory. The perturbation matrix elements for the energy of interaction between any two electron states are abbreviated by the letters J and K with subscripts s and p depending on whether both electrons are in an s state ($l = 0$) or one of them is in a p state ($l = 1$).

$$J_s = \langle 100, 200 \left| \frac{e^2}{r_{12}} \right| 100, 200 \rangle \equiv \int \psi_{100}^{(1)*}\psi_{200}^{(2)*}\frac{e^2}{r_{12}}\psi_{100}^{(1)}\psi_{200}^{(2)}d\tau_1 d\tau_2 \quad \textbf{(12.5)}$$

where (1) and (2) refer to each of the two electrons, $d\tau_1$ is the differential volume element $r_1^2 \sin\,\theta_1\,dr_1 d\theta_1 d\varphi_1$ over the three degrees of freedom available to the first electron, and similarly for $d\tau_2$. $\langle 100,200|$ and $|100,200\rangle$ are the products of two hydrogen-like wave functions derived in Section 8.6 involving the Laguerre polynomials. J_s is also written for

$$\langle 200, 100 \left| \frac{e^2}{r_{12}} \right| 200, 100 \rangle$$

which has an identical numerical value.

$$K_s = \langle 100, 200 \left| \frac{e^2}{r_{12}} \right| 200, 100 \rangle \equiv \int \psi_{100}^{(1)*}\psi_{200}^{(2)*}\frac{e^2}{r_{12}}\psi_{200}^{(1)}\psi_{100}^{(2)}d\tau_1 d\tau_2 \quad \textbf{(12.6)}$$

or

$$\langle 200, 100 \left| \frac{e^2}{r_{12}} \right| 100, 200 \rangle$$

which has the same numerical value. J_p, and also K_p, is written for six different matrix elements all having the same numerical result, one of which is

$$J_p = \langle 100, 211 \left| \frac{e^2}{r_{12}} \right| 100, 211 \rangle$$

also,

$$K_p = \langle 100, 211 \left| \frac{e^2}{r_{12}} \right| 211, 100 \rangle \quad \textbf{(12.7)}$$

The other J_p and K_p matrix elements differ from these only in having the two electrons interchanged or in the orientation of the total angular momentum in space (that is, in having $m = 0$ or -1 instead of $+1$). The matrix elements other than these 16 elements are zero. For example, consider the matrix element $\langle 100,200| e^2/r_{12} | 100,211 \rangle$: the $|211\rangle$ wave function is an odd function (see Appendix D); the other terms are all even, and thus, integrated over all space this matrix element must be zero. The J integrals are called *Coulomb integrals* since they give the potential energy contribution from the mutual repulsion of the two electrons. The K integrals are called *exchange integrals* since they result from the exchange of the two electrons in the wave function on the left as compared with the wave function on the right in the matrix element. Another name for the K integrals is *resonance integrals* since this energy contribution arises from the resonance of the two electrons between the states on the left and right of the matrix element.

By use of these matrix elements between the various hydrogen-like wave functions, the secular determinant Eq. (9.32) is written

$$\begin{vmatrix} (J_s - \Delta E) & K_s & 0 & 0 & 0 & 0 & 0 & 0 \\ K_s & (J_s - \Delta E) & 0 & 0 & 0 & 0 & 0 & 0 \\ 0 & 0 & (J_p - \Delta E) & K_p & 0 & 0 & 0 & 0 \\ 0 & 0 & K_p & (J_p - \Delta E) & 0 & 0 & 0 & 0 \\ 0 & 0 & 0 & 0 & (J_p - \Delta E) & K_p & 0 & 0 \\ 0 & 0 & 0 & 0 & K_p & (J_p - \Delta E) & 0 & 0 \\ 0 & 0 & 0 & 0 & 0 & 0 & J(_p - \Delta E) & K_p \\ 0 & 0 & 0 & 0 & 0 & 0 & K_p & (J_p - \Delta E) \end{vmatrix} = 0$$

(12.8)

where ΔE is the small energy shift due to the perturbation.

Equation (12.8) may be rewritten

$$[(\Delta E - J_s)^2 - K_s^2][\Delta E - J_p)^2 - K_p^2]^3 = 0 \tag{12.8a}$$

since all the J's and K's with the same subscript are equal. The solutions of (12.8a) are

$$\Delta E = J_s + K_s, \ J_s - K_s, \ J_p + K_p, \ J_p - K_p \tag{12.9}$$

with the last two terms still giving a triple degeneracy since the degeneracy in m has not been broken up. Since the energy does not depend on m, for convenience we may drop the magnetic quantum number from the bra and ket vectors for the rest of this section. The splitting up of the degeneracy between $|10, 20\rangle$ and $|10, 21\rangle$ (i.e., the 2s and 2p states, see Figure 12.1), corresponding to the difference between J_s and J_p is due to the greater penetration of lower l electrons within the electron cloud. The further splitting of these states, however, results from the resonance energy expressed by the exchange integrals. We have also discussed previously the raising of the ground state due to the mutual shielding of the positive nucleus by the electrons. Figure 12.1 illustrates the effect of these perturbation calculations.

Thus, we have found that the eightfold degenerate first excited level of the helium atom splits into four levels, of which the lower two are entirely nondegenerate due to the addition or subtraction of the exchange energy. If we choose new wave functions

$$\psi_S = \frac{1}{\sqrt{2}}(|100, 200\rangle + |200, 100\rangle)$$

$$\psi_A = \frac{1}{\sqrt{2}}(|100, 200\rangle - |200, 100\rangle) \tag{12.10}$$

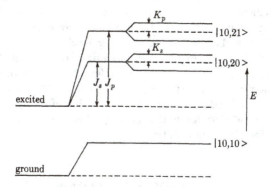

FIGURE 12.1 Illustration of the effect of the interaction energy of the two electrons on the lowest two unperturbed levels for the electrons in the helium atom. The two lines on the left represent the ground state energy and first excited state energy if the perturbation caused by the electron repulsion is neglected. Both levels are raised by the electron shielding of the nuclear electrostatic potential, but the excited level is split into several levels as a result of differences in shielding between s $(l = 0)$ and $p(l = 1)$ electrons and quantum mechanical resonance.

then K_s is Eq. (12.6) becomes

$$K_s = \int \psi_S^* \frac{e^2}{r_{12}} \psi_A \, d\tau$$

$$= \frac{1}{2}\Big(\langle 100, 200 \big| \frac{e^2}{r_{12}} \big| 100, 200 \rangle - \langle 100, 200 \big| \frac{e^2}{r_{12}} \big| 200, 100 \rangle$$

$$+ \langle 200, 100 \big| \frac{e^2}{r_{12}} \big| 100, 200 \rangle - \langle 200, 100 \big| \frac{e^2}{r_{12}} \big| 200, 100 \rangle \Big) = 0$$

$$(12.11)$$

Similarly, K_p also vanishes when the wave functions (12.10) are used. Similarly, the nonzero J integrals are found to be symmetric and antisymmetric.

$$J_{ss} = \langle 100, 200 \big| \frac{e^2}{r_{12}} \big| 100, 200 \rangle + \langle 100, 200 \big| \frac{e^2}{r_{12}} \big| 200, 100 \rangle$$

$$J_{sa} = \langle 100, 200 \big| \frac{e^2}{r_{12}} \big| 100, 200 \rangle - \langle 100, 200 \big| \frac{e^2}{r_{12}} \big| 200, 100 \rangle \quad (12.12)$$

ψ_S is a *symmetric* wave function since it does not change sign if the two electrons are interchanged. ψ_A, however, does change sign and is thus an *antisymmetric* wave function. ψ_A and ψ_S are seen to be the correct first-order wave functions since the off-diagonal elements K_s, K_p vanish if these wave functions are used. Whereas before inclusion of the interaction of the electrons only one energy level was predicted, now four closely spaced energy levels are predicted. The lowest level is due to the antisymmetric $2s$ state, the next to the symmetric $2s$ state, the third to the antisymmetric degenerate $2p$ level, and the highest to the symmetric $2p$ level.

In general, then, whenever a system comprised of two similar interacting members may have two exactly similar states, their interaction will cause a lowering and raising (i.e. a splitting) of the energy of these states compared to the energy calculated regarding just one member of the system at a time. The depression and elevation of the energy level is due to resonance between the two possible degenerate states.

12.3 COMPLEX ATOMS

As the number of electrons in the atomic system becomes larger, a perturbation calculation such as we have discussed for the helium atom rapidly becomes extremely difficult in practice to carry out. One way out of this difficulty was given by Hartree. In this approximation one assumes a system of the nucleus plus electrons wherein the Coulomb interactions between various electrons are accounted for by supposing that each electron moves independently in a central field $V_i(\mathbf{r}_i)$. This field consists of the field of the charged nucleus and a spherically symmetric field associated with the average spatial distribution of the remaining electrons. The Schrödinger equation for the total wave function of the electrons in the atomic system will then be of the form

$$\frac{-\hbar^2}{2m} \sum_{i=1}^{Z} \nabla_i^2 \psi_T + \sum_{i=1}^{Z} V_i(\mathbf{r}_i)\psi_T = E_T \psi_T \qquad (12.13)$$

Since the electrons are assumed to be noninteracting, Eq. (12.13) splits into Z one-particle Schrödinger equations. Therefore, the solution for ψ_T is

$$\psi_T = \phi_1(\mathbf{r}_1)\phi_2(\mathbf{r}_2) \dots \phi_Z(\mathbf{r}_Z)$$

where the $\phi_i(\mathbf{r}_i)$ are the solutions to the one-particle Schrödinger equations. The effects of the electron spin are ignored, except in that the Pauli exclusion principle is invoked in the assignment of quantum numbers to the single-particle wave functions so that no two electrons in one atom can have the same values for all four quantum numbers, n, l, m_l, and m_s.

At this point, neither the $V_i(\mathbf{r}_i)$ nor the $\phi_i(\mathbf{r}_i)$ are known so that in general one might expect the problem to be insolvable. However, reasonably accurate solutions can be obtained in the following way. First an educated guess for the $V_i(\mathbf{r}_i)$ is made and then the one-particle equations are solved.

The resulting ϕ_l are used to calculate the total charge density arising from the electrons in the atom. The potential, as seen by the ith electron, resulting from this charge density is calculated by means of Poisson's equation from electrostatics

$$\nabla^2 V'_i = -\rho_i$$

where ρ_i is the total charge density arising from all the electrons (except the ith) which produce the average potential V'_i. When V'_i is added to the potential of the charged nucleus one has the average potential seen by the ith electron. Ideally the average potentials calculated for all the electrons should then agree with the $V_i(\mathbf{r}_i)$ assumed initially in setting up the Schrödinger equation.

In general, these calculated potentials will differ from the original assumed $V_i(\mathbf{r}_i)$ and thus an improved choice for the $V_i(\mathbf{r}_i)$ is made and new ϕ_i are calculated, from which another set of $V_i(\mathbf{r}_i)$ can then be calculated. The iteration process is continued until the $V_i(\mathbf{r}_i)$ obtained from the final ϕ_i agree satisfactorily with $V_i(\mathbf{r}_i)$ used to calculate the final ϕ_i; the $V_i(\mathbf{r}_i)$ are then said to form a *self-consistent field*.

Note that the assumption of spherical symmetry for the charge density that any one electron sees should be quite good. We have already seen an example for the one-electron atom where, if all possible states for a given n are included, the corresponding electron density distribution is precisely spherically symmetric.

In the Hartree approximation one finds that the net potential any one electron sees is given by

$V(r) = -e^2 Z(r)/r$ where $Z(r) =$ effective charge seen by the electron.

For
$$r \longrightarrow 0 \quad Z(r) \longrightarrow Z$$
$$r \longrightarrow \infty \quad Z(r) \longrightarrow 1$$

In between, $Z(r)$ takes on intermediate values which can be calculated.

It should be emphasized that this has been a most elementary discussion and that actual calculations must include refinements such as spin-orbit interactions in order to get quantitative agreement with electron binding energies and other experimentally measurable quantities. Moreover, wave functions belonging to the same energy level are not orthogonal in the Hartree theory; this more precise refinement is embodied in the Hartree-Fock theory, which is beyond the scope of this book. In general, a product wave function can be made orthogonal by forming a Slater determinant wave function. This is a superposition of product wave functions of the individual electrons which makes use of the fact that the interchange of any two columns of a determinant (corresponding to the operation of exchanging two electrons) changes the sign of the determinant. With χ denoting the spin of the electron and φ the electron total space and spin state, the determinant is written

$$\psi = \text{Det} \begin{vmatrix} \varphi_1(\mathbf{r}_1, \chi_1) & \varphi_1(\mathbf{r}_2, \chi_2) \cdots \varphi_1(\mathbf{r}_n, \chi_n) \\ \varphi_2(\mathbf{r}_1, \chi_1) & \varphi_2(\mathbf{r}_2, \chi_2) \cdots \varphi_2(\mathbf{r}_n, \chi_n) \\ \cdot & \cdot & \cdot \\ \cdot & \cdot & \cdot \\ \cdot & \cdot & \cdot \\ \varphi_n(\mathbf{r}_1, \chi_1) & \varphi_n(\mathbf{r}_2, \chi_2) \cdots \varphi_n(\mathbf{r}_n, \mathbf{x}_n) \end{vmatrix} \qquad (12.14)$$

The fact that levels specified by antisymmetric space wave functions were found to have lower energies than levels with symmetric space wave functions in the discussion on the helium atom is particularly significant because the same result is found to hold for all atomic electron systems. This may be interpreted physically as resulting from the repulsive force between the electrons: The antisymmetric space wave functions overlap much less than the symmetric space wave functions, which makes the Coulomb repulsion between the two electrons less in the antisymmetric than in the symmetric space state. Thus, the Pauli exclusion principle leads, in general, to an indirect dependence of the electron energies on their spin state. Since electrons have half-integral spin, the total electronic wave function is required to be antisymmetric, and space-antisymmetric electron states are lower in energy than space-symmetric ones; therefore among degenerate space states the corresponding symmetric spin states will be lower in energy than the antisymmetric spin states. Another way of saying this is Hund's rule, *states of highest spin will have the lowest energy*, for the antisymmetric two-electron spin state is composed of electrons whose spins couple to zero, and the symmetric two-electron spin state is composed of electrons whose spins couple to one. In general, for n electrons, the state with z projection of spin $n\hbar/2$ has a total spin of $n\hbar/2$. This maximum value of z projection of spin, $n\hbar/2$, can occur only if all n electrons are lined up in one direction along the z axis with spin up, i.e., all electrons are in the same spin state α. This state is necessarily symmetric in spin, and multiplies a necessarily totally antisymmetric space wave function. Similarly, the state with z projection of spin $(n-1)\hbar/2$ has $n-1$ electrons in an α state and one electron in a β state. This state is composed *partly* of symmetric spin states ($\alpha \alpha \alpha \ldots (\alpha\beta + \beta\alpha)$) which are eigenfunctions of total spin $n\hbar/2$ and multiply a necessarily totally antisymmetric space wave function; however, this state also includes spin states which are eigenfunctions of total spin $(n-1)\hbar/2$, $\alpha \alpha \alpha \ldots (\alpha\beta - \beta\alpha)$. Hence these constituent spin states are spin-antisymmetric with respect to the exchange of two electrons and spin-symmetric with respect to any other electron pair. Since the space state must be symmetric with respect to the electron pair which is antisymmetric in spin, this state with total spin $(n-1)\hbar/2$ is higher in energy than the state of total spin $n\hbar/2$, in accord with Hund's rule. A state with total spin of $(n - 2)\hbar/2$ is of still higher energy, and so on for the other states. In summary, the space function must be symmetric in $(n/2) - (S/\hbar)$ pairs, where S is the total spin of the wave function. This phenomenon where the spin and energy are indirectly coupled is called *Russell-Saunders* coupling.

In Russell-Saunders coupling the electrostatic interaction dominates the spin-orbit coupling and, in general, the natural repulsion of the atomic electrons separates levels of different symmetry. In light atoms Russell-Saunders coupling is of primary importance. For the inner electrons of heavy atoms, on the other hand, the $\mathbf{L} \cdot \mathbf{S}$ or spin-orbit coupling is predominant. ($\mathbf{J}$ commutes with the Hamiltonian including the $\mathbf{L} \cdot \mathbf{S}$ term whereas $\mathbf{L}$ and $\mathbf{S}$ do not. Therefore $\mathbf{J}$ is a constant of the motion. The large size of the $\mathbf{L} \cdot \mathbf{S}$ term causes states of different J to be widely separated.) In this case neither the magnetic spin quantum numbers nor the magnetic orbital momentum quantum number l is unique for the widely separated levels of different total angular momentum, $j = |\mathbf{l} + \mathbf{s}|$. Hence the $|j, m\rangle$ states having total angular momentum $\mathbf{j}$ and $j_z \equiv m$, the magnetic quantum number, are the important states for these electrons. The weaker electrostatic interaction slightly separates levels of different symmetry as a weak perturbation. Thus the total angular momentum has a primary effect on the energy as a result of the $\mathbf{L} \cdot \mathbf{S}$ interaction. This situation is called *j-j coupling*, and in inner electrons of heavy atoms is the predominant effect.

In both R-S and *j-j* coupling the higher-order perturbation calculation is carried out in the representation in which the predominant perturbation is diagonal.

In discussing various atomic examples of perturbation theory we have always neglected the electronic interaction and the spin-orbit interaction on the electrons in zero-order approximation. Then in first-order approximation we used the hydrogen wave functions to estimate the change in energy levels and wave functions from either of these two perturbing effects. It is important to compare the magnitude of these perturbations with each other in order to know which perturbation should be used first in calculating energy levels and wave functions in higher order approximations. It is also important to compare the perturbations with the nuclear Coulombic potential for an indication of the reliability of the perturbation approximation.

Table 12.1 displays the magnitudes of the energy contributions of various atomic electron interactions. We note that neither Russell-Saunders nor *j-j* coupling is appropriate to the intermediate case of the outer electrons in heavy atoms. Successive perturbation treatment is not possible, so that both perturbations must be handled simultaneously, which is much more messy and difficult. Similar changes in the relative importance of these two physical effects occur in nuclear physics, as one goes from one end of the periodic table to the other.

TABLE 12.1

Magnitudes of the energies (in ev) of the various interactions involving atomic electrons.

Interaction effects	Light atoms	Valence electrons of heavy atoms	K or 1st Shell electrons of heavy atoms
Nuclear potential	~ 100–1000	~ 2–10	$\sim 20,000$–$100,000$
Electrostatic repulsion of electrons	~ 1	~ 1	~ 100
Spin-orbit coupling	$\sim 10^{-4}$–10^{-3}	~ 0.1–1.0	~ 3000

12.4 PERIODIC TABLE

We are now in a position to construct a table of the elements giving the quantum numbers appropriate to the various electrons and in particular to show how the periodic atomic structure arises.

Consider the one-electron atom, hydrogen. We have already seen that in the lowest energy level, the ground state, the electron will be described by the following quantum numbers, $n = 1$, $l = 0$, $m_l = 0$, $m_s = \pm\frac{1}{2}$.

If we choose $m_s = -\frac{1}{2}$ for hydrogen, then the next electron added to give the ground state of helium will have quantum numbers $n = 1$, $l = 0$, $m_l = 0$, $m_s = +\frac{1}{2}$ since the Pauli exclusion principle says that any given atomic state can be occupied by no more than one electron. When a third electron is added to give the ground state of lithium, it must necessarily go to the next energy level for which $n = 2$. Then l can be $= 0$ or ± 1. From the discussion in Section 12.2 we know that the shielding effect of the inner electrons will cause the $l = 0$ state to lie lower in energy than $l = 1$. Therefore, we can say that the third electron added to make a lithium atom will be specified by $n = 2$, $l = 0$, $m_l = 0$, $m_s = \pm\frac{1}{2}$.

Following this line of reasoning, we can start to build up the periodic table of the elements as shown in Table 12.2. The atomic shell structure is readily apparent if one examines the periodic table, and it is easy to see why families of elements should exhibit similar chemical properties and spectra. For instance, consider the alkali metals, lithium, sodium, potassium, rubidium, cesium, and francium. Each has one s electron outside a closed shell. Such an electron has a small binding energy and therefore all the alkali metals exhibit hydrogen-like spectra. Another interesting family is that of the noble gases, helium, neon, argon, krypton, xenon, and radon. Each of these elements has a filled outer shell of electrons and all are chemically inactive. (For many years it was thought that the noble gases were completely inert chemically, but recently a variety of interesting compounds of the four heaviest elements have been produced.) In general, it is the number and configuration of the outermost electrons which determine the chemical properties of the element.

The electron shells fill in sequence, as we might naively expect, up through argon which has the configuration $1s^2\,2s^2\,2p^6\,3s^2\,3p^6$. At this point we find a deviation from the expected ordering. Because of the various electron-electron interactions and the more penetrating orbit of $l = 0$ electrons, the $4s$ electrons are more tightly bound than are the $3d$ electrons. Consequently, after argon one $4s$ electron is added in potassium and a second in calcium. Following this the $3d$ shell is filled, starting with scandium, which has the configuration $1s^2\,2s^2\,2p^6\,3s^2\,3p^6\,4s^2\,3d$. Iron, cobalt, and nickel belong to this transition group and their magnetic properties are due to the presence of the partly filled d shell in each case. As more and more electrons are added the shells are formed to fill up on the average in the following order $1s^2\,2s^2\,2p^6\,3s^2\,3p^6\,4s^2\,3d^{10}\,4p^6\,5s^2\,4d^{10}\,5p^6\,6s^2\,4f^{14}\,5d^{10}\,6p^6\,7s^2\,5f^{14}\,6d^{10}$. The periodic table terminates at $Z \approx 100$ because these nuclei are too unstable to fission or radioactive decay.

TABLE 12.2

Periodic Table

n	l	m_l	m_s	element	configurations
1	0	0	$-\frac{1}{2}$	H	$1s$
1	0	0	$+\frac{1}{2}$	He	$1s^2$
2	0	0	$-\frac{1}{2}$	Li	$1s^2 2s$
2	0	0	$+\frac{1}{2}$	Be	$1s^2 2s^2$
2	1	-1	$-\frac{1}{2}$	B	$1s^2 2s^2 2p$
2	1	-1	$+\frac{1}{2}$	C	$1s^2 2s^2 2p^2$
2	1	0	$-\frac{1}{2}$	N	$1s^2 2s^2 2p^3$
2	1	0	$+\frac{1}{2}$	O	$1s^2 2s^2 2p^4$
2	1	1	$-\frac{1}{2}$	F	$1s^2 2s^2 2p^5$
2	1	1	$+\frac{1}{2}$	Ne	$1s^2 2s^2 2p^6$
3	0	0	$-\frac{1}{2}$	Na	$1s^2 2s^2 2p^6 3s$
3	0	0	$+\frac{1}{2}$	Mg	$1s^2 2s^2 2p^6 3s^2$

The 14 elements in which the $4f$ electrons are added one by one to fill the $4f$ shell are known as the rare earths. Since the orbits of the $4f$ electrons lie far inside those of the outer electrons and since the configurations of the outer electrons are similar for all the rare earths, the chemical properties of the rare earths are very nearly the same. Consequently, it is difficult to separate these elements by chemical means. A similar situation exists higher up in the periodic table when the $5f$ shell is being filled in elements such as protactinium and uranium.

12.5 SPECTROSCOPIC NOTATION—EQUIVALENT ELECTRONS

We have already seen in Section 11.7 that the properties of a single electron can be specified using standard spectroscopic notation. The problem now is to carry over this notation to the multi-electron case. As an example, we consider a specific case for an excited state of helium where one electron has quantum numbers $n = 1$, $l = 0$, $m_l = 0$ and the other has $n = 2$, $l = 1$, $m_l = 1$. The system of the two electrons has some angular momentum $\mathbf{J}$. (At this point we note that, customarily, in labeling a single electron wave function in a multi-electron system small letters are used, i.e., we may refer to a p electron ($l = 1$) with $\mathbf{j} = \mathbf{l} + \mathbf{s}$. For the total spin and angular momentum of a system of electrons capital letters are used, i.e., we may refer to a P state ($L = 1$) with $\mathbf{J} = \mathbf{L} + \mathbf{S}$.)

As we have seen in Section 12.4, there are two ways we can calculate $\mathbf{J}$ from the information given, either assuming Russell-Saunders or j-j coupling. In Russell-Saunders coupling it is appropriate to say that $\mathbf{J} = \mathbf{L} + \mathbf{S}$ where

$$\mathbf{L} = \sum_{i=1}^{z} \mathbf{l}_i$$

and

$$\mathbf{S} = \sum_{i=1}^{z} \mathbf{s}_i$$

Whereas, in the j-j coupling scheme, $\mathbf{J} = \sum_{i=1}^{z} \mathbf{j}_i$ with $\mathbf{j}_i = \mathbf{l}_i + \mathbf{s}_i$ for the ith electron.

Since the experimental evidence shows that the R-S coupling scheme describes the situation well for light- and medium-weight atoms, in future discussions it will always be assumed that the R-S coupling scheme is to be used unless specifically stated otherwise. (N.B. Because one adds the total l and s separately in Russell-Saunders coupling many authors refer to Russell-Saunders coupling as LS coupling in parallel to the j-j coupling terminology. This may be confusing since j-j coupling arises from the spin-orbit interaction.)

For our particular example of an excited state in helium, then, we have

$$l_1 = 0, l_2 = 1; L = 1$$
$$s_1 = \tfrac{1}{2}, s_2 = \tfrac{1}{2}; S = 1 \text{ or } 0$$

$$J = L + S = 2, 1, 0 \text{ for } S = 1, L = 1$$
$$= 1 \quad \text{for } S = 0, L = 1$$

Therefore, the two electrons could form a singlet-P-one state $^1P_1$ where the P shows that $L = 1$, and the subscript 1 shows that $J = 1$. The superscript 1 comes from the spin multiplicity: $2S + 1 = 1$ in this case since $S = 0$. On the other hand, the electrons could equally well form $^3P_2$, $^3P_1$, or $^3P_0$ states because for $L = 1$ and $S = 1$, J can be 2, 1, or 0.

The general procedure for nonequivalent electrons (electrons having different n or l quantum numbers) is to find all possible values of the total orbital angular momentum $\mathbf{L}$ and add them vectorially to all possible values of the total spin angular momentum $\mathbf{S}$ of the system. This can very quickly lead to a large multiplicity of states.

The situation is changed when we are dealing with equivalent electrons because then the Pauli exclusion principle limits the number of acceptable combinations. The ground state of the helium atom has already furnished a good example of this; there we saw that two equivalent s electrons could combine to give only a $^1S_0$ configuration. Another instructive example is the case of two equivalent p electrons. Perhaps the most direct method of finding the allowed configurations is to make a schematic representation of the possible combinations such as is shown in Table 12.3, where the arrows show electrons with spins pointing up or down. In the table all possible

TABLE 12.3

Two Equivalent p Electrons

m_l			$\sum m_l = M_L$	$\sum m_s = M_S$	Terms
$+1$	0	-1			
↑↓			2	0	^1D
↑	↑		1	1	^3P
↑	↓		1	0	^1D
↓	↓		1	-1	^3P
↓	↑		1	0	^3P
↑		↑	0	1	^3P
↑		↓	0	0	^1D
↓		↓	0	-1	^3P
↓		↑	0	0	^3P
	↑	↑	-1	1	^3P
	↑	↓	-1	0	^1D
	↓	↓	-1	-1	^3P
	↓	↑	-1	0	^3P
	↑↓		0	0	^1S
		↓↑	-2	0	^1D

combinations of the electrons consistent with the exclusion principle are included. Then the M_L and M_S are obtained in each case. Looking at the largest value of $M_L = 2$ one sees that it must belong to an $L = 2$, or D, state. The corresponding $M_S = 0$, so the state must be a $^1D_2$ which has a fivefold multiplicity since M_J can $= 2, 1, 0, -1, -2$. Five terms compatible with this designation are so labeled. Of the remaining terms, the highest value for M_L is 1; this must belong to a ^3P state since the corresponding $M_S = 1$. The various terms are $^3P_2$ with $M_J = 2, 1, 0, -1, -2$, $^3P_1$ with $M_J = 1, 0, -1$, and $^3P_0$ with $M_J = 0$. Once these terms are specified, there remains only one combination left for which M_L and $M_S = 0$. This then must correspond to a $^1S_0$ state.

From these considerations we see that two equivalent p electrons can combine to give $^1D_2$, $^3P_{2,1,0}$ and $^1S_0$ states. Various other combinations of equivalent electrons can be treated similarly. However, for more than two electrons and higher values of L the method can become quite tedious and the problem can be solved more readily by following a more sophisticated approach utilizing group theory.

It is worthwhile noting at this time that for the case where an electron shell is more than half filled, a simplification results if one makes use of the fact that all closed electron shells have a total angular momentum equal to zero, i.e., both L and S must equal 0. As an illustration, if we have a p^5 configuration it must be equivalent to one p electron because if one more p electron is added the shell becomes filled. Similarly, any allowed combination for a p^4 case must be equivalent to a p^2 situation.

TABLE 12.4

Possible terms of Equivalent Electrons with Russell-Saunders Coupling.
The numbers in parentheses give the number of distinct states having the designated term.

Electron Configuration	R-S Terms
s^2, p^6	^1S
p^1, p^5	^2P
p^2, p^4	$^1S, {}^1D, {}^3P$
p^3	$^2P, {}^2D, {}^4S$
d^1, d^9	^2D
d^2, d^8	$^1S, {}^1D, {}^1G, {}^3P, {}^3F$
d^3, d^7	$^2P, {}^2D(2), {}^2F, {}^2G, {}^2H, {}^4P, {}^4F$
d^4, d^6	$^1S(2), {}^1D(2), {}^1F, {}^1G(2), {}^1I, {}^3P(2), {}^3D, {}^3F(2), {}^3G, {}^3H, {}^5D$
d^5	$^2S, {}^2P, {}^2D(3), {}^2F(2), {}^2G(2), {}^2H, {}^2I, {}^4P, {}^4D, {}^4F, {}^4G, {}^6S$

The foregoing discussion also makes it clear that only electrons outside a closed shell can join in such a way as to give nonzero total angular momentum. These electrons outside a closed shell are often referred to as the *optically active* electrons because they are usually the ones involved in transitions in which visible or near visible light quanta are emitted or absorbed. Table 12.4 shows possible allowed terms for various combinations of equivalent electrons.

We are now in a position to find the spectroscopic terms describing the ground state configuration for some of the lighter elements. (See Table 12.5.) In the event there is more than one possible term to choose from we have already seen that when two electrons are involved, giving rise to triplet and singlet spin states, the triplet spin state will lie at lower energy. This is an example of Hund's general rule, which, as the reader will recall, states that of all possible terms the one having the largest spin multiplicity will have the lowest energy. Another general rule is that, for cases where the electron shell is less than half filled, of all the possibilities the term having the smallest value for J will have the lowest energy. This ordering is reversed for cases when the shell is more than half filled.

TABLE 12.5

Spectroscopic Terms for Ground States of Light Elements

Element	Electron configuration	Possible terms	Ground state term
H	$1s$	$^2S_{1/2}$	$^2S_{1/2}$
He	$1s^2$	$^1S_0$	$^1S_0$
Li	$1s^22s$	$^2S_{1/2}$	$^2S_{1/2}$
Be	$1s^22s^2$	$^1S_0$	$^1S_0$
B	$1s^22s^22p$	$^2P_{3/2}, {}^2P_{1/2}$	$^2P_{1/2}$
C	$1s^22s^22p^2$	$^1S_0, {}^1D_2, {}^3P_{2,1,0}$	$^3P_0$
N	$1s^22s^22p^3$	$^4S_{3/2}{}^2D_{5/2,3/2}{}^2P_{3/2,1/2}$	$^4S_{3/2}$
O	$1s^22s^22p^4$	$^1S_0, {}^1D_2, {}^3P_{2,1,0}$	$^3P_0$
F	$1s^22s^22p^5$	$^2P_{3/2,1/2}$	$^2P_{3/2}$
Ne	$1s^22s^22p^6$	$^1S_0$	$^1S_0$

Problem 12.2: Find the total number of possible levels for a two-electron system with $l_1 = 3$, $s_1 = \frac{1}{2}$, $l_2 = 1$, $s_2 = \frac{1}{2}$.

Problem 12.3: Find all the allowed terms arising from a $3p^24p$ electron configuration.

Problem 12.4: Find the possible terms for a nd^2 electron configuration and compare your results with the terms listed in Table 12.4.

12.6 ENERGY LEVELS

In considering the possible excited levels for the various elements, one must realize that the Pauli exclusion principle will not prevent the occupation of any of the unoccupied antisymmetric states of higher energy. Therefore, all possible quantum numbers among the higher energy levels are available. Also, in general only one electron is excited; i.e., for a $1s^2 2s^2 2p^2$ ground state configuration the first five electrons remain in the original state and the other $2p$ electron is excited so that a possible excited state could be represented as $1s^2 2s^2 2p 3d$. In this case, the $2p$ and $3d$ electrons would be considered as the optically active electrons and their angular moments would couple to give various possible states. To find these terms we need all possible values of L and S.

$$L = 3, 2, \text{ or } 1$$

$$S = 0 \text{ or } 1$$

The possible terms are

$$^3F_{4,3,2} \quad ^3D_{3,2,1} \quad ^3P_{2,1,0} \quad ^1F_3 \quad ^1D_2 \quad ^1P_1$$

To see how the various terms will be split we must include the Coulomb effects and the spin-orbit interaction. In Section 11.5 the spin-orbit interaction has already been discussed for the hydrogen atom. Looking at this work we see that the only change needed for the results there to be valid also for multi-electron atoms is to use the **S** and **L** found by the Russell-Saunders coupling of the optically active electrons. The expression for the spin-orbit interaction energy becomes

$$\overline{\Delta E}_{s \cdot l} = \iint \psi^*_{nljm_j}(\Delta E_{s \cdot l}) \, \psi_{nljm_j} d\tau \tag{12.15}$$

$$= \frac{\hbar^2 Z e^2}{4 m_e^2 c^2} \overline{\frac{1}{r} \frac{dV(r)}{dr}} [J(J+1) - L(L+1) - S(S+1)]$$

and $\overline{\Delta E}_{s \cdot l} = 0$ for either $L = 0$ or $S = 0$.

The ψ_{nljm_j} which might be used here to carry out the calculation would be the appropriate combinations of the $\psi_{nlm_lm_s}$ found in the Hartree approximation. An actual calculation would be somewhat difficult, but could be carried out.

In Section 11.10 we discussed hyperfine structure arising from the interaction of the nuclear magnetic moment with the orbital and electron spin magnetic moments for the one-electron atom and we introduced an $\mathbf{I} \cdot \mathbf{J}$ interaction in analogy with the $\mathbf{S} \cdot \mathbf{L}$ interaction between the electron magnetic moment and the magnetic moment due to the orbital angular momentum. Here again we can carry over the same conclusions to the case of the multi-electron atom as long as R-S coupling applies. The total angular momentum of the system is given by

$$F = I + J = I + L + S$$

and Eq. (11.34) is valid. Moreover, the effect of an external magnetic field will still be in accord with the discussion at the end of Section 11.10, as illustrated by Figure 11.14. Since Eq. (11.34) is similar to Eq. (12.15), a similar *interval rule* holds for the relative separation between lines of a multiplet (see Problem 12.5).

In certain cases this interval rule appears not to be valid, but the discrepancies can be shown to be due to a nonspherical distribution of charge in the nucleus which produces a nuclear electric quadrupole moment. The

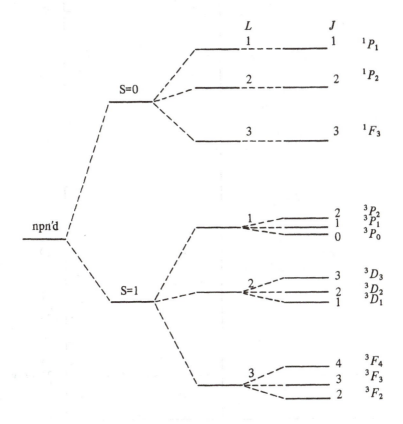

FIGURE 12.2 Illustrating the splitting of levels produced by two optically active electrons with Russell-Saunders coupling.

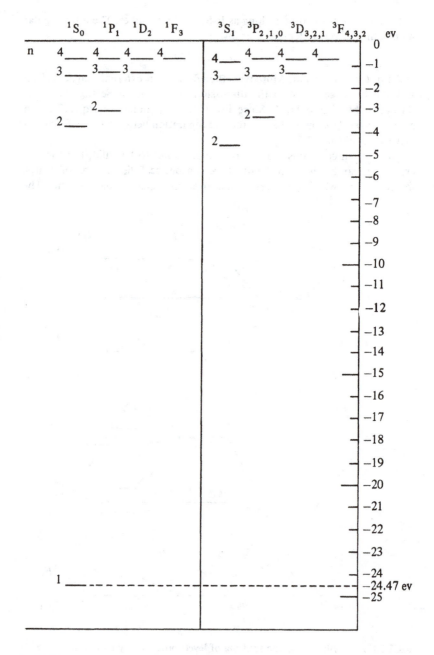

FIGURE 12.3 Energy-level diagram for helium.

interaction of the electric quadrupole moment with the gradient of the electric field at the nucleus (related to the average charge distribution of the electronic states involved) gives rise to small changes in the energies of the multiplet levels.

Problem 12.5: In an atom which obeys Russell-Saunders coupling the relative separations between adjacent levels of a particular multiplet are given by the *Landé interval rule*

$$(\overline{\Delta E}_{s \cdot l})_J - (\overline{\Delta E}_{s \cdot l})_{J-1} \propto J$$

Justify this rule on the basis of Eq. (12.15) and then apply it to assign the quantum numbers S, L, and J to the levels of a particular multiplet in an atom where the relative separations between adjacent levels of the multiplet are $4:3:2:1$.

Taking into account the various interactions between the two optically active electrons, we find that the levels arising from a *npn'd* combination would be split as shown in Figure 12.2. Here we have made use of the fact that for most cases (small and medium Z) the effects of the Coulomb interactions are quite large compared to the spin-orbit interactions.

Figure 12.3 shows an energy-level diagram for He. Notice that it is divided into separate sections for the singlet and triplet states. This is convenient because the selection rules show that transitions between singlet and triplet spin states are forbidden. The selection rules which are obeyed for allowed transitions between levels in atoms with R-S coupling are:

1. Only transitions can take place which involve a change in the n and l quantum numbers of a single electron. Two or more electrons cannot simultaneously make transitions between subshells.
2. $\Delta l = \pm 1$ where l is the orbital quantum number of the electron undergoing the transition.
3. $\Delta S = 0$ shows there are no transitions between states of different spin multiplicity.
4. $\Delta L = 0, \pm 1$.
5. $\Delta J = 0, \pm 1$ (but not $J = 0 \rightarrow J = 0$).

12.7 j–j COUPLING

It has already been mentioned in Section 12.3 that Russell-Saunders coupling is found to be valid for light- and medium-weight atoms where the Coulomb effects are larger than the spin-orbit interactions. However, for atoms of large Z, the spin-orbit interaction becomes comparable to and even greater than the effects of the residual Coulomb interactions. When this happens, the

R-S coupling scheme is no longer valid and *j-j* coupling must be used. In this case

$$\mathbf{j}_i = \mathbf{l}_i + \mathbf{s}_i$$

and $\mathbf{J} = \mathbf{j}_1 + \mathbf{j}_2 + \cdots + \mathbf{j}_3$ gives the total angular momentum of the system. Of course, for closed shells the total $\mathbf{J} = 0$ as before so that the net total angular momentum is still due to the optically active outer electrons.

If we take again as an example $2p$ and $3d$ electrons, the splitting of the various terms would now be as is shown in Figure 12.4.

There are a few general statements that should be made about the *j-j* coupling scheme. The exclusion principle still holds and there will be a direct correspondence between states in the R-S system and those of the *j-j* system. Therefore, one could still use the R-S labeling with *j-j* coupling, but this practice would not be very useful because it would not give the correct positions of the terms and the R-S selection rules for transitions would not be valid.

In general, calculations for *j-j* coupling are more difficult to carry out. The selection rules for this case turn out to be:

$$\Delta J_i = 0, \pm 1 \text{ between subshells}$$

$$\Delta J_i = 0 \text{ for other electrons}$$

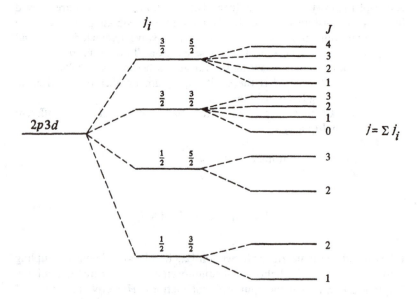

FIGURE 12.4 Illustrating the splitting of levels produced by two optically active electrons with *j-j* coupling.

It is instructive to consider what happens to low-lying states in several atoms all with the same electron configuration, but progressing from R-S coupling to *j-j* coupling. This is shown in Figure 12.5. Again the reader is reminded that most atoms obey R-S coupling rather well. There are practically no cases of pure *j-j* coupling (In nuclear physics, however, excited states of heavy nuclei come close to it.)

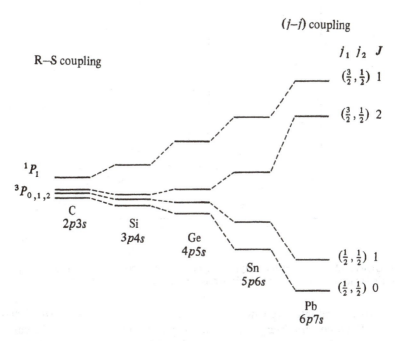

FIGURE 12.5 Illustrating the transition from RS to *j-j* coupling. From H.E. White, *Introduction to Atomic Spectra*, New York: McGraw-Hill Book Co., 1934.

12.8 X RAYS

If a beam of electrons is accelerated through a potential difference V_0 in an X-ray tube, then electromagnetic radiation will be emitted from the target. A typical spectrum would appear as shown in Figure (12.6).

The kinetic energy acquired by an electron in moving through a potential difference V_0 is just eV_0 where e is the electronic charge. When an electron of energy eV_0 hits the target, it is always found that the most energetic X rays have a limiting, or cutoff, frequency v_0 where $hv_0 = eV_0$. The X-ray spectrum

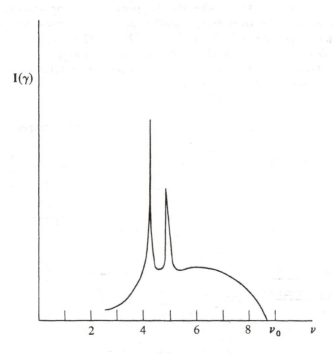

FIGURE 12.6 Illustrating a typical spectrum of electromagnetic radiation produced when a beam of energetic electrons strikes a metal target in an X-ray tube.

can be divided into two categories, *bremsstrahlung* and *characteristic* radiation. The *bremsstrahlung* produces the smooth continuous part of the curve in Figure 12.6 and arises from the sudden deceleration of the electrons in the target material. It is a well-known result that whenever a free charged particle undergoes acceleration, electromagnetic radiation will be emitted. The rate of emission of radiation from a beam of charged particles scattered by target atoms is found to be $\propto \cdot Z^2 z^2 / m^2$, where Z = charge of the scattered particle, z the charge of the target particle, and m is the mass of the scattered particle. This explains why bremsstrahlung production is important chiefly in the case of electrons. It is interesting to note that in synchrotrons (machines used for producing energetic electrons) the electrons are constrained to move in circular orbits by a magnetic guide field and the electromagnetic radiation, emitted because the electrons are continually undergoing centripetal acceleration, can appear as visible light.

The other kind of X ray is the *characteristic* radiation represented by sharp spikes in the drawing. As we shall see, it is directly related to transitions between various atomic energy levels; consequently, it is of more interest to us here than bremsstrahlung. The characteristic X rays arise in the target material of the X-ray tube because the incoming electrons can produce vacancies in the inner atomic shells. As these vacancies are filled by electrons from higher-lying levels, photons of characteristic energies are emitted.

We see, therefore, that for a given X-ray tube and accelerating voltage, the extent of the continuous bremsstrahlung background is determined by the accelerating voltage, while the characteristic X-ray peaks are to be identified with the specific kind of atoms making up the target material. Of course, characteristic X rays do not have to be produced by external electron bombardment; they will result any time a vacancy occurs in an inner electron shell due to photoelectric absorption, electron capture in radioactive nuclei, or any other means. Electron capture occurs for certain radioactive atomic species wherein the nucleus captures one of the K-shell electrons. Such an event is conveniently detected by observing the characteristic X ray emitted when the vacancy in the K shell is filled. (By convention, the energy shells corresponding to successive values of the principal quantum number $n = 1$, 2, 3, 4, 5 . . . are labeled K, L, M, N, O . . . .)

From our study of atomic energy levels we already have sufficient information to discuss characteristic X rays in some detail. For instance, suppose that a K-shell vacancy has been produced in a krypton atom ($Z = 36$). Then we know that any electrons in a higher level can make a transition to the unoccupied level, so long as the various selection rules are obeyed. For $Z = 36$ the electron configuration will be $1s^2 2s^2 2p^6 3s^2 3p^6 4s^2 3d^{10} 4p^6$. Therefore, the K, L, and M shells will be filled and eight electrons will be in the N shell. This is illustrated in Figure 12.7.

By convention the various possible X rays are labeled according to the shell in which the vacancy is being filled. For the case of a vacancy in the K shell we would have K_α, K_β, K_σ, or K_γ X rays emitted, depending on which higher level supplied the electron to fill the hole. If a K_α X ray were emitted, this would mean a vacancy had occurred in the L shell and so we should expect an L X ray as well. In practice, especially for high Z atoms, the process of filling holes in an inner shell can be quite complicated and produces a variety of different characteristic X rays. As an added illustration we should consider that each electron shell will consist of various separate levels, which will of course add to the complexity of the characteristic X-ray patterns observed. For R-S coupling the sublevels would be labeled as shown in Figure 12.7. All possible allowed transitions among them in general would be expected to take place.

Problem 12.6: Find the spectrum resulting from the transition from a $^4F_{3/2}$ to a $^4D_{5/2}$ state in a weak magnetic field. If $B = 15 \times 10^4$ gauss, calculate the Zeeman splitting of the levels in each state. Express the separations in ev and also in terms of the frequency shift. (Note that as long as R-S coupling is valid the discussion of Section 11.9 applies here also.)

Problem 12.7: Determine the hyperfine structure pattern that would result from a $^6P_{7/2}$ to $^6S_{5/2}$ transition in manganese, assuming that the

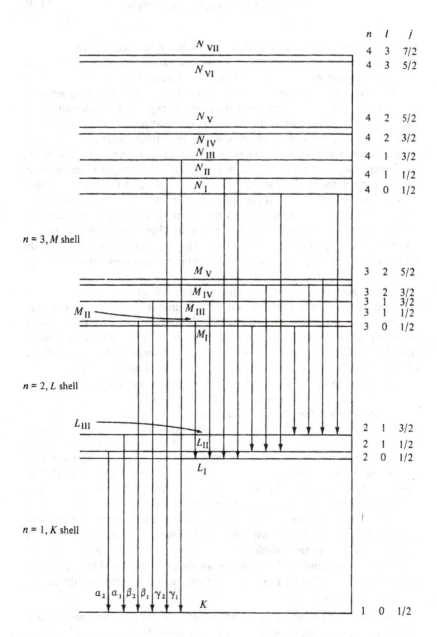

FIGURE 12.7 Illustrating possible X rays to be obtained from the krypton atom. The Roman numeral subscripts are used to identify the various subshells which depend on the value of *j* for electrons in any particular energy shell denoted by the principal quantum number *n*.

nuclear spin of manganese is 5/2. Show the Zeeman splitting to be expected for one hyperfine level of each of the $^6P_{7/2}$ and $^6S_{5/2}$ states.

Problem 12.8: By means of vector diagrams illustrate the terms arising from an s and a d electron for both R-S and j-j coupling.

diagrams

BIBLIOGRAPHY

Herzberg, Gerhard, *Atomic Spectra and Atomic Structure*. Englewood Cliffs, N.J.: Prentice-Hall, Inc., 1944.

Kuhn, H.G., *Atomic Spectra*. New York: Academic Press, Inc., 1963.

Eisberg, R. M., *Fundamentals of Modern Physics*. New York: John Wiley & Sons, Inc., 1961.

Leighton, R. B., *Principles of Modern Physics*. New York: McGraw-Hill Book Co., 1959.

Pauling, L. and E. B. Wilson, *Introduction to Quantum Mechanics*. New York: McGraw-Hill Book Co., 1935.

XIII

MOLECULAR
STRUCTURE

Since molecules consist of atoms held together by relatively low binding energies ($\approx$ few ev) we may well expect that a knowledge of atomic electron systems can help in understanding molecular systems. In this discussion, we shall deal mainly with the simplest of molecules, namely those containing only two atoms, diatomic molecules. The molecular binding holding the atoms together can be either of two types, ionic binding (heteropolar) or covalent binding (homopolar).

13.1 IONIC BINDING

As an illustration of ionic binding let us consider the NaCl molecule. For sodium $Z = 11$ and the electron configuration for the sodium atom is $1s^2 2s^2 2p^6 3s$. Furthermore, the binding energy of the $3s$ electron is -5.1 ev. For chlorine with $Z = 17$ the electron configuration is $1s^2 2s^2 2p^6 3s^2 3p^5$. The chlorine atom lacks one electron for filling the $3p$ subshell. Since the filled p subshell is a particularly stable configuration, the lack of the electron acts as though there were a "hole" present with an effective net binding energy of $+3.8$ ev. If the $3s$ electron is transferred from the sodium atom to the $3p$ shell of the chlorine atom, it would take $5.1 - 3.8 = 1.3$ ev of energy, or, in other words, the system of Na^+ and Cl^- would have a net gain of 1.3 ev. However there will also be a Coulomb attraction between the two ions. The Coulomb potential energy is $-e^2/r_0$ so that the net binding energy for the system will be

$$BE = +1.3 \text{ ev} - \frac{e^2}{r_0}$$

Experimentally one finds a binding energy for $NaCl = -4.24$ ev. Thus $r_0 = 2.36$ Å, a value which is in good agreement with results of independent measurements using X rays.

The NaCl molecule cannot collapse because when the atoms get too close the wave functions overlap. At this point the electron densities get distorted, partially as a consequence of the Pauli exclusion principle, and there is a net repulsion between the electron clouds and between the positively charged nuclei. The r_0 is the value for which the total potential energy of the system is a minimum. This is shown schematically in Figure 13.1.

From the foregoing picture, we might expect the molecule to vibrate about the equilibrium distance r_0 if one of the atoms should be hit in some way. These vibrations do occur and give rise to characteristic vibrational spectra as we shall see later.

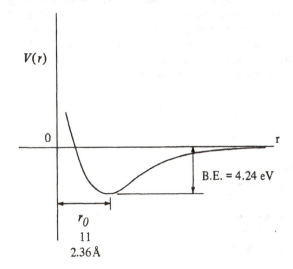

FIGURE 13.1 Illustrating the potential energy for ionic binding of the NaCl molecule as a function of the separation distance between the ions. The value of 2.36 Å for r_0 at the minimum of the curve is found from the experimentally observed binding energy of 4.24 ev.

Because the NaCl molecule consists of a positively charged Na ion together with a negatively charged Cl ion, we also expect, and find, that NaCl has a permanent electric dipole moment. A permanent electric dipole moment is characteristic of all diatomic molecules with ionic binding.

13.2 COVALENT BINDING—H_2^+ MOLECULE

In ionic binding we have seen that the individual atoms either gain or lose electrons and become effectively ions. In covalent binding the electrons are shared between the atoms and one should not think in terms of gaining or losing electrons. Again we will choose a particularly simple system to illustrate; in this case it is H_2^+, two protons sharing an electron.

For this system we can easily write down the Hamiltonian

$$H = \frac{P_1^2}{2M} + \frac{P_2^2}{2M} + \frac{p^2}{2m} + \frac{e^2}{r} - \frac{e^2}{r_1} - \frac{e^2}{r_2}$$

where the r's are shown in Figure 13.2, P_1 and P_2 are the momenta of protons 1 and 2 respectively and p is the electron momentum. M is the mass of a proton and m the electron mass. By using the Hamiltonian we could write the Schrödinger equation, although it could not be solved in closed form since this is a three-body problem. Therefore, it is helpful to use the adiabatic approximation in which the motion of the protons is neglected. This is a reasonable procedure because the protons are so much more massive than the electron that their relative motion will necessarily be much smaller than that of the electron. In the adiabatic approximation we ignore the $P_1^2/2M$, $P_2^2/2M$ and e^2/r and the Hamiltonian becomes

$$H = \frac{p^2}{2m} - \frac{e^2}{r_1} - \frac{e^2}{r_2} \tag{13.1}$$

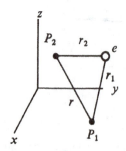

FIGURE 13.2   The H_2^+ molecule.

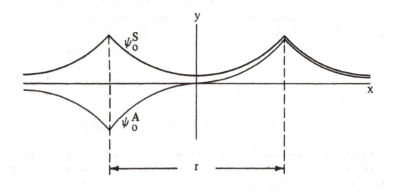

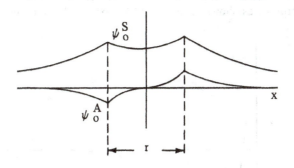

FIGURE 13.3 Illustrating the form of the symmetric and antisymmetric wave function of the H_2^+ molecule for two different proton separation distances.

When r is large the electron can be thought of as belonging to one proton or the other, but as r becomes small this distinction is not possible and we must consider the effects of exchange degeneracy. In this case, we ought then to take the total wave function as being either symmetric or antisymmetric. For convenience we take both protons to be lying on the x axis so that under an exchange it is only the x coordinate that varies. Then we write

$$\psi_n^s(xyz) = \frac{1}{\sqrt{2}}[U_n(x_1, y, z) + U_n(x_2, y, z)] \qquad (13.2)$$

or

$$\psi_n^A(xyz) = \frac{1}{\sqrt{2}}[U_n(x_1, y, z) - U_n(x_2, y, z)]$$

Further, for the ground state U_{100} we choose the x axis so that we have U_{100} $(x00)$. Both ψ_0^s and ψ_0^A are shown in Figure 13.3 for different values of r. Looking at these plots, displayed in Figure 13.3, we see that

$$\text{As } r \longrightarrow 0, \text{ at } x = 0 \qquad \psi_0^s(0) \longrightarrow \text{maximum}$$
$$\psi_0^A(0) \longrightarrow 0$$

As the protons come reasonably close together the electron has the greatest probability for being found between them. It acts to give a net attraction until the Coulomb repulsion between the protons overrides the attraction to give a net repulsion. Therefore, we again have a situation where there will be a minimum in the potential energy for a certain equilibrium separation of the protons as shown in Figure 13.4. For the H_2^+ molecule $r_0 = 2.65$ Å

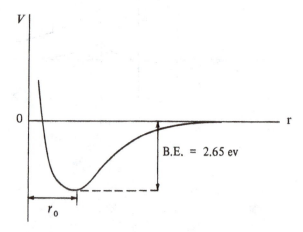

FIGURE 13.4 Illustrating the potential energy for covalent binding of the H_2^+ molecule as a function of the separation distance between the protons. The binding energy and separation distance at the minimum point have been obtained from experiment.

and the binding energy $= -2.65$ ev. Since the binding energy is negative the H_2^+ system is stable; it is interesting to note that H_2^+ molecules can readily be produced in the laboratory.

13.3 THE HYDROGEN MOLECULE (HEITLER—LONDON TREATMENT)

The preceding discussion on the H_2^+ molecule has been made quite qualitative. Taking the H_2 molecule as an example we will now show how a much more quantitative treatment can be carried out. In the diatomic hydrogen molecule the two electrons can alternate in orbiting around one proton or the other; thus the hydrogen molecule ground state furnishes us with another simple example of resonance. (See Section 12.2.) In general, the hydrogen-hydrogen bond is composed of two types of states, one in which both electrons surround one proton forming an ionic bond and the other type in which each proton has an electron close to it. This latter type of state, corresponding to a perturbation of the wave functions for infinitely separated hydrogen atoms, is the one we will assume to predominate (covalent binding). Figure 13.5 illustrates the change in electron potential and consequent change in wave functions as the two atoms are brought close together. A continuous wave function and derivative cannot be realized by joining the two infinite-separation wave functions when the two atoms are joined. Only by lowering the energy eigenvalue or the system (or raising it to a new first excited level) can a smooth fit for the total wave function be made.

To begin with we consider the two atoms well separated as compared to the distance of the electrons from the closest protons. The initial wave function will then consist of the product of the ordinary ground state hydrogen wave functions for each atom, $\psi_A(r_1)\psi_B(r_2)$, and will yield resonance or exchange energy. Let r_{1A} represent the separation of electron 1 from proton A, and similarly r_{2B} for electron 2 and proton B. The complete wave equation is then

$$\left[-\frac{\hbar^2}{2m}(\nabla_1^2 + \nabla_2^2) - \frac{e^2}{r_{1A}} - \frac{e^2}{r_{2B}} - \frac{e^2}{r_{1B}} - \frac{e^2}{r_{2A}} + \frac{e^2}{r_{12}} + \frac{e^2}{r_{AB}}\right]\psi = E\psi$$

The unperturbed function is $\psi = \psi_A(\mathbf{r}_1)\psi_B(\mathbf{r}_2)$ where $\psi_A(\mathbf{r}_1)$ is a solution to

$$\left(-\frac{\hbar^2}{2m}\nabla_1^2 - \frac{e^2}{r_{1A}}\right)\psi_A(\mathbf{r}_1) = E_1\psi_A(\mathbf{r}_1)$$

$\psi_A(\mathbf{r}_1)$ is the associated Laguerre and spherical harmonic function of Chapter VIII, and similarly for $\psi_B(\mathbf{r}_2)$. Let

$$H' = -\frac{e^2}{r_{1B}} - \frac{e^2}{r_{2A}} + \frac{e^2}{r_{12}} + \frac{e^2}{r_{AB}}$$

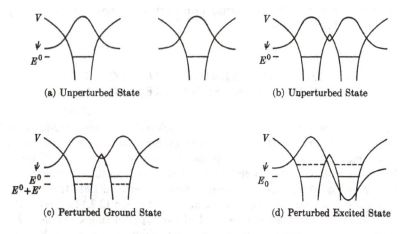

(a) Unperturbed State   (b) Unperturbed State

(c) Perturbed Ground State   (d) Perturbed Excited State

FIGURE 13.5 Illustration of the behavior of the potential energy, the energy levels, and the complete two-electron wave functions for two hydrogen atoms (a) infinitely separated, (b) brought together with the separated wave functions drawn (unchanged from a), (c) brought together and showing the corrected, smoothly joining wave functions, (d) brought together and showing the correct wave function for the first excited state. The dashed lines in (c) and (d) are for the correct energy levels appropriate to the wave functions shown in (c) and (d). The straight solid horizontal lines on all four figures show the ground energy level of the infinitely separated atoms, for comparison.

Then the degenerate perturbation integrals are

$$J = \int\int \psi_A^*(\mathbf{r}_1)\psi_B^*(\mathbf{r}_2)\, H'\, \psi_A(\mathbf{r}_1)\psi_B(\mathbf{r}_2)\, d\tau_1 d\tau_2$$

$$K = \int\int \psi_A^*(\mathbf{r}_1)\psi_B^*(\mathbf{r}_2)\, H'\, \psi_A(\mathbf{r}_2)\psi_B(\mathbf{r}_1)\, d\tau_1 d\tau_2$$

$$+ \int\int \psi_A^*(\mathbf{r}_1)\psi_B^*(\mathbf{r}_2)\, H^0\, \psi_A(\mathbf{r}_2)\psi_B(\mathbf{r}_1)\, d\tau_1 d\tau_2 \qquad (13.3)$$

where $d\tau_1$ and $d\tau_2$ are the differential spherical volume elements

$$d\tau_1 = r_1^2 \sin\theta_1\, dr_1\, d\theta_1\, d\varphi_1$$

The second term in the exchange integral K, which we will label K_0, must be included because our initial wave functions are not orthogonal. Since

$$H^0\, \psi_A(\mathbf{r}_2)\psi_B(\mathbf{r}_1) = E^0\, \psi_A(\mathbf{r}_2)\psi_B(\mathbf{r}_1) = 2E_1\, \psi_A(\mathbf{r}_2)\psi_B(\mathbf{r}_1)$$

we have

$$K_0 = 2E_1 \int\int \psi_A^*(\mathbf{r}_1)\psi_B^*(\mathbf{r}_2)\psi_A(\mathbf{r}_2)\psi_B(\mathbf{r}_1)\, d\tau_1 d\tau_2$$

The contribution of e^2/r_{AB} to J is simply e^2/r_{AB}, since the wave functions are normalized and e^2/r_{AB} may be taken outside the integration over the electron coordinates. Similarly, the contribution of the e^2/r_{AB} term to K is simply $[(e^2/r_{AB})/2E_1]K_0$. The contribution of (e^2/r_{1B}) to J would be

$$\int \psi_A^*(\mathbf{r}_1)\left(\frac{-e^2}{r_{1B}}\right)\psi_A(\mathbf{r}_1)\,d\tau_1$$

since $\psi_B(\mathbf{r}_2)$ is normalized and r_{1B} is not a function of the position of the second electron. The contribution of (e^2/r_{1B}) to K would be similar.

We thus obtain the secular determinant

$$\begin{vmatrix} (J - \Delta E) & K \\ K & (J - \Delta E) \end{vmatrix} = 0 \qquad (13.4)$$

from which the shifts in energy level caused by the resonance energy are found to be

$$\Delta E = J + K, \quad J - K$$

The Coulomb effects between all the charges roughly cancel out, leaving the exchange effects which cause chemical binding, as illustrated in Figure 13.6.

It is found that the symmetric wave function

$$\psi_S = \frac{1}{\sqrt{2 + K_0/E_1}} [\psi_A(\mathbf{r}_1)\psi_B(\mathbf{r}_2) + \psi_A(\mathbf{r}_2)\psi_B(\mathbf{r}_1)] \qquad (13.5)$$

is the correct zero-order wave function which with its antisymmetric twin causes the perturbation matrix to be diagonal. Equation (13.5) is the symmetric wave function whose eigenvalue is approximately $E_0 - K$, the lower of the two energy eigenvalues including resonance. This is in contrast to the case

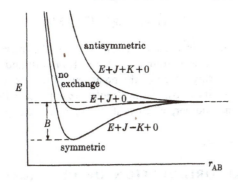

FIGURE 13.6 Illustration of effect of exchange or resonance energy on the binding energy B of the hydrogen molecule. The middle curve includes all effects $J + 0$ other than the resonance of the two electrons where 0 represents all interactions not specifically calculated in the text.

of helium where both electrons are in the same atom, and the antisymmetric wave function was found to be the lower of the two energy eigenvalues including resonance. The attractive bonding force of the hydrogen molecule has been shown to be due to the possibility of two electrons being in a symmetric state (i.e., being interchangeable in the space wave function representation). Since no more than two electrons can be in one symmetric state at the same time, the degeneracy in this case can be only twofold.

The symmetric wave function Eq. (13.5) yields a probability of finding both electrons between the two protons larger than the probability of this event for the two atoms placed a distance r_{AB} apart and no resonance effects taken into account. That is, $|\psi_A(\mathbf{r}_1)\psi_B(\mathbf{r}_2)|^2$ gives a probability of this event smaller than that given by the absolute square of the symmetric wave function Eq. (13.5) (see Figure 13.6). The antisymmetric wave function, on the other hand, gives a smaller electron probability density in this region, compared to the nonresonance value. Thus, despite the mutual repulsion of the two electrons, the attraction of the two protons for the (symmetric state) negative cloud causes the hydrogen molecule to be a stable configuration. The mutual repulsion of the less screened protons in the antisymmetric electronic state results in an unstable configuration. ΔE, J, and K are functions of internuclear distance as shown in Figure 13.6.

From the discussion it is evident why covalent binding does not give rise to a permanent electric dipole moment as was the case for ionic binding. Further, it is also reasonable to expect that negative ions such as H_2^- would be producible in the laboratory.

Problem 13.1: If two dipoles with moments $e\mathbf{1}_1$ and $e\mathbf{1}_2$ (e being the electronic charge and $\mathbf{1}_1$ and $\mathbf{1}_2$ being the charge separation in the two dipoles) are a large distance r apart, their interaction energy is

$$V = \frac{e^2}{r_3}[\mathbf{1}_1\cdot\mathbf{1}_2 - 3(\mathbf{1}_1\cdot\hat{\mathbf{r}})(\mathbf{1}_2\cdot\hat{\mathbf{r}})]$$

where $\hat{\mathbf{r}}$ is a unit vector in the $\mathbf{r}$ direction. What is the interaction energy between an excited hydrogen atom ($n = 2$, $l = 1$, $m = 0$) and an unexcited hydrogen atom r cm away? Assume the polar axis of the excited atom to be parallel to $\mathbf{r}$. Note that if two oscillating dipoles are in phase with one another they will attract one another, whereas if out of phase they will repel one another.

13.4 SPATIAL ORIENTATION OF CHEMICAL BONDS

We have already observed (Chapter XI) that electrons in various subshells tend to have definite angular distributions in space. Of course, when the shell is filled the total angular distribution becomes isotropic. However, the outer or valence electrons in the unfilled shells will tend to keep their geometrical distribution in forming molecular structures.

As an example, we consider the common water molecule H_2O. Oxygen has $Z = 8$ and its electron configuration is $1s^2 2s^2 2p^4$. The electrons in the p shell tend to be distributed along orthogonal directions. Therefore, we might naively expect to find the H_2O molecule appearing as is shown in the diagram with a pair of electrons being shared between each hydrogen atom (proton) and the oxygen atom. The excess p electrons simply form a pair in themselves with opposite spins along the third orthogonal direction, as shown in Figure 13.7.

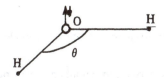

FIGURE 13.7 Illustrating the spatial orientation of the hydrogen and oxygen atoms in the water molecule.

This simple picture agrees surprisingly well with the form of the H_2O molecule as determined experimentally. The only real difference is that the angle θ between the hydrogen bonds is $\approx 100°$ instead of $90°$ and the increases in the angle can readily be explained by the Coulomb repulsion between the two protons.

Another very interesting and common example is given by carbon for which Z is 6 and the electron configuration is $1s^2 2s^2 2p^2$. In computing the distribution of the outer 4 electrons using the appropriate combination of ^1s and ^3p wave functions, the electrons are found to be situated at vertices of a regular tetrahedron. This characteristic structure is maintained in carbon bonding both for compounds and in the structure of diamond crystals.

13.5 DIATOMIC MOLECULES

We have seen that the attractive potential energy curves for both ionic and covalent binding have a minimum corresponding to the equilibrium distance of separation of the atoms and we have already speculated that, if somehow the system were displaced from this equilibrium position, it might vibrate with simple harmonic motion, much like two balls held together by a spring. We might also expect that the simple diatomic molecule would appear to be like a dumbbell, so that it could rotate about its center of mass. Let us assume that a diatomic molecule can be considered from these simple classical viewpoints and see what the consequences are.

First we consider the dumbbell model where the two atoms are considered to be a fixed distance apart as illustrated in Figure 13.8.

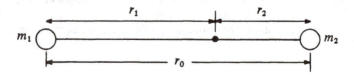

FIGURE 13.8 Schematic diagram of the dumbbell model of a diatomic molecule.

Classically, we could have rotations about an axis through the center of mass where the energy of rotation is given by

$$E = \frac{1}{2}\mathscr{I}\omega^2 = \frac{P_A^2}{2\mathscr{I}} \tag{13.6}$$

$\mathscr{I}$ = moment of inertia about a perpendicular axis through the center of mass where $P_A = \mathscr{I}\omega$ = angular momentum and of course E, ω, or P_A could have any values $\geqslant 0$. However, the molecule is a microscopic system and so we should expect to need the use of quantum mechanics. For r_0 fixed, the Schrödinger equation becomes:

$$\frac{\hbar^2}{2\mathscr{I}}\left[\frac{1}{\sin\theta}\frac{\partial}{\partial\theta}\left(\sin\theta\frac{\partial}{\partial\theta}\right) + \frac{1}{\sin^2\theta}\frac{\partial}{\partial\phi}\right]\psi(\theta,\phi) = E\psi(\theta,\phi) \tag{13.7}$$

We have already considered this type of equation (Section 8.4) and found the solutions to be

$$\psi_{lm} = Y_{lm}(\theta,\phi)$$

$$E_l = \frac{\hbar^2}{2\mathscr{I}}l(l+1) \tag{13.8}$$

By convention, in dealing with the diatomic molecules we take $l = J$.

The rotational energy levels of the system are shown in Figure 13.9. Just as in the atomic case, we expect allowed transitions between the levels only if there is a changing dipole moment associated with the system. We can get a changing electric dipole moment, of course, for a permanent electric dipole which is rotating. Therefore, we should expect to get allowed transitions only for molecules consisting of unlike atoms since for them the centers of positive and negative charge do not coincide, thereby giving rise to permanent electric dipole moments.

If the calculations were carried through, we should find the selection rules for allowed transitions to be similar to the atomic case:

$$\Delta J = \pm 1$$
$$\Delta m = 0, \pm 1$$

The emission spectrum for this system should be particularly simple since

$$h\nu = E_{J+1} - E_J = \frac{\hbar^2}{2\mathscr{I}}\{(J+1)(J+2) - J(J+1)\}$$
$$= \frac{\hbar^2}{\mathscr{I}}(J+1) \tag{13.9}$$

where $J = 0, 1, 2, 3, \ldots$, the term value of the lower state. As is also shown in Figure 13.9, the emission spectrum of the rigid rotator consists therefore of a series of equidistant lines. The same would hold true for the absorption spectrum as well.

Next we consider the harmonic oscillator approximation in which the two atoms are pictured as being held together by a spring, so that the poten-

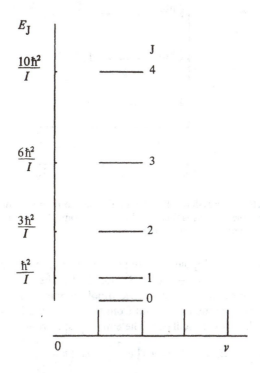

FIGURE 13.9 Energy-level diagram of the rigid rotator. As shown at the bottom, the emission or absorption spectrum would consist of equally spaced lines.

tial energy becomes $V = \frac{1}{2}kx^2$ (the one-dimensional harmonic oscillator potential). The solution to this case has also been already given (Section 6.3) and we know the energy levels are given by

$$E_v = \hbar\omega\left(v + \frac{1}{2}\right) \text{ where } \omega = \sqrt{\frac{k}{m_r}} \qquad k = \text{spring constant}$$

$$v = 0, 1, 2, 3 \ldots \qquad\qquad m_r = \text{reduced mass}$$

Figure 13.10 shows these vibrational energy levels.

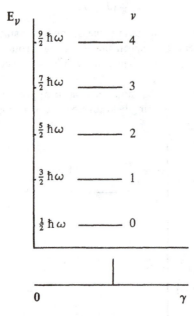

FIGURE 13.10 Energy-level diagram of the harmonic oscillator approximation to the dipole molecule. As is illustrated at the bottom, the emission or absorption spectrum would consist of a single line.

Allowed electric dipole transitions between the levels are governed by the selection rules $\Delta v = \pm 1$. (See Section 10.5.) The selection rule results from the assumption that the electric dipole moment varies linearly with the distance between molecules and from the use of the known wave functions for the harmonic oscillator. The emission spectrum is given by

$$h\nu_v = E_{v+1} - E_v = \hbar\omega\left(v\frac{3}{2}\right) - \hbar\omega\left(v + \frac{1}{2}\right) = \hbar\omega$$

and consists of a single frequency.

$$\nu_v = \frac{1}{2\pi}\sqrt{\frac{k}{m_r}} \qquad\qquad (13.10)$$

Of course, for a real diatomic molecule, one cannot completely separate the rotational and vibrational motions as we have done in order to simplify the calculations. That there will be interactions between the rotational and vibrational motion is easily seen by considering the two atoms as being held together by a harmonic or spring force. Then, as the molecule rotates about an axis perpendicular to the line between the atoms, the atoms will tend to move farther apart, thereby giving a different effective vibrational potential for different rotational states. However, for low rotational states, the effect is small and reasonable agreement with experimental results can be obtained by separating the two motions.

For a more rigorous treatment, one could start with the Hamiltonian to be used in the Schrödinger equation for two atoms of mass M_1 and M_2 separated by a distance r

$$H = \frac{P_1^2}{2M_1} + \frac{P_2^2}{2M_2} + V_n(r)$$ (13.11)

where $V_n(r)$ is the interaction potential for any given electronic state of the system. From Chapter VIII we have already seen that so long as $V_n(r)$ is only a function of r, one can go to center-of-mass coordinates and separate out the angular dependence. Therefore, both the total orbital angular momentum of the system and its z component will be quantized and only the solution for the radial part of the motion will differ from the solution for the one-electron atom.

A detailed solution of the radial equation is beyond the scope of this book, but it is instructive to consider how the problem can be treated. If the angular motion is considered as being due to the nuclear motion (ignoring angular momentum contributions from the electrons), then the radial wave equation becomes of the form

$$\frac{1}{r^2}\frac{d}{dr}\left(r^2\frac{dR}{dr}\right) + \frac{2m_r}{\hbar^2}\left[E_s - V_n(r) - \hbar^2\frac{J(J+1)}{2m_r r^2}\right]R = 0$$ (13.12)

where m_r = reduced mass and J = rotational quantum number. If we let $\chi(r) = rR(r)$ then

$$\frac{-\hbar^2}{2m_r}\frac{d^2\chi}{dr^2} + \left[V_n(r) + \frac{\hbar^2 J(J+1)}{2m_r r^2} - E_s\right]\chi = 0$$ (13.13)

But this is the same equation one gets for a one-dimensional system consisting of a particle moving on a line under the influence of a potential

$$U_n(r) = V_n(r) + \frac{\hbar^2 J(J+1)}{2m_r r^2}$$ (13.14)

Considering the form of $V_n(r)$, it is readily seen that $U_n(r)$ should have the form shown in Figure 13.11. For a given value of J, and near its minimum point, $U_n(r)$ is approximately the same as the harmonic oscillator potential. Note that as J increases, the minimum of the potential $U_n(r)$ becomes less pronounced and finally disappears. Physically we see that the molecule

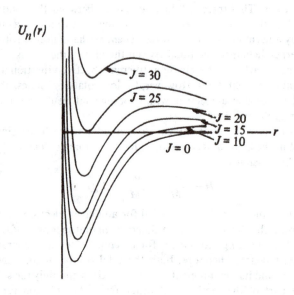

FIGURE 13.11 Illustrating the effective potential for the diatomic molecule showing the dependence on J as well as r.

stretches as it rotates, and at high enough rotational speeds the stretching becomes so great the atoms will fly apart.

The radial equation cannot be solved exactly, but good results can be obtained using approximation methods. When this is done, it is found that the major feature of the nuclear motion is the "harmonic" vibration with the rotational effects appearing as a fine structure upon each vibrational energy level. The energy levels of the system are found to be given by

$$E_{nkv} = V_n + \hbar\omega_n(v + \tfrac{1}{2}) + \frac{\hbar^2 J(J + 1)}{2m_r r_J^2} \qquad (13.15)$$

Here V_n represents the energy of the electronic state and r_J is the atomic separation for the Jth rotational state. Transitions between these electronic states usually occur in the visible region of the spectrum. The second term corresponds to vibrational (harmonic oscillator) energy levels and gives rise to transitions in the near infrared region, while the last term (rotational states) produces spectral lines in the far infrared. Since these energy regions are far removed from one another, they can be treated separately as illustrated by Figure 13.12.

At low temperatures, the vast majority of the molecules will be in the lowest electronic and vibrational states and transitions will generally take place between various rotational levels. At higher temperatures where excitation to higher vibrational states is frequent, the rotational states are also observed as "fine structure" in the observed vibrational spectrum.

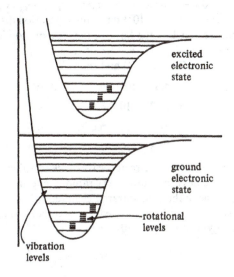

FIGURE 13.12 Energy-level diagram for a typical diatomic molecule, showing electronic, vibrational, and rotational levels.

13.6 COMPARISON WITH EXPERIMENT—HCl

If infrared light is passed through a cell containing HCl gas, many absorption lines will appear. In the far infrared (longer wavelengths) there will be a series of equally spaced lines resulting from transitions between rotational levels, while in the near infrared one sees a "single" intense line arising from transitions between vibrational levels. With improved spectral resolution this "single" line will be observed to be made up of many closely spaced lines due to simultaneous transitions in many rotational states. From the experimentally observed value of 20.68 cm^{-1} for the separation of the equally spaced lines in the far infrared, we can determine the moment of inertia of the HCl molecule and the separation of the hydrogen and chlorine atoms. Since the frequency equals the wave number times the speed of light, from Eq. (13.9) we obtain

$$hc \times 20.68 \text{ cm}^{-1} = \frac{\hbar^2}{\mathscr{I}}$$

and

$$\mathscr{I} = 2.70 \times 10^{-40} \text{ gm-cm}^2.$$

Then using the reduced mass $m_r = m_H m_{Cl}/m_H + m_{Cl} = 1.63 \times 10^{-24}$ gm, we find $r = \sqrt{\mathscr{I}/m_r} = 1.29 \times 10^{-8}$ cm. This separation distance is just what one would expect for HCl on the basis of the known atomic (and molecular) radii. Also we note from Eqs. (13.6, 13.8) that the frequencies of rotation are given by $\nu = (1/2\pi)(\hbar/\mathscr{I})\sqrt{J(J + 1)}$ and thus

$$\nu = 8.8 \times 10^{11} \text{ sec}^{-1} \qquad J = 1$$
$$= 15.2 \times 10^{11} \text{ sec}^{-1} \qquad = 2$$
$$= 21.6 \times 10^{11} \text{ sec}^{-1} \qquad = 3$$
$$= 27.8 \times 10^{11} \text{ sec}^{-1} \qquad = 4$$

The periods of rotation are the reciprocals of these values, and we see that the individual molecules are rotating very rapidly indeed.

The "single" line in the near infrared centered at 2885.9 cm^{-1} represents the vibrationa spectrum and can be used to obtain an estimate of the molecular binding energy. From Eq. 13.10

$$\nu_v = 8.66 \times 10^{13} \text{ sec}^{-1} = \frac{1}{2\pi} \sqrt{\frac{k}{m_r}}$$

and $k = 4.95 \times 10^5$ dynes/cm. Of course the range of r for which this value of k is valid is restricted to small displacements from the equilibrium radius of 1.29×10^{-8} cm.

It is interesting to note that the ratios of the "single" line vibrational frequency to the frequencies of the rotational spectrum are $\approx 10^2$ and the corresponding ratios of vibrational to rotational energies are also $\approx 10^2$.

13.7 EFFECTS OF NUCLEAR SPIN

When nuclear spins are included, the total wave function for a molecule can be written (to a good approximation) as:

$$\Psi_T = \Psi_{el}\Psi_{vib}\Psi_{rot}\Psi_{spin}$$

where

Ψ_{el} describes the electronic state

Ψ_{vib} describes the vibrational motion

ψ_{rot} describes the rotational motion

ψ_{spin} describes the state of the nuclear spins

Nuclear spins will generally have small effect on the molecular properties except for special cases when the molecules are composed of identical nuclei. In this situation the spins can indirectly exert a very great influence. To see how this comes about let us consider the case of ortho- and para-hydrogen, first explained by Dennison. Since each proton has a spin of 1/2, the H_2 molecule can be formed with the proton spins parallel (*ortho-hydrogen*) and antiparallel (*para-hydrogen*). In cases such as this involving identical particles, it has long been known that the system must obey Fermi-Dirac statistics (total wave function antisymmetric and the Pauli exclusion principle follows) if the nuclear spins are half-integer and Bose-Einstein statistics (total wave function symmetric) if the nuclear spins are integer. Consequently, for the H_2 molecule the total wave function must be antisymmetric.

At reasonably low temperatures where there is little excitation energy available, we expect to find the H_2 molecule in the electronic ground state (symmetric) and also in the vibrational ground state (symmetric). Therefore, the symmetry of ψ_T will depend on the relative symmetries of the nuclear spin and the rotational states. We recall from section 12.1 that if one has two spin 1/2 particles, 3 symmetric and 1 antisymmetric spin functions can be formed. Also, from the discussion in Section 8.2 for the angular momentum eigenfunctions we see that rotational states with even values of J have even symmetry and rotational states with odd J are antisymmetric. It follows that H_2 molecules in the ground electronic and vibrational states will have symmetric spin states matched with antisymmetric rotational states and antisymmetric spin states with symmetric rotational states. At room temperatures where many rotational states of the molecule are formed one expects and finds a statistical distribution of

3/4 ortho-hydrogen molecules (3 symmetric spin functions)
1/4 para-hydrogen molecules (1 antisymmetric spin function)

However, at very low temperatures only the $J = 0$ rotational band will be occupied. In this case the spin wave function must be antisymmetric and we should expect to find just para-hydrogen present. In practice, to achieve this result one must overcome the obstacle that the spins of the nuclei interact only very slightly with each other; it is very difficult to change the nuclear spin state from triplet to singlet, just as in the case of transitions between atomic energy levels where transitions between different electron spin states are forbidden. The difficulty can be overcome by cooling the hydrogen in the presence of activated charcoal. The charcoal acts as a catalyst and allows the hydrogen to recombine into the most stable form, para-hydrogen. If the charcoal is removed and the para-hydrogen is heated to room temperature, it will maintain its identity for several weeks. This can be verified by measuring physical properties such as molecular spectra, heat capacity, and heat conductivity. Here we have a striking example of how important the Pauli exclusion principle can be when one has homonuclear molecules.

The preceding discussion can easily be generalized to include homonuclear molecules composed of nuclei with arbitrary spin. In this situation the total wave function is still expressed as

$$\psi_T = \psi_{el}\psi_{vib}\psi_{rot}\psi_{spin}$$

where for each nucleus

I integral ψ_T must be symmetric
I half integral ψ_T must be antisymmetric

To a very good approximation, the spins of the nuclei do not interact with each other so that we use the results for noninteracting particles where each nucleus can have $2I + 1$ spin states.

Assuming *an even electronic ground state configuration and an even vibrational ground state*, the space symmetry part of the molecular wave function is then determined by the orbital angular momentum: even J—even parity, odd J—odd parity. For possible nuclear spins we have

integer I for each nucleus	even integers for total I are associated with even-parity orbital states
	odd integers for total I are associated with odd-parity orbital states
1/2 integer I for each nucleus	symmetric spin states are associated with odd-parity orbital states
	asymmetric spin states are associated with even-parity orbital states.

It is interesting to note that for the special case of molecules having nuclei with zero spin there can be only even rotational levels.

The requirement that the nuclear spin state be unchanged during a transition between different levels limits the number of transitions that can occur. For instance, allowed transitions between rotational states will be restricted to those cases in which $\Delta J = 0, \pm 2$. For a given-type molecule the relative transition intensities will be proportional to the relative number of molecules present in the various states, i.e., the relative intensities will be proportional to the statistical weights of the levels. The ratio of the statistical weights between symmetric and antisymmetric spin states is given by

$$\frac{N_s}{N_A} = \frac{I + 1}{I}$$

where N_s = number of symmetric spin states
                    N_A = number of antisymmetric spin states

This relationship can easily be verified for cases where I is small. If I

for each of the nuclei $= 0$ then as we have already seen $N_A = 0$ and $N_s/N_A = \infty$ as predicted. Similarly from Section 12.1 we know that with $I = \frac{1}{2}$ for each nucleus there will be 3 symmetric spin states and 1 antisymmetric spin state. So $N_s/N_A = 3$.

We could continue using the approach of Section 12.1 to write down the possible symmetric and antisymmetric spin combinations for molecules with nuclei having $I \geqslant 1$ to find N_s and N_A. However, it is perhaps more convenient in general to find N_s and N_A by using the following scheme. We note that the total nuclear spin is given by $\mathbf{I}_T = \mathbf{I} + \mathbf{I}$ and then

Total Spin	Number of States	
$I_T = 2I$	$= 2(2I) + 1$	all even symmetry
$= 2I - 1$	$= 2(2I - 1) + 1$	all odd symmetry
$= 2I - 2$	$= 2(2I - 2) + 1$	all even symmetry
$= 2I - 3$	$= 2(2I - 3) + 1$	all odd symmetry
......	......	etc. through
		$I = 0$ state

To summarize, we see that we may expect to find the following results for homonuclear diatomic molecules:

(a) Nuclear spin wave function will be constant, i.e., unchanged by transitions.

(b) Space symmetry character of the system will be unchanged by transitions.

(c) There will be transitions between states of even J, and between states of odd J so only definite types of transitions will be allowed $\Delta J = 0, \pm 2$, etc.

(d) Once in a state of even or odd J the system will remain so even after collisions.

(e) Therefore, it makes sense to talk about two distinct types of diatomic molecules

ortho (symmetric nuclear spin wave function) greater statistical weight
para (antisymmetric nuclear spin wave function) lesser statistical weight

13.8   THERMAL DISTRIBUTION OF VIBRATIONAL AND ROTATIONAL LEVELS

If we consider a gas of diatomic molecules in thermal equilibrium, then the relative populations of the vibrational and rotational levels will be governed by the Maxwell-Boltzmann distribution law. The relative number of molecules in a particular vibrational energy state $E_v = (v + 1/2)\hbar\omega$ will be

$$\frac{N_v}{N_{\text{total}}} = \frac{e^{-E_v/kT}}{\sum\limits_{v=0}^{\infty} e^{-E_v/kT}} = \frac{e^{-(v+1/2)\hbar\omega/kT}}{\sum\limits_{v=0}^{\infty} e^{-(v+1/2)\hbar\omega/kT}}$$ (13.16)

$$= (1 - e^{-\hbar\omega/kT}) e^{-(v\hbar\omega/kT)}$$

Under ordinary conditions the ratio of E_v/kT is large for most diatomic molecules so that the great majority of the molecules will be found in the lowest vibrational energy state with $v = 0$. Therefore, at room temperatures and below, particularly for light molecules, it is reasonable to assume the molecules are in the ground vibrational state. For rotational levels, the situation is quite different because, as we have seen, the energy differences between low-lying rotational levels are much smaller than those between low-lying vibrational levels.

In considering the populations of the rotational levels, we must take into account the fact that each level specified by a given J is $2J + 1$ degenerate. Therefore, a statistical weighting factor of $2J + 1$ must be included in specifying the relative number of molecules occupying the Jth energy level. We have then, just considering rotational energy levels,

$$N_J \propto (2J + 1) \exp[-E_J/kT] = (2J + 1) \exp\left[-\frac{\hbar^2}{2\mathscr{I}} \frac{J(J+1)}{kT}\right]$$ (13.17)

where the normalization is such that

$$\sum_{J=0}^{\infty} N_J \approx \int_{J=0}^{\infty} N_J \, dJ = N_{\text{total}}$$

For a rotational level belonging to a particular vibrational level occupied by N_v molecules N_J is given as

$$N_J = N_v \frac{\hbar^2}{2\mathscr{I}} \frac{(2J + 1)}{kT} \exp\left[-\frac{\hbar^2}{2\mathscr{I}} \frac{J(J+1)}{kT}\right]$$ (13.18)

where N_v may be found by using Eq. (13.16).

Since the separation between rotational energy levels is so small we should expect to find many rotational levels occupied at room temperature and the rotational level with the maximum population can readily be obtained in the usual way by finding dN_J/dJ and setting it $= 0$.

$$\frac{dN_J}{dJ} \propto 2\exp\left[-\frac{\hbar^2}{kT}J(J+1)\right] - (2J + 1)\frac{\hbar^2}{2\mathscr{I}}\frac{(2J+1)}{kT}\exp\left[-\frac{\hbar^2}{kT}J(J+1)\right]$$

and setting $dN_J/dJ = 0$ gives

$$J_{\text{max}} = \frac{\mathscr{I}kT}{\hbar} - \frac{1}{2}$$

Of course, since there are only a finite number of levels occupied, J_{max} will not necessarily be integer, in which case one merely chooses the nearest integer value for J_{max}.

In any given case when one is calculating N_J, an additional weighting factor representing the degeneracy of the nuclear spin wave functions must also be added. We have discussed the question of nuclear spin wave functions in Section 13.6. The inclusion of the statistical weight factor g_N for the number of nuclear spin wave functions changes Eq. (13.18) to

$$N_J = g_N N_v \frac{\hbar^2}{2\mathscr{I}} \frac{(2J+1)}{kT} \exp\left[-\frac{\hbar^2}{2\mathscr{I}} \frac{J(J+1)}{kT}\right] \qquad \textbf{(13.18a)}$$

It is interesting to observe that the measurement of intensity ratios can be used to measure the nuclear spins. In particular if $I = 0$ for each nucleus of a homonuclear diatomic molecule, then every other line will be missing in the rotational energy-level spectra. Figure 13.13 shows the degeneracies of several rotational levels for the homonuclear diatomic molecules, H_2 and D_2. Since the proton has half-integral spin, H_2 molecules obey Fermi-Dirac

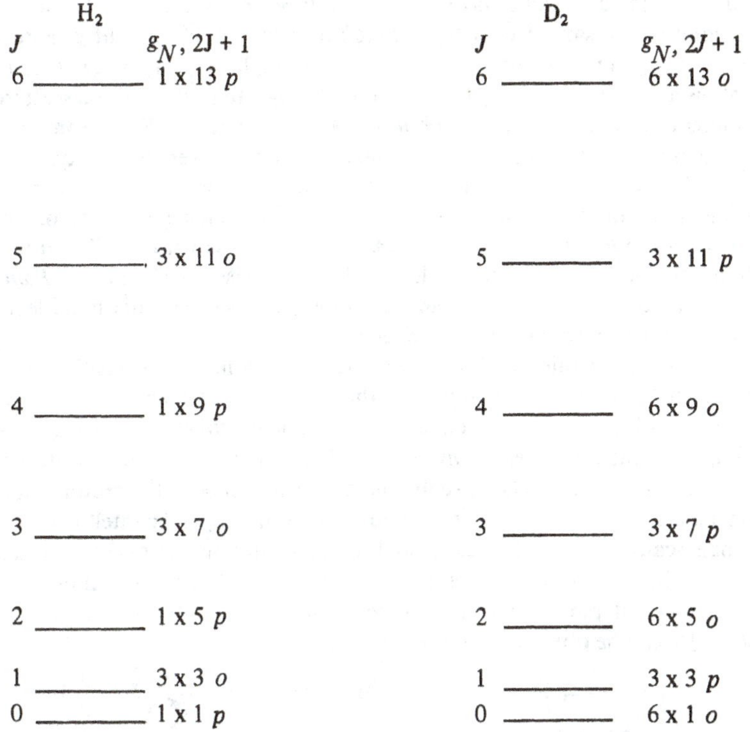

H_2				D_2	
J	$g_N, 2J+1$			J	$g_N, 2J+1$
6	1 x 13 p			6	6 x 13 o
5	3 x 11 o			5	3 x 11 p
4	1 x 9 p			4	6 x 9 o
3	3 x 7 o			3	3 x 7 p
2	1 x 5 p			2	6 x 5 o
1	3 x 3 o			1	3 x 3 p
0	1 x 1 p			0	6 x 1 o

FIGURE 13.13 Level diagrams showing nuclear spin and rotational statistical weights for the ground state and low-lying rotational levels of H_2 and D_2. The ortho- and para-states are indicated by o and p respectively.

statistics and their total wave function must be antisymmetric. On the other hand, the deuteron has one unit of spin and the D_2 molecule must obey Bose-Einstein statistics with a symmetric total wave function. Both the H_2 and D_2 are assumed to be in the electronic and vibrational ground state. The degeneracy of each level is given by the product of g_N and $2J + 1$. Notice that the ortho states are the states having the greatest spin degeneracy in each case.

13.9 RAMAN SPECTRA

As we have seen in Section 13.3, a homonuclear diatomic molecule cannot have a permanent electric dipole moment. However, the presence of an external electric field can cause an induced dipole moment which will be proportional to the electric field. Under these conditions the varying electric field of an incident light wave should produce an induced dipole moment changing at the same frequency. Therefore, when incident light quanta of frequency v_0 are incident on the system we might expect to see scattered photons also of frequency v_0 (Rayleigh scattering). In practice if the scattered photons are examined closely, photons of energy $hv_0 \pm \Delta E$ can sometimes be seen as well as those of energy hv_0. Since the molecule represents a quantum mechanical system the ΔE's must correspond to energy differences between various states of the system. Scattering for which the incident photon of energy hv_0 goes to $hv_0 \pm \Delta E$ is called *Raman scattering*. When the induced transitions are between vibrational levels they give rise to *vibrational Raman spectra* as distinguished from induced transitions between rotational levels, which produce *rotational Raman Spectra*.

The selection rule for Raman scattering involving vibrational levels can be shown to be $\Delta v = \pm 1$ and thus the scattered spectrum would consist of a central line with energy equal to the incident radiation (Rayleigh scattering) and satellite lines of $hv_0 \pm hv$ where hv is the energy difference between vibrational levels. (Usually the molecule will be in the ground vibrational state, in which case there would only be the $hv_0 - hv$ satellite line for Raman scattering.) With rotational Raman scattering the selection rule is $\Delta J = 0, \pm 2$. For $\Delta J = 0$ the emitted radiation would be just equal in energy to the incident radiation (Rayleigh scattering). The shifts in energy when $\Delta J = \pm 2$ can be obtained using Eq. (13.8).

$$E_{J+2} - E_J = \frac{\hbar^2}{2\mathscr{I}}[(J + 2)(J + 3) - J(J + 1)] = \frac{2\hbar^2}{\mathscr{I}}\left(J + \frac{3}{2}\right)$$

(Note that ΔJ is defined as $J' \rightarrow J''$ where J' is the upper state and J'' the lower state. Consequently, ΔJ is always $+$ and we need to consider only $\Delta J = 2$. From this we see

$$J \rightarrow J + 2 \quad \text{results in a shift to longer wavelengths}$$
$$J + 2 \rightarrow J \quad \text{results in a shift to shorter wavelengths}$$

From the foregoing we see that the rotational Raman spectrum consists of the central line plus a series of other lines of greater and lesser energy.

Problem 13.2: Experimentally it is found that the equally spaced lines of the rotational spectrum of the hydrogen fluoride molecule are 41.1 cm^{-1} apart. From this information calculate the separation distance of the hydrogen and fluoride atoms. What would be the expected line separations for the rotational spectrum for molecules of deuterium fluoride?

Problem 13.3: Find the rotational energy quanta for the first four rotational levels of HCl and compare them with kT at room temperature. Do the same for the vibrational energy quantum for the first excited vibrational level.

Problem 13.4: Consider the molecule Na_2. The separation distance is 1.5 Å and the spin of the Na^{23} nucleus is 3/2. Assume the molecule is in its ground electronic state (symmetric) and ground vibrational state, then draw the first 5 rotational energy levels showing the degeneracies of each level and calculate the energy difference in ev between the two lowest-lying levels. If the force constant (for vibrational motion) is 5×10^5 dynes/cm, what is the energy separation between the two lowest vibrational levels?

BIBLIOGRAPHY

Herzberg, Gerhard, *Spectra of Diatomic Molecules*. New York: D. Van Nostrand Company, Inc., 1950.

King, Gerald W., *Spectroscopy and Molecular Structure*. New York: Holt, Rinehart and Winston, Inc., 1964.

Pauling, L. and E. B. Wilson, *Introduction to Quantum Mechanics*. New York: McGraw-Hill Book Co., 1935.

XIV

FREE ELECTRON
AND BAND THEORY
OF METALS

In Chapter VI we saw the solution of the wave equation for the simple one-dimensional square well. One of the best and simplest applications of a square well in one dimension is the application to the problem of determining the possible energy levels available to electrons in metals. The simple one-dimensional Cartesian coordinate system is sufficient for a qualitative analysis and the long wavelength of the electrons compared to the distance between atoms causes the electronic behavior to be independent in many respects of the exact shape of the potential. Hence, we should expect a simple square well potential to yield a good qualitative picture of electrons in metals.

14.1 FREE ELECTRON THEORY

We start by assuming that a metal can be represented as a simple square well having the dimensions of the piece of metal. The electron motion in any one of the x, y, or z directions is assumed independent of motion in the other two directions. Therefore, the problem can be idealized by studying each motion separately, thus reducing our study to motion in one dimension.

Assume the piece of metal has an x dimension $= d$ with N atoms spaced a distance a apart. Then in the x direction the Schrödinger equation for the electron motion is

$$-\frac{\hbar^2}{2m}\frac{\partial^2}{\partial x^2}\psi = E\psi \qquad (14.1)$$

and

$$\psi_k = Ce^{+lkx}$$

$$E = \frac{\hbar^2}{2m}k^2$$

The solution, similar to the radiation problem of Chapter I, consists of plane waves whose wave numbers ($|k| = 1/\lambda = p/\hbar$) must be proportional to an integral number of the reciprocal of the length of the sides of the piece of metal. Thus $k = n\pi/d$, $n = \pm1, \pm2, \pm3 \ldots$. Since $d = Na$, $|k|$ must be equal to or less than π/a in order that the electrons be represented by unique wave functions.

Figure 14.1 illustrates the problems encountered if one attempts to describe the positions of the N electrons having N degrees of freedom in their motion about the N atoms of the crystal by more than the N waves with wave number $k < \pi/a$. The very concept of wave motion implies a continuous rather than a discrete medium, and therefore the wavelengths must be equal to or greater than, but not less than, the discreteness of the medium (twice the atomic separations), in order for a wave description to be valid. Therefore, $k = n\pi/Na$ where n is an integer equal to or less than N.

Note that, as shown in Figure 14.1, if k were greater than π/a, λ would be less than twice the distance between atoms and the basic treatment of the solid as a continuous medium would be invalid. This limitation on the magnitude of the wave number implies a limitation on the energy of the electron. Since

$$\lambda = \frac{h}{p} = \frac{h}{\sqrt{2mE}}$$

and

$$E = \frac{h^2}{2m\lambda^2} \leqslant \frac{h^2}{8ma^2}$$

If we choose $a = 2 \times 10^{-8}$ cm, then $E \leqslant 10$ ev in agreement with more detailed studies from which it is known that the maximum kinetic energy of free electrons in metals is usually of the order of a few ev.

The energy difference between any two levels n and $n + 1$ is given by

$$E_{n+1} - E_n = \frac{\hbar^2}{2m}\left(\frac{\pi}{d}\right)^2[(n+1)^2 - n^2] = \frac{\hbar^2}{2m}\left(\frac{\pi}{d}\right)^2(2n+1)$$

The maximum difference ΔE is readily estimated to be $\approx 10^{-14}$ ev, which is very much smaller than the average kinetic energy of thermal motion at usual temperatures.

The wave function [Eq. (14.1)] which we took to be the solution of the wave equation has been written in a form suitable to satisfy a periodic boundary condition. By this we mean that a solution is chosen such that $\psi_k(x) = \psi_k(x + d)$, where $d =$ the x dimension of the metal.

$$\psi_k(x) = Ce^{ikx} = C \exp\left\{\frac{i2n\pi}{d}x\right\}$$

$$\psi_k(x + d) = Ce^{ik(x+d)} = C \exp\left\{i\frac{2n\pi}{d}x + i\,2n\pi\right\} = \psi_k(x)$$

From the normalization condition

$$\int_0^d \psi_k^*\psi_k\,dx = 1$$

we find

$$C = 1/\sqrt{d}$$

The $\psi_k(x)$ form an orthonormal set of eigenfunctions. To show this we consider

$$\int_0^d \psi_k^*\psi_{k'}\,dx = \frac{1}{d}\int_0^d e^{i(k-k')x}\,dx$$

$$= \frac{1}{d}\frac{1}{i(k-k')}[e^{i(k-k')d} - 1]$$

But $(k - k')d = 2\pi(n - n') =$ an even multiple of π and therefore

$$\int_0^d \psi_k^*\psi_{k'}\,dx = 0 \qquad \text{if } k \neq k'$$

or

$$\int_0^d \psi_k^*\psi_{k'}\,dx = \delta_{k,k'}$$

This form for ψ really represents a solution in the form of traveling waves as can be readily seen by including the time dependence of the wave function

$$\Psi_k(x, t) = \frac{1}{\sqrt{d}}\exp\left\{i\,kx - \frac{i}{h}Et\right\}$$

If we had chosen to satisfy the boundary condition that requires the wave function to go to zero at the surface [$\psi(0) = 0$ and $\psi(d) = 0$] the solutions of the wave equation would take the form

$$\psi = \sqrt{\frac{2}{d}}\sin\frac{n\pi}{d}x \qquad n = \text{an integer}$$

which represents a set of standing waves as can readily be seen from the probability density

$$\psi^*\psi = \frac{2}{d}\sin^2\frac{n\pi x}{d}$$

This is to be contrasted to the probability density obtained from the traveling wave solutions where

$$\psi^*\psi = \frac{1}{d}$$

In the simple free electron model under discussion either type boundary condition is satisfactory. However, when the effect of the crystal lattice is being considered then it turns out that the periodic boundary condition is more useful.

For a three-dimensional piece of metal the frequencies or energies are determined by the condition that $H\psi = -(\hbar^2/2m)\nabla^2\psi = E\psi$. Therefore,

$$E = \frac{\hbar^2\pi^2}{2m}\left[\left(\frac{n_1}{d_1}\right)^2 + \left(\frac{n_2}{d_2}\right) + \left(\frac{n_3}{d_3}\right)^2\right] \tag{14.2}$$

where the subscripts refer to the three sides of the metal. (Note that since the Schrödinger wave equation is first-order in time, E appears to the first power, rather than the frequency to the second power as for radiation in Chapter I.) Thus, in close analogy to radiation in a box, the total number of electronic wave functions up to an energy E is

$$\int_0^E N(E)dE = \frac{\pi}{6}d_1 d_2 d_3 \left(\frac{2mE}{\hbar^2\pi^2}\right)^{3/2}$$

$$= \frac{V}{6\pi^2}\left(\frac{2m}{\hbar^2}\right)^{3/2} E^{3/2}$$

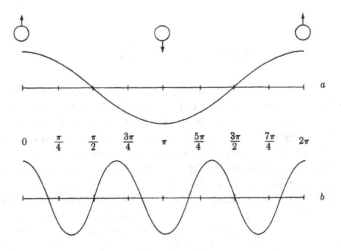

FIGURE 14.1 (a) Illustration of the shortest wavelength that may be used to describe the motion of single-valence electrons indicated above (a). If shorter wavelengths were permitted, the motion could be described by the wave (b) and an infinite number of higher-frequency waves.

and the number of electronic wave functions having an energy between E and dE is given by

$$N(E)dE = \frac{V}{4\pi^2}\left(\frac{2m}{\hbar^2}\right)^{3/2}\sqrt{E}\,dE \qquad (14.3)$$

where V is the volume of the piece of metal.

The Pauli exclusion principle requires that only one electron at a time can occupy a single state described by a given wave function and spin polarization. (Since electrons may have spin with vector up or down, there are two directions of polarization just as for light, and when we include spin the total number of states is double the number given in Eq. (14.3).) The total number of electrons in the lowest total wave function is obtained by integrating Eq. (14.3) up to a maximum energy E_{max}, with the result

$$N = 2\frac{V}{4\pi^2}\left(\frac{2m}{\hbar}\right)^{3/2}\left(\frac{2}{3}E_{max}^{3/2}\right) \qquad (14.4)$$

or

$$E_{max} = \frac{(3\pi^2)^{2/3}\hbar^2}{2m}\left(\frac{N}{V}\right)^{2/3} = 36.1\left(\frac{N}{V}\right)^{2/3}$$

in units of electron volts if V is in cubic angstroms. This maximum energy which electrons in the ground state of a metal may have is called the *Fermi energy*. An electron having this energy is in the *Fermi level*. Hence even in a metal at zero degrees absolute temperature, with all electrons in the lowest possible states, some electrons will have kinetic energies of several electron volts.

14.2 BAND THEORY

Solids are, in actuality, more complicated than they have been represented in the foregoing. They consist of a lattice of atomic nuclei; hence, in place of a square well of constant depth, the electron potential well is a periodic function of distance with a period given by the atomic separation in the lattice. It is still possible, however, to simplify the problem by assuming the variables separable (which in actuality they are not) and then considering the wave motion in just one dimension. The Kronig-Penney potential energy model in one dimension for electrons in a crystal is illustrated in Figure 14.2. The crystal atoms form a periodic barrier of height V_0 and widths b spaced a distance a apart. Due to the periodicity of the potential the wave function must have a coefficient periodic in x. The x-dependent part of the wave function in Cartesian coordinates is made up of plane waves e^{ikx} (where k, an integral multiple of π/a, is less than twice the total number of atoms N along a side of the crystal) times a periodic function of period a. Let

$$\psi = u_k(x)e^{ikx} \qquad (14.5)$$

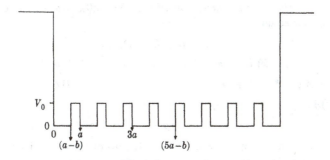

FIGURE 14.2 Illustration of the Kronig-Penney step-function barrier potential function representing the atoms in a crystal lattice.

where $u_k(x)$ is a general periodic function called a *Bloch wave function* for periodic potentials of any shape in a crystal.

Substitution of the Kronig-Penney potential into Eq. (4.8) yields time-independent standing wave solutions which are:

For $(n - 1)a \leqslant x \leqslant na - b$, where n is an integer $0 < n < N$:

$$\psi_k = A\, e^{i\alpha x} + B\, e^{-i\alpha x}$$

$$u_k = A\, e^{i(\alpha - k)x} + B\, e^{-i(\alpha + k)x} \qquad (14.6)$$

where

$$\alpha = p_1/\hbar = \sqrt{2mE/\hbar^2}$$

For $na - b \leqslant x \leqslant na$ the electron potential energy is increased by V_0 over the electron potential energy for $(n - 1)a \leqslant x \leqslant na - b$, and the electron wave number changes from $\sqrt{2mE/\hbar^2}$ to $i\sqrt{2m(V_0 - E)/\hbar^2}$. For $na - b \leqslant x \leqslant na$ and $V_0 > E$:

$$\psi_k = C\, e^{\beta x} + D\, e^{-\beta x}$$

$$u_k = C\, e^{(\beta - ik)x} + D\, e^{-(\beta + ik)x} \qquad (14.7)$$

where

$$\beta = iq_2/\hbar = \sqrt{2m(V_0 - E)/\hbar^2}$$

Equations (14.6, 14.7) are customarily found as solutions to the complete wave equation in which the Bloch wave function Eq. (14.5) has been substituted, which is

$$\frac{d^2 u_k}{dx^2} + 2ik\frac{du_k}{dx} + \left[\frac{2m}{\hbar^2}(E - V) - k^2\right]u_k = 0 \qquad (14.8)$$

The conditions requiring that u and its first derivative be continuous in general at $x = (n - 1)a$ and $x = na - b$ and in particular at $x = 0$ and $x =$

a, and the particular periodicity condition $u(a) = u(0)$, $\psi(a - b) = \psi(-b)$, yield the four equations

$$A + B = C + D$$

$$i(\alpha - k)A - i(\alpha + k)B = (\beta - ik)C - (\beta + ik)D$$

$$A\,e^{i(\alpha-k)(a-b)} + B\,e^{-i(\alpha+k)(a-b)} = C\,e^{-(\beta-ik)b} + D\,e^{(\beta+ik)b}$$

$$i(\alpha - k)A\,e^{i(\alpha-k)(a-b)} - i(\alpha + k)B e^{i(\alpha+k)(a-b)}$$
$$= (\beta - ik)Ce^{-(\beta-ik)b} - (\beta + ik)D\,e^{(\beta+ik)b}$$

Either A, B, C, and D are all simultaneously zero or the determinant of the coefficients of these four simultaneous algebraic equations must vanish. Setting the determinant equal to zero, going through a great deal of algebraic manipulation and inserting $\sin \alpha(a - b)$ for $(1/2i)(e^{i\alpha(a-b)} - e^{-i\alpha(a-b)})$, $\cos \alpha(a - b)$ for $\frac{1}{2}(e^{i\alpha(a-b)} + e^{-i\alpha(a-b)})$, $\sinh \beta b$ for $\frac{1}{2}(e^{\beta b} - e^{-\beta b})$, and $\cosh \beta b$ for $\frac{1}{2}(e^{\beta b} + e^{-\beta b})$, we find

$$\frac{\beta^2 - \alpha^2}{2\alpha\beta} \sinh \beta b \sin \alpha(a - b) + \cosh \beta b \cos \alpha(a - b) = \cos ka \qquad (14.9)$$

For simplicity, we may replace our potential barrier by a delta function so constructed that

$$\lim_{\substack{b \to 0 \\ \beta \to \infty}} \beta^2\, ab = 2P$$

where P is some constant, and use $\sin \alpha(a - b) = \sin \alpha a$, $\cos \alpha(a - b) = \cos \alpha a$, $\cosh \beta b = 1$, and $[(\beta^2 - \alpha^2)/2\alpha\beta] \sinh \beta b = (\beta/2\alpha)\,\beta b$, so that Eq. (14.9) becomes

$$(\beta/2\alpha)(\beta b) \sin \alpha a + \cos \alpha a = \cos ka$$

whence

$$[(P \sin \alpha a)/\alpha a] + \cos \alpha a = \cos ka \qquad (14.10)$$

The left-hand side of Eq. (14.10) must be less than or equal to one in order that k be real and not give rise to rapidly decaying exponential wave functions. A plot of the left-hand side of (14.10) for $P = 3\pi/2$ as a function of αa is shown in Figure 14.3. The shaded regions are for those values of αa for which solutions do not exist. Figure 14.4 illustrates the allowed values of $\alpha^2 a^2 = 2mEa^2/\hbar^2$ in Eq. (14.10) as a function of ka. Note the energy gaps occurring at integral multiples of $\pi = ka$, separating the bands of allowed energy levels. Since $E \sim k^2$, the allowed energies have a roughly parabolic dependence on ka. Portions of the three lowest eigenfunctions which occur in a Kronig-Penney potential are shown approximately in Figure 6.8. The exact wave functions in this potential are slightly shifted from those in the figure since the wave function in Eq. (14.7) includes an exponentially increasing term.

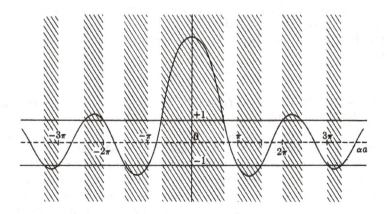

FIGURE 14.3 Graph of $[(P \sin \alpha a)/\alpha a] + \cos \alpha a$ as a function of αa. The dashed lines $+1$ and -1 are the maximum values the function can have in possible solutions, hence the shaded areas are unallowed values for αa.

14.3 EFFECTIVE MASS

From this qualitative discussion we can see that an electron in a solid should not be expected to act like a free electron in the presence of an external electric field. However, it turns out that in general the motion of the electron will be like that expected for the free electron if we introduce an "effective" mass into the equations of motion. To see how this comes about let us consider first the case of the free electron.

In Section 2.4 it is shown from wave mechanics that the particle velocity is equal to the group velocity of the wave packet representing the particle

$$v = \frac{d\omega}{dk}$$

where ω is the angular frequency corresponding to the de Broglie wavelength of the electron having an energy E.

Then since $E = \hbar\omega$, $v = (1/\hbar)(dE/dk)$.

For the special case of a free electron we know

$$E = \frac{p^2}{2m} = \frac{\hbar^2 k^2}{2m}$$

and $v = \hbar k/m = p/m$ as expected. However, for a solid in the band theory, E is no longer strictly proportional to k^2 and so we should expect results different from those in the case of the free electron. For instance, suppose an external electric field F is applied to a solid and acts on an electron initially specified by wave number k. Then, if the electron moves a distance dx in a time dt

$$dE = eFdx = eFvdt = \frac{eF}{\hbar}\frac{dE}{dk}dt$$

or $dk/dt = eF/\hbar$ since $dE = (dE/dk)dk$. This will produce an acceleration $a = dv/dt = (1/\hbar)(d^2E/dk^2)(dk/dt) = (eF/\hbar^2)(d^2E/dk^2)$ Compared to the free electron case where $a = eF/m$, we see that the electron in the solid behaves similarly if we assign to it an effective mass given by $m^* = \hbar^2/d^2E/dk^2$.

At this point it is convenient to refer to Figure 14.4a, which shows the allowed values for the electron energy as a function of ka. Within a given energy band, the energy is a periodic function of k since $\cos ka = \cos (k + 2\pi n/a)a$ in Eq. (14.10), if $n =$ an integer. Therefore, for convenience we choose a *reduced wave vector* such that $-\pi/a \leqslant k \leqslant \pi/a$. A plot such as that of Figure 14.4a now appears as shown in Figure 14.4b. Referring to the plot of E versus the reduced wave vector, it is easy to see that E, dE/dk and d^2E/dk^2 vary with k.

Since $m^* \propto (dE^2/dk^2)^{-1}$, it too will vary. This variation has been verified experimentally and the concept of an effective mass has proved to be useful in discussing phenomena such as the Hall effect and conductivity of metals, among others.

The concept of allowed energy bands separated by forbidden energy regions (gaps) can be used to explain qualitatively the existence of conductors (metals), insulators, and intrinsic semiconductors. In a metal, the electrons about half-fill the conduction band. An applied electric field causes the electron to accelerate in the direction of the field between collisions with the lattice vibrations. In this way a net flow of electricity takes place in the metal and we say it is a conductor. Certain other substances, however, have energy bands which are either completely filled or completely empty at ordinary temperatures. An electric field applied in this case cannot accelerate the electrons, for acceleration would raise the energy of the electron, which is not permitted unless the energy is raised discontinuously so that the electron jumps to an unfilled energy band. (The Pauli exclusion principle does not allow the electron to enter a state which is already occupied.) Under these conditions we say that the material is a nonconductor or insulator.

The intrinsic semiconductor arises from the interesting situation where a filled band is separated from an unfilled band lying above it with only a small energy gap. At absolute zero this material is an insulator. However, with rising temperatures some electrons gain enough thermal energy to cross the gap into the unfilled band. There they could move under the influence of an applied electric field and so the material now has become a conductor. It is interesting to note that the conductivity of a semiconductor increases with increasing

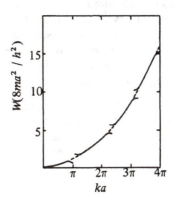

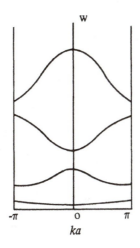

FIGURE 14.4 (a) Graph of energy as a function of wave number for the Kronig-Penney potential. (b) The same except that here the energy is plotted against the reduced wave number.

temperature as more electrons are then able to move into the unfilled band. This is in sharp contrast to the behavior in metals where increasing the temperature decreases the conductivity. Although we have greatly simplified the discussion of electrons in a solid to one dimension, nevertheless the results obtained in this way are in good qualitative agreement with those of more rigorous calculations in three dimensions.

Problem 14.1: Derive Eq. (14.3) by the methods given in Chapter I.

Problem 14.2: Assume a cube of side d. Then, assuming periodic boundary conditions, find: (a) all the wave functions for the first three distinct energy levels, (b) the energy for each of the levels, (c) the degeneracy of each level. Repeat the calculations using the boundary condition requiring the wave function to go to zero on the surface.

BIBLIOGRAPHY

Mott, N. F. and H. Jones, *The Theory of the Properties of Metals and Alloys.* New York: Dover Publications, Inc., 1958.

Kittel, C., *Introduction to Solid State Physics.* New York: John Wiley & Sons, Inc., 1966.

XV

ELASTIC
SCATTERING
THEORY

Elastic scattering theory treats collisions between particles in which no internal excitation of the particles takes place. Inelastic scattering, on the other hand, occurs when the same kind of particle is present in the scattered beam as in the incident beam but the total translational kinetic energy of the system is less, the missing energy going into excitation of internal degrees of freedom of one or both particles. For example, if two atoms collide and part of their translational kinetic energy is lost to electronic excitation or ionization, the scattering is inelastic. Inelastic scattering theory is somewhat more complicated and is outside the scope of this book, so that the internal energy of the system or the complete eigenfunction of the colliding systems' constituent parts will be ignored. A complex, many-component particle such as an atom will be regarded as a single particle.

A cross section for a given scattering event is defined as the fraction of the number of *incident-particles-per-unit-area* for which the scattering event happens. For example, suppose a group of boys threw 900 baseballs at random through an opening nine square feet in area, producing a beam of 100 balls/square foot. Suppose that out of every 100 balls/foot² 50 balls hit a target on the other side of the opening. The cross section would be given by

$$\frac{50 \text{ balls}}{100 \text{ balls/ft}^2} = \frac{1}{2} \text{ft}^2$$

This is the fraction of the incident beam of 100 balls/ft² which hit the target on the average. We see that a cross section is given in units of area and gives the probability that a beam of one ball per square foot will hit the target.

Suppose that 15 balls cracked the target. The cross section for cracking the target would be

$$\frac{15 \text{ balls}}{100 \text{ balls/ft}^2} = 0.15 \text{ ft}^2$$

It is clear that there are different kinds of cross sections. We will be concerned solely with the probability of elastically scattering particles into a differential element of solid angle $d\omega$ (the differential elastic scattering cross section) and the integral of this cross section over all angles (the total elastic scattering cross section).

Figure 15.1 illustrates the scattering of a beam of particles by a scattering center. The problem raised is: What is the probability that a particle will be deflected or absorbed and it or a like particle reemitted into a particular differential element of solid angle $d\omega$ if the incident beam has unit intensity (1 particle per cm²-sec)? The dimension of this probability is, as we have just seen, the dimension of area used to express the incident beam intensity. The scattering cross section is normally measured in square centimeters, barns, or millibarns where 1 barn $= 10^{-24}$ cm². (The name "barn" is said to have been used first in the Manhattan Project during the Second World War, 10^{-24} cm² being as huge as a barn compared with most nuclear scattering cross sections.)

To reiterate, in general the ratio of total number of scattered particles to the number of particles per unit area in the incident beam is called the total *scattering cross section*. Similarly, the ratio of the number of particles scattered into the differential element of solid angle $d\omega$ to the number of particles per unit area in the incident beam is called the *differential scattering cross section*. The number of scattered particles having an energy between E and $E + dE$ is also of interest, but this is for *inelastic* scattering and outside the scope of this chapter.

The subject of scattering theory is the relationship of scattering cross sections to the potential energy between the interacting particles. For example if the wave functions of the electrons in an atom are known, the electrostatic potential energy of an incident ion moving near this atom can ideally be approximated everywhere as a function of space by ignoring the changes in atomic wave functions introduced by the ion; scattering theory then enables us to predict exactly what the differential and total scattering cross section of these atoms for the particular incident ions will be. Conversely, detailed

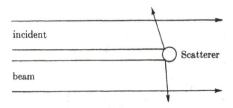

FIGURE 15.1   Illustration of the scattering of a beam of particles by a scattering center.

measurements of particle scattering from atomic nuclei as a function of scattering angle and energy of incident and scattered particle may be interpreted by use of scattering theory to give us a great deal of information about the potential energy of the scattered particle in the force field of atomic nuclei. This information is essential to a knowledge of nuclear forces.

15.1   THE PATH INTEGRAL FORMULATION OF THE SCATTERING PROBLEM

Before developing the conventional techniques for scattering problems, in order to obtain a better insight into the scattering process we will first analyze a simple scattering problem by the path integral method of Chapter IV. That is, we will compute the phase of the exponential in Eq. (4.8) for each path through the scattering center and then integrate over the projected area of the scattering center. As an illustrative example let us take the case of a neutron incident on a completely absorbing nucleus of radius a cm which after absorbing reemits the neutron at angle θ (i.e., the nucleus elastically scatters the neutron). The fraction of the wave stopped by the circular obstruction in a transparent wall may be analyzed in terms of a calculation which treats the nucleus as if it were a two-dimensional opening of a cm radius in an infinite opaque plane perpendicular to the momentum of the incident neutron (see Figure 15.2). The path length x for a path through the point (ρ, φ) in the opening is given by $r - \rho \sin \theta \cos \varphi$, where r is the path length of a ray passing through the center of the opening from the opening to a point r cm away. θ is the scattering angle, and φ is measured from the plane containing the incident and diffracted wave propagation vector. If in Eqs. (4.5, 4.8) we neglect the constant, energy-dependent part, we find

$$\psi(\theta) = Ce^{(i/\hbar)pr} \int_0^a \int_0^{2\pi} \exp\left\{-(i/\hbar)p\rho \sin \theta \cos \varphi\right\} \rho \, d\rho \, d\varphi \qquad \textbf{(15.1)}$$

where C is a normalization constant.

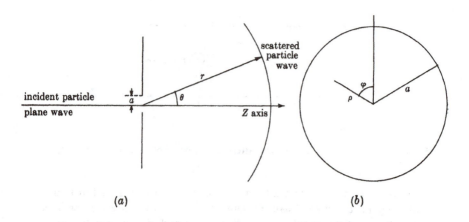

(a)   (b)

FIGURE 15.2 (a) Illustration of scattering, showing scattered particle waves. The incident neutron wave is diffracted through the "circular opening" into angle θ. (b) Diagram of circular opening, in plane of the opening, showing a path position ρ, φ at the opening. φ is measured from the plane containing the incident and diffracted wave propagation vectors.

Since the integral representation of the Bessel function of order n is

$$J_n(x) = \frac{i^{-n}}{2\pi} \int_0^{2\pi} e^{ix \cos \varphi} e^{in\varphi} d\varphi \qquad (15.2)$$

after integrating over φ we are left with

$$\psi(\theta) = Ce^{(i/\hbar)pr} 2\pi \int_0^a J_0\left(\frac{p\rho \sin \theta}{\hbar}\right) \rho \, d\rho$$

and since

$$xJ_0(x) = \frac{d}{dx}[xJ_1(x)]$$

therefore

$$\psi(\theta) = Ce^{(i/\hbar)pr} \frac{2\pi\hbar a}{p \sin \theta} J_1\left(\frac{ap \sin \theta}{\hbar}\right) \qquad (15.3)$$

which yields the amplitude of the scattered wave at angle θ.

The total intensity at angle θ and distance r beyond the nucleus is found by adding the amplitude of this scattered wave to the amplitude of the incident plane wave $e^{(i/\hbar)pz}$ at the point θ, r and taking the square of the absolute value of this sum. Thus, the probability of finding a neutron at a differential solid angle $d\Omega$ at angle θ is given by

$$\sigma(\theta) \, d\Omega = |e^{(i/\hbar)pz} + \psi(\theta)|^2 \, d\Omega \qquad (15.4)$$

In order to evaluate this expression explicitly, $e^{(i/\hbar)pz}$ must be expanded in spherical harmonics, as is done later in this chapter.

When the neutron is not re-emitted with the same energy as the incident neutron (inelastic scattering) or is not re-emitted at all, the absorbed wave still interferes with the transmitted wave, however, so that elastic scattering is still present. The elastic scattering in terms of our $\psi(\theta)$ just calculated assumes a particularly simple form by application of Babinet's theorem for physical optics, i.e.: The diffraction patterns produced by two complementary screens (the complementary screen is opaque where the other screen is transparent and vice versa) are identical except for the central spot, which corresponds to the diffraction angle zero. Hence, the elastic scattering cross section for a totally absorbing nucleus is the same an for an opaque screen with a transparent opening of the same size.

$$\sigma(\theta)d\Omega \; = |\psi(\theta)|^2 d\Omega = \left[\frac{CahJ_1(ap\sin\theta/\hbar)}{p\sin\theta} \right]^2 d\Omega \qquad (15.5)$$

The total elastic scattering cross section when neutrons are absorbed by the nucleus is the fraction of the incident beam intensity which is incident on the nucleus, namely πa^2. By evaluating the integral of $\sigma(\theta)$ over all angles and setting the result equal to πa^2, we determine that $C = 2\sqrt{\pi p/h}$ so that

$$\sigma(\theta)\, d\Omega = 4\pi a^2 \csc^2\theta J_1^2 \left(\frac{ap\sin\theta}{\hbar} \right) d\Omega \qquad (15.6)$$

This is so-called shadow scattering. The absorption of the neutron causes a loss in the incident wave, a "shadow," and the interference of this shadow with the incident plane wave causes a diffraction pattern which is a function of the angle θ. If the neutron is re-emitted with the same energy, the total scattering cross section will be $2\pi a^2$, as we shall see in Section 15.9.

15.2 GENERAL FORMULATION OF SCATTERING PROBLEMS

The incident beam of particles is represented by a plane wave e^{ikz} propagated along the z axis, where the particle wave number k is the free particle momentum divided by the particle mass m. Figure 15.2(a) illustrates the physical situation. At very large distances beyond the scattering center the incident particle spatial wave function will be composed of two parts, the initial plane wave and a radially diverging outgoing scattered wave. That is, for r very much larger than the range of the particle potential in the force field of the scattering center,

$$\psi \sim e^{ikz} + (i/r)e^{ikr}f(\theta) \qquad (15.7)$$

Actually all we need say is that ψ is made up of the original plane wave plus a small contribution from the scattered wave representing the few particles which are scattered. Since, however, the scattered wave, which will be our prime concern, will have a radially diverging form, it is most convenient to take as the unknown function to be computed $f(\theta)$, the coefficient of the radially diverging wave, rather than the entire expression for the scattered wave. We are interested in the current of particles scattered into the differential element of solid angle $d\omega$, given by

$$v\,|\,\psi(\theta)\,|^2\,r^2\,d\omega = v\,|\,f(\theta)\,|^2\,d\omega \qquad (15.8)$$

where the incident particle current is v for a single particle. The differential scattering cross section of the scattering center $\sigma(\theta)\,d\omega$ is thus

$$\sigma(\theta)\,d\omega = |\,f(\theta)\,|^2\,d\omega \qquad (15.9)$$

where $f(\theta)$ has the dimensions of a length. For inelastic scattering in which the particle energy and speed changes, Eq. (15.9) would be multiplied on the right-hand side by v_{fin}/v_{init}.

15.3 THE BORN APPROXIMATION

The Born approximation is the most effective, simplest, and most generally applied method of computing scattering problems. It is essentially an application of time-dependent or time-independent perturbation theory to the scattering of a plane wave with the initial and final wave functions both approximately plane waves far from the scattering center. We have already touched on its application to inelastic atomic scattering in Chapter X. The primary condition for the validity of this approximation is that the perturbing potential energy V between particle and target be much less than the kinetic energy of the scattered particle E, so that that the amplitude of the perturbed wave $\psi' = (e^{ikr}/r)f(\theta)$ will be much less than that of the incident plane wave $\psi^0 = e^{ik_0 z}$.

The time-independent Schrödinger equation for the two-particle interaction may be written for the space-dependent wave function alone since the energy of the particles is a constant in the center-of-mass system; it is

$$\nabla^2\psi + \frac{2m}{\hbar^2}[E - V(r)]\psi = 0 \qquad (15.10)$$

where m is the reduced mass of the system, $m_1 m_2/(m_1 + m_2)$. We assume $V \ll E$ and $\psi' \ll \psi^0$, where

$$\psi = \psi^0 + \psi' \qquad (15.11)$$

and $\psi^0 = e^{ik_0 z}$ is the plane wave solution of the homogeneous differential equation

$$\nabla^2\psi^0 + \frac{2m}{\hbar^2}E\psi^0 = 0 \qquad (15.12)$$

Then first-order time-independent perturbation theory, in which $V\psi'$ is neglected, yields

$$\nabla^2\psi' + \frac{2m}{\hbar^2}E\psi' = \frac{2m}{\hbar^2}V(r)\psi^0 \tag{15.13}$$

The right-hand side of (15.13) is merely a known function of r; the left-hand side is the homogeneous part of a differential equation in ψ'.

In the theory of electromagnetism one obtains an equation identical in form to (15.13) for the electrostatic potential φ in a medium with charge density ρ (note that both φ and ρ are the Fourier transforms from time to $k = \omega$),

$$\nabla^2\varphi + k^2\varphi = -4\pi\rho$$

whose solution by use of Green's theorem gives one the Lienard-Wiechart retarded potential $\varphi(\mathbf{r})$ at the point $\mathbf{r}$ in terms of the source charge density $\rho(\mathbf{r}')$ at a different point $\mathbf{r}'$,

$$\varphi(\mathbf{r}) = -\int \frac{\rho(\mathbf{r}')}{|\mathbf{r} - \mathbf{r}'|}e^{ik|\mathbf{r}-\mathbf{r}'|}d\tau'$$

where the absolute value of the vector difference $\mathbf{r} - \mathbf{r}'$ is its length. We could take over this result complete and apply it to Eq. (15.13), but particularly for students who are unfamiliar with potential theory we will run through an elementary derivation. Let

$$k_0^2 = \frac{2m}{\hbar^2}E = \frac{p_0^2}{\hbar^2}, \qquad U = \frac{2m}{\hbar^2}V \tag{15.14}$$

then Eq. (15.13) becomes

$$(\nabla^2 + k_0^2)\psi' = Ue^{ik_0 z} \tag{15.15}$$

It is convenient to work with the Fourier transform of ψ',

$$A(\mathbf{k}) = \frac{1}{(2\pi)^{3/2}}\int \psi'(\mathbf{r})e^{-i\mathbf{k}\cdot\mathbf{r}}d\tau \tag{15.16a}$$

and the inverse transform,

$$\psi'(\mathbf{r}) = \frac{1}{(2\pi)^{3/2}}\int A(\mathbf{k})e^{i\mathbf{k}\cdot\mathbf{r}}d\mathbf{k} \tag{15.16b}$$

where $d\mathbf{k}$ is the three-dimensional volume element in momentum space. Further,

$$\nabla^2\psi'(\mathbf{r}) = \nabla^2\frac{1}{(2\pi)^{3/2}}\int A(\mathbf{k})e^{i\mathbf{k}\cdot\mathbf{r}}d\mathbf{k} = -\frac{1}{(2\pi)^{3/2}}\int k^2 A(\mathbf{k})e^{i\mathbf{k}\cdot\mathbf{r}}d\mathbf{k} \tag{15.17}$$

Equation (15.15) then becomes

$$\frac{1}{(2\pi)^{3/2}}\int (-k^2 + k_0^2)A(\mathbf{k})e^{i\mathbf{k}\cdot\mathbf{r}}d\mathbf{k} = Ue^{ik_0 z} \tag{15.18}$$

We now multiply both sides of Eq. (15.18) by a particle plane wave function of particular wave number k', $e^{-i\mathbf{k}'\cdot\mathbf{r}}$, and write the integral of both sides over coordinate space,

$$\frac{1}{(2\pi)^{3/2}}\iint A(\mathbf{k})e^{-i\mathbf{k}'\cdot\mathbf{r}}e^{i\mathbf{k}\cdot\mathbf{r}}(-k^2+k_0^2)d\mathbf{k}d\tau = \int U(\mathbf{r})e^{ik_0 z}e^{-i\mathbf{k}'\cdot\mathbf{r}}d\tau \qquad (15.19)$$

By making use of the first of the following mathematical identities resulting from the orthogonality of the wave functions

$$\frac{1}{(2\pi)^{3/2}}\int e^{-i\mathbf{k}\cdot\mathbf{r}}e^{i\mathbf{k}'\cdot\mathbf{r}}d\tau = (2\pi)^{3/2}\delta(\mathbf{k}-\mathbf{k}')$$

$$\frac{1}{(2\pi)^{3/2}}\int e^{-i\mathbf{k}\cdot\mathbf{r}}e^{i\mathbf{k}\cdot\mathbf{r}'}d\mathbf{k} = (2\pi)^{3/2}\delta(\mathbf{r}-\mathbf{r}') \qquad (15.20)$$

we find that Eq. (15.19) becomes

$$(2\pi)^{3/2}A(\mathbf{k}')(-k'^2+k_0^2) = \int U(\mathbf{r})e^{i(\mathbf{k}_0-\mathbf{k}')\cdot\mathbf{r}}\,d\tau$$

and Eq. (15.16b) becomes

$$\psi'(\mathbf{r}') = \frac{1}{(2\pi)^3}\int\left\{\frac{\int U(\mathbf{r})e^{i(\mathbf{k}_0-\mathbf{k}')\cdot\mathbf{r}}d\tau}{k_0^2-k'^2}\right\}e^{i\mathbf{k}'\cdot\mathbf{r}}d\mathbf{k}' \qquad (15.21)$$

Reversing the order of integration in Eq. (15.21) and integrating over $\mathbf{k}'$, we obtain

$$\psi'(\mathbf{r}') = -\frac{1}{4\pi}\int\frac{e^{ik_0|\mathbf{r}'-\mathbf{r}|}}{|\mathbf{r}'-\mathbf{r}|}U(\mathbf{r})e^{ik_0 r}d\tau \qquad (15.22)$$

which is the expression we sought for ψ' analogous to the Lienard-Wiechart retarded potential in electromagnetic wave theory. In the Born approximation we are interested in the wave function ψ' only in the region far beyond where the potential energy is nonzero, that is, for $r' \gg r$. For this situation a convenient and useful approximation is to let

$$|\mathbf{r}'-\mathbf{r}| = r' - \hat{\mathbf{r}'}\cdot\mathbf{r} \qquad (15.23)$$

where $\hat{\mathbf{r}'}$ is the unit vector in the direction of $\mathbf{r}'$ (see Figure 15.3). The approximation results from neglecting terms of order r^2/r'^2 and higher in the expansion of the law of cosines:

$$|\mathbf{r}'-\mathbf{r}| = \sqrt{r'^2+r^2-2\mathbf{r}'\cdot\mathbf{r}} \simeq r'\left(1-\frac{\mathbf{r}'\cdot\mathbf{r}}{r'^2}\right)$$

In the center-of-mass system of coordinates, the momentum of the elastically scattered particle $\mathbf{k}$ is equal in magnitude to the momentum of the incident particle and, of course, it has the direction $\hat{\mathbf{r}'}$. That is,

$$\mathbf{k} = k_0\hat{\mathbf{r}'} \qquad (15.24)$$

so that from (15.23)

$$k_0|\mathbf{r}'-\mathbf{r}| \simeq k_0 r' - \mathbf{k}\cdot\mathbf{r} \qquad (15.25)$$

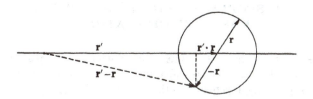

FIGURE 15.3 Illustration of the approximation $|r' - r| = r' - \hat{r}' \cdot r$ which is obtained from the law of cosines for $r \ll r'$. r' is the position of the scattered particle after being scattered in the region $r \sim 0$.

Therefore, ψ' of (15.22) may be written approximately as

$$\psi' = -\frac{1}{4\pi} \frac{e^{ik_0 r'}}{r'} \int U(r) e^{i(k_0 - k) \cdot r} d\tau \qquad (15.26)$$

$$= \frac{e^{ik_0 r}}{r'} f(\theta)$$

where $|r' - r|$ in the denominator has been put equal to r',

$$f(\theta) = -\frac{1}{2\pi} \frac{m}{\hbar^2} \int V(r) e^{(i/\hbar) q \cdot r} d\tau \qquad (15.27)$$

and

$$q = \hbar(k_0 - k) = (p_0 - p) \qquad (15.28)$$

where p is the momentum of the particle after scattering. q is thus the momentum transfer of the scattered particle, that is, the change in momentum of the particle due to the collision. *Note that $f(\theta)$ is the Fourier transform of the potential energy of the particle interaction.* Thus the result far from the scattering center (15.27) depends only on q and is completely independent of any other property of the particle. If $V(r) = V(r)$, a central (angle-independent) potential, $f(\theta)$ is a function of q only, that is, $f(\theta)$ is a function of $|p_0 - p|$. Since $p = p_0$, the change in momentum is

$$|p_0 - p| = 2p \sin (\theta/2) \qquad (15.29)$$

The particle mass used in the equations throughout this section is the reduced mass $m_1 m_2/(m_1 + m_2)$. If neutrons are scattered from free protons, for example, the reduced mass is one-half the nucleon mass. If the protons are bound chemically in an experimental target of polyethylene, however, the effective mass of the target is infinite, i.e., $m_t \gg m_n$ for low energy neutron scattering experiments, and the reduced mass is the neutron mass. This difference of a factor of two in $f(\theta)$ becomes a factor of four in $\sigma(\theta)$ and results in considerably less scattering of neutrons from free protons than from hydrogen compounds.

15.4 SIMPLE APPLICATIONS OF THE BORN APPROXIMATION

1. SQUARE WELL   If the particle interaction potential is a square well and angle independent, which is a reasonable approximation to low-energy nuclear scattering, Eq. (15.27), together with the general result $f(q) = f(\theta)$ from Eqs. (15.28) and (15.29), yields

$$f(\theta) = -\frac{m}{\hbar^2} \int_0^a \int_0^\pi V_0 e^{(i/\hbar)qr\cos\varphi} r^2 \sin\varphi \, d\varphi \, dr$$

$$= -2\frac{mV_0}{\hbar q} \int \left(\sin\frac{qr}{\hbar}\right) r \, dr$$

$$= \frac{2mV_0\hbar}{q^3} \left[\frac{qa}{\hbar}\cos\frac{qa}{\hbar} - \sin\frac{qa}{\hbar}\right] \tag{15.30}$$

Although of course low-energy nuclear scattering does not meet the validity condition of the Born approximation ($V' \ll E$), Eq. (15.30) is useful in giving a rough idea of the scattering of medium-energy neutrons. Since $q = 2p$ $\sin(\theta/2)$ from Eq. (15.29), $f(\theta)$ in Eq. (15.30) may be expressed directly in terms of θ, but it is easier to leave the result in this simple form. Note that $f(\theta)$ is a rapidly decreasing function of q in Eq. (15.30) and is close to a minimum at $\theta_{cm} = \pi$ representing backward scattering in the center-of-mass system. The differential scattering cross section $|f(\theta)|^2$ for a square well potential is shown as a function of q in Figure 15.4. For large values of $(qa/\hbar)$, $|f(\theta)|^2 \sim q^{-4} \cos^2(qa/\hbar)$.

2. A COULOMB POTENTIAL   It is unnecessary to give a quantum mechanical treatment to the Coulomb potential for dissimilar particles of medium or low energy, because, in contrast to the short-range interaction potentials considered thus far, the Coulomb potential, at distances greater than the classical distance of closest approach within the Coulomb and centripetal barriers, changes insignificantly within a particle wavelength. Very little scattering interaction occurs in these classically forbidden regions of space. Therefore the classical calculation by Rutherford is valid. It is always illuminating, however, to see how one may obtain classical results by quantum mechanical theory, and a Coulomb potential provides a nice illustration of the scattering to be expected for long-range, $1/r$-dependent potentials. Equation (15.27), the Born approximation for this potential, yields the result

$$f(\theta) = -\frac{mZze^2}{2\pi\hbar^2} \int_0^\infty e^{(i/\hbar)\mathbf{q}\cdot\mathbf{r}}\frac{1}{r} d\tau = \frac{-2mZze^2}{q^2} \tag{15.31}$$

whence, by (15.9) and (15.29),

$$\sigma(\theta)d\omega = \left(\frac{mZze^2}{2p^2}\right)^2 \frac{1}{\sin^4(\theta/2)} d\omega \tag{15.32}$$

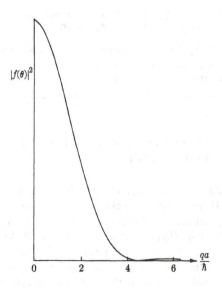

FIGURE 15.4 Graph of $|f(\theta)|^2$ for a square well potential as a function of $qa/\hbar$.

which is the classical Rutherford formula, in which z is the charge on the incident ion or electron and Z the atomic number of the target atom. Note that in general, $f(\theta)$ is a monotonically decreasing function of the momentum exchange q, depending inversely on q^2, even for a rather long-range potential. There are interesting quantum mechanical effects occurring in the Coulombic collision of energetic, indistinguishable particles, which will be treated later in this chapter.

3. CONSTANT POTENTIAL If $V = V_0$, constant over all space, the momentum transform Eq. (15.27) is a delta function of q, $\delta(q)$; thus $f(\theta) = f(q) = 0$ unless $q = 0$, meaning that therefore $q = 0$ in this example, i.e., no momentum transfer takes place: the incident plane wave is transmitted undeflected, as would be expected in a force-free ($\partial V/\partial x = 0$) situation.

4. BILLIARD BALL COLLISION The interaction between two small hard spheres is essentially a delta function of space. The potential is a square well of nearly infinite height and almost negligible extent. The momentum transform Eq. (15.27) is a constant independent of q and therefore independent of scattering angle θ. Isotropic scattering is approached in a square well at low energies where $V \gg E$ and the particle wavelength $\lambda \gg a$, the range of the potential. However, in this case for which $V \gg E$, the Born approximation

becomes rough. The isotropy is an invariable result of $\lambda \gg a$ (see Figure 15.4 for which $ga/\hbar = (2a/\lambdabar) \sin \theta/2 \ll 1$ for all θ).

Problem 15.1: Why do all instances of inelastic scattering (scattering in which the scattered particle loses energy) include elastic scattering? When elastic scattering is taken into account, what is the maximum total cross section in terms of the geometrical cross section of the target?

15.5 QUALITATIVE FEATURES OF THE BORN APPROXIMATION

One can see from these simple problems, and from the discussion of wave packets in Chapter II (particularly the example of the Fourier transform of a Gaussian function), that the differential scattering cross section is, in general, a decreasing function of q. That is, scattering is most likely to occur in directions requiring small change in momentum. Also, since $q = 2p \sin (\theta/2)$ for central potentials, the scattering may be expected to become more pronounced in the forward direction with increasing energy. These are very striking and fundamental features of scattering due to simple, central field interactions. The latter feature is so common and characteristic of all scattering observations that an exception is an exciting and vitally informative discovery in itself. For example, the experimental observation that neutrons scatter backward from protons slightly more than they scatter forward proved that the nuclear force is not a simple central force field. Subsequently this field was found to be best represented by a Serber potential which exchanges half the particles. That is, the nuclear potential is represented most simply as $[\frac{1}{2}(1 + P)]V(r)$ where P is the parity operator and $V(r)$ is a simple central potential. The effect of the parity operator is that an incident neutron emerges as a proton (presumably corresponding to absorption of a positively charged meson from the proton or loss of a negative meson to the proton). In forward exchange scattering, an incident neutron would cause a proton to be scattered at small angles and a neutron at large angeles.

15.6 ATOMIC SCATTERING BY THE BORN APPROXIMATION

The potential between an atomic scatterer and an incident ion or electron is that between a negatively screened nuclear positive charge and the charge of the incident particle, and may be written by crudely approximating the atom with its many electrons as a simple potential due to only one particle

$$V(r) = ze\left(\frac{Ze}{r} - e\varphi\right) \tag{15.33}$$

where z and Z are as defined for (15.31) and $-e\varphi$ is the potential resulting from the presence of the target atom electrons. From Poisson's equation in electromagnetic theory, we obtain a differential equation relating the potential to the charge density from which the potential may be computed,

$$\nabla^2(-e\varphi) = -4\pi\rho = -4\pi(-e)n(\mathbf{r})$$

where $n(\mathbf{r})$ is the electron-number density (electrons/cc). The nuclear potential is trivial and we have already obtained it, Eq. (15.31). In order to compute the scattering cross section of an incident charge off an atom we need the atomic shape or the form of the distribution of the atomic electrons. To be specific, we need the atom form factor $F(q)$, which is defined by the integral taken over all space

$$F(q) = \frac{q^2}{4\pi} \int \varphi(r)e^{i\mathbf{q}\cdot\mathbf{r}}d\tau \tag{15.34}$$

Let $\chi = -e^{i\mathbf{q}\cdot\mathbf{r}}$; then $\nabla^2\chi = q^2 e^{i\mathbf{q}\cdot\mathbf{r}}$. We now use Green's theorem,

$$\int \nabla^2\chi d\tau = \int \nabla\cdot(\nabla\chi)d\tau = \int \nabla\chi\cdot d\mathbf{A} = \int \frac{\partial\chi}{\partial n}dA$$

where dA is a differential element of surface area and n is in the direction normal to the surface. Since the volume integral is taken over all space, the surface integral over the surface which encloses this volume is that of a sphere of infinite radius. Integrating $F(q)$ twice by parts, we obtain

$$4\pi F(q) = \int \varphi\nabla^2\chi d\tau = \int \left(\varphi\frac{\partial\chi}{\partial n} - \chi\frac{\partial\varphi}{\partial n}\right)dA + \int \chi\nabla^2\varphi d\tau$$

The surface integral is zero in the limit $r = \infty$, owing to the radial dependences of χ and φ, the atomic potential which must be evaluated on the surface of infinite radius. Therefore

$$F(q) = \int n(r)e^{i\mathbf{q}\cdot\mathbf{r}}d\tau \tag{15.35}$$

$F(q)$ may be calculated by use of various atomic electron-density approximations such as the simple, inverse power of r-dependent Fermi-Thomas approximation (see Mott and Snedden, Slater, or Pauling and Wilson cited in bibliography of this chapter). In this way we obtain the result

$$f(\theta) = 2me^2 z \frac{Z - F(q)}{q^2}$$

whence

$$\sigma(\theta) = \left(\frac{me^2z}{2p^2}\right)^2 \frac{|Z - F(q)|^2}{\sin^4(\theta/2)} d\omega \tag{15.36}$$

If q is large, corresponding to large-angle scattering and close approach to the nucleus, Eq. (15.35) gives a nearly zero result due to phase cancellation in the integral over space. Hence if $qa \gg 1$ (where a is the atomic radius), simple Rutherford scattering results. The atomic radius for all atoms is $a \sim a_0 Z^{-1/3}$

where a_0 is the Bohr radius of the hydrogen atom, 0.53 Å. For an alpha particle of 5 Mev, incident on an atom of radius a, for example, $qa \sim 10^4$. As $q \rightarrow 0$,

$$Z - F(q) \rightarrow q^2$$

even though Z alone is a constant. Hence the scattering remains finite. Without screening by the atomic electrons, the range of the Coulomb potential is infinite and $\sigma(\theta)$ would diverge as $\theta \rightarrow 0$ ($q \rightarrow 0$).

Problem 15.2: If alpha particles are scattered from a lithium nucleus of radius 10^{-12} cm, what momentum exchange will occur at 30° if the initial alpha particle energy is 5.3 Mev in the center-of-mass system? in the laboratory system?

Problem 15.3: Calculate alpha particle scattering by a lead atom. The atomic potential may be approximated as $(zZe^2/r)e^{-r/r_0}$ where $r_0 \sim$ 1 Å.

Problem 15.4: Use the Born approximation to calculate neutron-proton scattering if their interaction potential is $V_0(e^{-(\hbar/mc)r})/r$ where m is the mass of the π meson (273 electron masses). Calculate the scattering resulting if the potential were $V_0 e^{-\hbar r/mc}$.

15.7 SCATTERING BY IDENTICAL PARTICLES

If a particle is scattered into angle θ by an identical target particle, which simultaneously scatters into angle $\pi - \theta$, the event cannot be distinguished from the similar event in which the target particle emerges at angle θ (see Figure 15.5). Therefore, one might think that the total expected differential scattering would be not $\sigma(\theta) = |f(\theta)|^2$ but (neglecting spin interactions)

$$\sigma(\theta) \, d\omega = |f(\theta) + f(\pi - \theta)|^2 \, d\omega \qquad (15.37)$$

This result is incomplete, however, if the two particles each have odd half-integral spin, inasmuch as then they must be in antisymmetric states. There

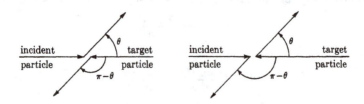

FIGURE 15.5 Illustration of two scattering events indistinguishable in the center-of-mass coordinate system.

are two possibilities: Either the particles are in antisymmetric spin states and Eq. (15.37), which represents scattering in a symmetric space state, is correct; or the particles are in a symmetric spin state and the scattering is given by the antisymmetric space function

$$\sigma(\theta) \, d\omega = |f(\theta) - f(\pi - \theta)|^2 \, d\omega \qquad (15.38)$$

Similar considerations apply to particles of integer spin which must be in a combined spin-space symmetric state. The wave amplitudes for the two situations of (15.37, 15.38) cannot both be present to interfere with one another; for since the spin states are quantized, the complete wave function for the two particles in any individual scattering event must be either completely symmetric or antisymmetric in spin. Hence, the total space state must be correspondingly antisymmetric or symmetric, and interference cannot occur between states of different space symmetry. (If the particle spin symmetry is changed in the interaction, however, the space symmetry of the final states may differ from the space symmetry of the initial states. All the foregoing is valid only if there is no spin interaction.)

If electrons whose spins are random are scattered, there will be three times as many scatterings in the symmetric spin states $\alpha\alpha$, $\beta\beta$, and $(\alpha\beta + \alpha\beta)$ as in the antisymmetric state $(\alpha\beta - \beta\alpha)$. Therefore, assuming no electron spin interaction, the total scattering will be given by

$$\sigma(\theta) \, d\omega = \{\tfrac{3}{4} | f(\theta) - f(\pi - \theta)|^2 + \tfrac{1}{4} | f(\theta) + f(\pi - \theta)|^2\} \, d\omega \qquad (15.39)$$

Mott has derived electron-electron scattering exactly and obtained an additional factor in the interference term:

$$\sigma(\theta) d\omega = \left(\frac{e^2}{mv^2}\right)^2 \left\{ \frac{1}{\sin^4 \theta/2} + \frac{1}{\cos^4 \theta/2} - \frac{\cos\left[\frac{2e^2}{\hbar v} \log \tan \theta/2\right]}{\sin^2 \theta/2 \cos^2 \theta/2} \right\} d\omega$$

$$(15.40)$$

15.8 THE METHOD OF PARTIAL WAVES

Though convenient and simple to apply, the Born approximation is invalid for many problems of physical interest. Relativistic effects invalidate the assumption of static potentials at high energies; and at low energies it is incorrect to treat the potential energy of interaction as a small perturbation on the total energy of the interacting particles. The exact general treatment of the collision problem at low and medium energies where the static potential applies is based on phase shifts of the radial wave function. Since we are concerned only with central (angle-independent) potential in this introductory account of scattering theory, we will be concerned primarily with the radial wave function and its wave equation. At large distances from the scattering center the radial wave equation is sinusoidal independently of whether or not a scattering potential exists. If a scattering center is present, however, the phase of the argument of the sinusoidal function is shifted by a constant amount over that for a free particle.

In a central potential where $V(\mathbf{r}) = V(r)$, the wave equation is easily transformed by the substitution

$$\psi_l(\mathbf{r}) = \frac{u_l(r)}{r} Y_l^m(\theta, \varphi) \tag{15.41}$$

to

$$\frac{d^2 u_l}{dr^2} + \frac{2m}{\hbar^2}[E - V(r)]u_l - \frac{l(l+1)}{r^2}u_l = 0 \tag{15.42}$$

where m is, as usual, the reduced particle mass $m_1 m_2/(m_1 + m_2)$, and the Y_l^m are, as before, the spherical harmonic wave functions with quantum numbers l and m. The solutions to Eq. (15.42) are subject to the usual boundary conditions: the ψ's are finite at the origin and zero at infinity so that

$$u(0) = 0, \quad u(\infty) = 0 \tag{15.43}$$

The effects of spin and symmetry on the scattering, discussed in the preceding section, may be included by writing the free-particle wave functions for odd or even l, whichever is appropriate. That is, in (15.42) one could make $V = 0$ for odd or even l, whichever is appropriate. Since the spatial state is specified by $(-1)^l$ as antisymmetric for $(-1)^l = -1$ and symmetric for $(-1)^l = +1$, protons in symmetric spin states, for example, would experience no interaction with one another in even l states.

The expected asymptotic behavior of ψ far beyond the scattering center (see Section 15.2) is a sum of a plane wave and a spherically outgoing wave,

$$\psi(r, \theta, \varphi) \longrightarrow A\left[e^{ikz} + \frac{e^{ikr}}{r}f(\theta)\right] \tag{15.44}$$

Hence, we expect that at large r, $u_l(r)$ will assume the form

$$u_l(r) = v_l(r)e^{\pm ikr} \tag{15.45}$$

where $k = p/\hbar = \sqrt{2mE/\hbar^2}$, and $v_l(r)$ is a slowly varying function of r as we will now show. Substituting (15.45) into (15.42) we obtain

$$v_l'' \pm 2ikv_l' = \left[\frac{2m}{\hbar^2}V(r) + \frac{l(l+1)}{r^2}\right]v_l \tag{15.46}$$

If v_l is slowly varying, v_l'' may be neglected and the rest of Eq. (15.46) may be solved by quadrature:

$$\pm 2ik \log v_l = \int^r \left[\frac{2m}{\hbar^2}V(r) + \frac{l(l+1)}{r^2}\right]dr \tag{15.47}$$

Since $V(r)$ approaches zero as r goes to infinity, v_l does indeed approach a constant for large r, thus justifying our assumption that v_l is slowly varying at large r. Therefore, the solutions to Eq. (15.42) at large particle separations become

$$u_l(r) = A'_l \sin(kr + \delta'_l) \qquad (15.48)$$

where δ'_l is a constant phase factor in the solution—one of the two arbitrary constants appearing in the solutions of all second-order differential equations. Since these constants are determined by the boundary conditions, we might expect, as is in fact the case, that δ'_l depends on the scattering potential; furthermore, the scattering cross section may be put in terms of δ'_l. A direct connection between the scattering cross section and the scattering potential does not exist in the exact theory. *The connection may be made only through the phase shift of the radial wave function δ_l.*

The complete wave function may be expanded in terms of the orthonormal set of radial and angular wave functions,

$$\psi = \sum_{l=0}^{\infty} \frac{u_l(r)}{r} P_l(\cos\theta) \qquad (15.49)$$

where $P_l(\cos\theta)$ is the Legendre polynomial of order l.

A slight modification of the theory is necessary in problems where $V(r)$ decreases with r faster than does $l(l+1)/r^2$. In this case the solution in the intermediate region where $V(r)$ is zero but $l(l+1)/r^2$ is appreciable is the solution to

$$\frac{d^2 u_l}{dr^2} + k^2 u_l - \frac{l(l+1)}{r^2} u_l = 0 \qquad (15.50)$$

namely,

$$u_l(r) = \sqrt{\frac{\pi r}{2k}} A_l[(\cos\delta_l)J_{l+1/2}(kr) - (\sin\delta_l)N_{l+1/2}(kr)] \qquad (15.51)$$

which is the radial solution to the spherical square well potential discussed in Chapter VIII. $A_l \cos\delta_l$ and $A_l \sin\delta_l$ are two undetermined constants which are part of the solution to any second-order differential equation; δ_l must be fixed by the boundary conditions. For large r, the asymptotic form of the spheroidal Bessel and Neumann functions may be used, so that $u_l(r)$ becomes

$$u_l(r) = k^{-1} A_l \sin\left(kr - \frac{l\pi}{2} + \delta_l\right) \qquad (15.52)$$

which again is the free-particle wave function just as Eq. (15.48) is, with the difference that here the phase of the wave has been shifted by $[-(l\pi/2) + \delta_l]$ intsead of by the phase shift computed by neglecting the centrifugal potential, δ'_l. This solution (Eqs. 15.48, 15.52) includes both the incident plane wave and the scattered spherical wave. The complete solution requires that one take the sum of $u_l(r)$ over all l. The scattering cross section is found from the spherical wave contribution. To evaluate the scattering cross section brought about by the scattering potential we must compute how much of (15.52) is due to the scattering potential. That is, we are interested only in the difference between the sum over l of (15.41) with (15.52) and the free-particle wave functions. To find this difference we must first express the free-particle plane wave function e^{ikz} in terms of functions similar to (15.41) and (15.52).

The plane wave e^{ikz} may be expanded in terms of the complete orthonormal set of radial Bessel functions of order $\frac{1}{2}$ and the Legendre polynomials. The expansion is

$$e^{ikz} = e^{ikr\cos\theta} = \sum_{l=0}^{\infty} (2l + 1)i^l \sqrt{\frac{\pi}{2kr}} J_{l+1/2}(kr) P_l(\cos\theta) \qquad (15.53)$$

The asymptotic form of the Bessel function may be used at large r so that

$$e^{ikz} = \sum_{l=0}^{\infty} (2l + 1)i^l \frac{1}{kr} \sin\left(kr - \frac{l\pi}{2}\right) P_l(\cos\theta) \qquad (15.53a)$$

Therefore, by substituting Eqs. (15.52) and (15.53a) into Eq. (15.44), we obtain

$$\sum_{l=0}^{\infty} (2l + 1)i^l \frac{1}{kr} \sin\left(kr - \frac{l\pi}{2}\right) P_l(\cos\theta) + \frac{1}{r} f(\theta) e^{ikr}$$

$$= \sum_{l=0}^{\infty} A_l \frac{1}{kr} \sin\left(kr - \frac{l\pi}{2} + \delta_l\right) P_l(\cos\theta) \qquad (15.54)$$

which is the desired relation between $f(\theta)$ and A_l, δ_l we sought. Because of the term in e^{ikr}, it is useful to write the sine functions in terms of e^{+ikr} and e^{-ikr} so that each term in Eq. (15.54) is multiplied by either e^{ikr} or e^{-ikr} ($\sin kr = (e^{ikr} - e^{-ikr})/2i$). The equation is true for all values of k and large r, so that the equation must be true independently for the coefficients of e^{+ikr} and e^{-ikr}. Equating the coefficients of e^{+ikr}, we obtain

$$2ikf(\theta) + \sum_{l=0}^{\infty} (2l + 1)i^l e^{-il\pi/2} P_l(\cos\theta) = \sum_{l=0}^{\infty} A_l e^{i\delta_l - (il\pi/2)} P_l(\cos\theta) \qquad (15.55)$$

and equating the coefficients of e^{-ikr} we obtain

$$\sum_{l=0}^{\infty} (2l + 1)i^l e^{il\pi/2} P_l(\cos\theta) = \sum_{l=0}^{\infty} A_l e^{-i\delta_l + (il\pi/2)} P_l(\cos\theta) \qquad (15.56)$$

Multiplying both sides of Eq. (15.56) by the particular Legendre polynomial $P_n(\cos\theta)$ and integrating over all angles, we find that because of the orthogonality of the P_l's,

$$A_n = (2n + 1)i^n e^{i\delta_n} \qquad (15.57)$$

Substituting this result into Eq. (15.55), we may rewrite Eqs. (15.7) and (15.9) as

$$f(\theta) = \frac{1}{2ik} \sum_{l=0}^{\infty} \{-(2l + 1)i^l e^{-il\pi/2} P_l(\cos\theta) + (2l + 1)i^l e^{i\delta_l} e^{i\delta_l - (il\pi/2)} P_l(\cos\theta)\}$$

$$= \frac{1}{2ik} \sum_{l=0}^{\infty} (2l + 1)(e^{2i\delta_l} - 1) P_l(\cos\theta) \qquad (15.58)$$

and

$$\sigma(\theta) = |f(\theta)|^2 = \frac{1}{k^2} \left| \sum_{l=0}^{\infty} (2l+1)e^{i\delta_l}(\sin \delta_l)P_l(\cos \theta) \right|^2 \qquad (15.59)$$

which is the desired connection we have sought between the phase shift and the scattering cross section. Equation (15.59) is Lord Rayleigh's formula for wave diffraction, which was rederived for quantum mechanics by H. Faxen and J. Holtzmark. Note that waves of different orbital momenta interfere with one another. This is an exact result for $\sigma(\theta)$; thus the scattering problem is reduced to finding the appropriate phase shift for each angular momentum. δ_l is the phase shift in the radial wave function (15.52) for a wave in a potential relative to a wave in no potential. (It is *not* a space angle related to some direction in the scattering process!) δ_l is just a way of expressing the constants in the general solution which must be fitted to particular boundary conditions.

The total scattering cross section is obtained by integrating Eq. (15.59) over all angles (note that the interference terms integrate to zero because of the orthogonality of the $P_l(\cos \theta)$):

$$\sigma = 2\pi \int_0^\pi \sigma(\theta) \sin \theta \, d\theta = \frac{4\pi}{k^2} \sum_{l=0}^{\infty} (2l+1) \sin^2 \delta_l$$

$$= 4\pi\lambda^2 \sum_{l=0}^{\infty} (2l+1) \sin^2 \delta_l \qquad (15.60)$$

In order to calculate the scattering cross section, the solution to the wave equation within the range of the scattering potential u_l must be calculated and then fitted to the free-particle wave function (15.48) or (15.52) by an appropriate choice of δ_l. For example, if necessary, u_l could be integrated numerically for $r < a$, where a is the range of the potential, and then at $r = a$, u_l and its first derivative could be matched to Eq. (15.52) and its derivative, thus furnishing the boundary conditions needed to fix δ_l.

The radial wave function $u_l(r)$ at small r is an exponential function decreasing as $r \rightarrow 0$ for repulsive potentials, including the centripetal potential for which $u_l(r)$ behaves as r^{l+1} near $r = 0$; at large r, $u_l(r)$ behaves as $\sin[kr - (l\pi/2) + \delta_l]$. The various functions for a typical problem are illustrated in Figure 15.6 for $l = 0$. At very low energies, for which $k \rightarrow 0$ and $\lambda \rightarrow \infty$, the intercept of the tangent to the slope of the wave function at the edge of the potential ($r = a$), b, illustrated in Figure 15.6, is approximately equal to δ_0/k and is a constant independent of E or k since $\delta_0 \sim k$. The terms for $l \neq 0$ in the summation over l in the cross section (15.59) are negligible and do not contribute to the sum for $k \ll a^{-1}$, since the penetration through the angular momentum barrier $\hbar^2 l(l+1)/r^2$ is small, as will be discussed in the following sections. For this very important low-energy case, only δ_0 need be considered, that is, $\delta_0 = kb$ and approaches zero with k. Thus,

$$\sigma \xrightarrow[k \rightarrow 0]{\text{lim}} \frac{4\pi}{k^2} k^2 b^2 = 4\pi b^2 \qquad (15.61)$$

b has the dimensions of a length and is called the *Fermi scattering length*.

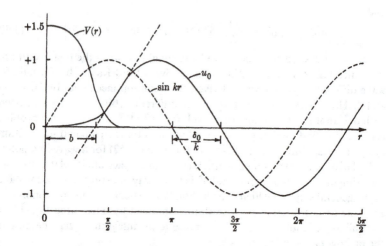

FIGURE 15.6 The wave functions of a free particle and the wave function $u_l(r)$ of a particle influenced by a repulsive potential $V(r)$. $V(r)$ and the radial function u_l for $l = 0$ are shown in solid lines. The dashed sine curve is for the undisturbed free-particle wave function, $\sin kr$. The problem illustrated here has a negative phase shift $-\delta_l$, although conceivably for other potentials or higher energies (and therefore short free-particle wavelengths compared to the range of the potential well) the phase shift could just as well be positive.

15.9 SCATTERING BY PARTICLE WAVES OF HIGH ORBITAL ANGULAR MOMENTUM

For large angular momentum quantum numbers we would expect a classical result. For this reason and for greater physical insight it is worthwhile to examine the classical theory of scattering.

Classically, the total scattering cross section of a small incident particle would be πa^2 where a is the range of the interaction between incident particle and target object. The angular momentum of the incident particle with respect to the center of coordinates located in the target particle would be $\mathbf{r} \times \mathbf{p}$. If angular momentum were to be conserved, no particle with linear momentum p could be deflected which had an angular momentum $l\hbar$ greater than ap. That is, classically, for deflection in the potential field of the target particle we must have $l\hbar \leqslant ap(l < a/\lambda)$. Similarly, in quantum mechanics the particle wave functions for the radial wave equation for large l,

$$\frac{\partial^2 u_l}{\partial r^2} + \frac{2m}{\hbar^2}[E - V(r)]u_l - \frac{l(l+1)}{r^2}u_l = 0 \qquad (15.62)$$

tail off exponentially at small r and behave as r^{l+1} for very small r, and hence, although there may be some penetration into the potential region and therefore scattering of particles with large l, it is very small. That is, if $V(r) \ll E$ for $r > a$, then $\delta_l \ll 1$ for all $l \gg a/\lambda$. Nevertheless, although the scattering for $l \gg a/\lambda$ is small, it is of interest and the foregoing suggests a simple perturbation theory calculation for it.

The radial wave equation for a free particle in spherical coordinates is Eq. (15.50) and that for a particle in a potential is Eq. (15.42). Writing v_l for u_l as the free-particle wave function (see Eq. 15.50), we have

$$\frac{d^2v_l}{dr^2} + \frac{2m}{\hbar^2}Ev_l - \frac{l(l+1)}{r^2}v_l = 0 \qquad (15.63)$$

$$\frac{d^2u_l}{dr^2} + \frac{2m}{\hbar^2}Eu_l - \frac{2m}{\hbar^2}V(r)u_l - \frac{l(l+1)}{r^2}u_l = 0 \qquad (15.64)$$

Multiplying Eq. (15.63) by u_l and Eq. (15.64) by $-v_l$ and adding the results, we obtain

$$u_l\frac{d^2v_l}{dr^2} - v_l\frac{d^2u_l}{dr^2} + \frac{2m}{\hbar^2}V(r)v_lu_l = 0 \qquad (15.65)$$

Integrating this equation over r we obtain

$$\left[u_l\frac{dv_l}{dr} - v_l\frac{du_l}{dr}\right]_0^R + \frac{2m}{\hbar^2}\int_0^R V(r)u_lv_ldr = 0 \qquad (15.65a)$$

If the upper limit $R \gg a$, where R is taken arbitrarily far out from a, then the left-hand integrated term is independent of R, as we shall see. Evaluation at the lower limit gives 0, inasmuch as u_l and v_l are both zero at $r = 0$. At the upper limit, $v_l = \sin[kR - (\pi l/2)]$ and $u_l = \sin[kR - (\pi l/2) + \delta_l]$. The integrated expression in (15.65a) therefore becomes

$$\sin\left(kR - \frac{\pi l}{2} + \delta_l\right)k\cos\left(kR - \frac{\pi l}{2}\right) - \sin\left(kR - \frac{\pi l}{2}\right)k\cos\left(kR - \frac{\pi l}{2} + \delta_l\right)$$

$$= k\sin\left[\left(kR - \frac{\pi l}{2} + \delta_l\right) - \left(kR - \frac{\pi l}{2}\right)\right]$$

$$= k\sin\delta_l \qquad (15.66)$$

Therefore, substituting in (15.65a) gives

$$k\sin\delta_l = -\frac{2m}{\hbar^2}\int_0^a u_lv_lV(r)dr \qquad (15.67)$$

If δ_l is very small so that $v_l \sim u_l$, then the unknown δ_l in u_l may be neglected, so that Eq. (15.67) becomes

$$k\sin\delta_l = -\frac{2m}{\hbar^2}\int_0^a v_l^2V(r)dr \qquad (15.68)$$

which is just the result that would be obtained from perturbation theory.

If $a \gg k^{-1}$ and $V(r)$ is slowly varying, the average value of v_l^2, $\frac{1}{2}$, may be used in Eq. (15.68). At large r,

$$\int v_l^2 V(r) dr \sim \tfrac{1}{2} \int V dr$$

and unless $V < (\text{const.}/r)$ at large r, the integral will diverge. Hence, this method will not work for the Coulomb potential, which has been solved exactly by other means.

For particle energies sufficiently high that several l would contribute classically to the scattering cross section (i.e., $ap/l \gg \hbar$, and the potential range is several wavelengths in radius), the total scattering cross section assumes a particularly simple form. In the classical limit, δ_l is assumed large for $l < a/\lambda$ and negligible for $l > a/\lambda$. Thus the total cross section, Eq. (15.60), becomes

$$\sigma = 4\pi\lambda^2 \sum_{l=0}^{a/\lambda} (2l+1)\sin^2 \delta_l$$

$$\simeq 4\pi\lambda^2 \sum_{l=0}^{a/\lambda} (2l+1)\tfrac{1}{2}$$

$$= 2\pi a^2 \qquad\qquad\qquad (15.60a)$$

The geometrical cross section is πa^2. Where does the extra πa^2 come from? It is the shadow scattering—the interference of the absorbed wave with the transmitted wave—and it occurs in optics in diffraction phenomena also. Thus πa^2 from the reemitted wave plus another πa^2 due to shadow scattering add up to $2\pi a^2$.

Problem 15.5: Calculate the phase shifts δ_0 for a square well with the following conditions:

(a) ka arbitrary and V_0 very small but positive (that is, an attractive potential).
(b) ka arbitrary and V_0 very small but repulsive.
(c) What is the precise expression for the condition on V_0 which justifies the approximations used in (a) and (b)? What is V_0 less than?
(d) How would you solve this problem if V_0 were -10 Mev (repulsive!)? if E were 100 kev and a were 10^{-12} cm? for a neutron being scattered from an infinitely heavy nucleus? For the latter two conditions include $l > 0$ and obtain $\sigma(\theta)$ within $\sim 5\%$.

Problem 15.6: Calculate the scattering cross section for low-energy particles ($E = 0.200$ Mev) incident on a square well $V_0 = -20$ Mev, $a = 10^{-12}$ cm. How much does the $l = 1$ wave contribute to the differential cross section if $E = 0.200$ Mev?

Problem 15.7: Calculate the scattering cross section for low-energy particles incident on a square well as a function of particle energy and the depth of the well. Assume that only s-wave ($l = 0$) scattering occurs at this energy. For what values of V_0 and E is the scattering a maximum, and why? If $\delta_0 = 180°$, what is the scattering cross section? (This is essentially the Ramsauer effect discussed in Section 5.7.)

Problem 15.8: (a) Calculate approximately the contribution of the $l = 1, 2,$ and 3 states to the scattering of 10 Mev neutrons by protons if their interaction potential is given by $V = -V_0 e^{-\hbar/mc}$, where m is the mass of the π meson (273 electron masses). Ignore any dependence of spin on nuclear forces. (b) Further suppose $\delta_0 = (\pi/2)$, $\delta_1 = (\pi/2)$, and $\delta_l = 0$ for $l > 1$. What will be the fractional contributions of the two waves to the scattering? (c) How may δ_0 and δ_1 be separately determined experimentally?

Problem 15.9: Calculate the differential cross section for low-energy neutron-neutron scattering, assuming a square well attractive force. Neglect waves with $l > 2$.

Problem 15.10: Calculate the differential scattering cross section for low-energy alpha particles incident on helium nuclei, assuming a square well attractive force. Neglect waves with $l > 2$.

15.10 NEUTRON-PROTON SCATTERING

An interesting and illustrative example of the usefulness of the method of partial waves is furnished by neutron-proton scattering. We recall that in Section 6.2 the problem of finding the ground state binding energy of the deuteron was considered under the assumption that the potential describing the nuclear interaction was given by a square well. There we found that the information obtained about the nuclear force from a knowledge of the binding energy was incomplete in that the detailed shape of the nuclear potential could not be determined. Further information concerning the nuclear force can be obtained from neutron-proton scattering experiments.

Before discussing the neutron-proton system in an unbound state (scattering), it is useful to summarize the properties of the deuteron as determined from the results of various experiments:

Binding Energy	$= 2.2256 \pm 0.0015$ Mev
Spin	$I = 1$
Magnetic Moment	$\mu_d = 0.85739\ \mu_N$
Electric Quadrupole Moment	$= 2.82 \times 10^{-27}$ cm^2

Because $I = 1$, we know the deuteron is in a triplet spin state and it is reasonable to expect that the ground state of the deuteron is mainly a $^3S_1$ state. (The spectroscopic notation of nuclear physics is patterned after the notation used in atomic physics.) This choice is also supported by the value of the deuteron magnetic moment, which is very close to what one obtains by simply adding the individual magnetic moments of a neutron and proton having parallel spins. To wit

$$\mu_p = 2.7927 \ \mu_N$$
$$\mu_n = -1.9135 \ \mu_N$$
$$\text{Sum} = 0.8892 \ \mu_N \approx \mu_d$$

(The difference can be attributed to a slight admixture of the $^3D_1$ state.)

The absence of a bound $^1S_0$ state shows the nuclear forces are spin-dependent. The fact that the deuteron has a nonvanishing electric quadrupole moment provides evidence that the probability density is not spherically symmetric and hence the nuclear force must also have a noncentral component. However, since the deuteron ground state is mainly $^3S_1$, we should expect the noncentral component to be small compared with the central part.

With the foregoing information in mind, we now return to the problem of neutron-proton scattering. Since the de Broglie wavelength of incident neutrons is much greater than the range of the nuclear forces for energies up to 10 Mev or so, we expect that most interactions will take place with neutrons having zero angular momentum with respect to the scattering center. Consequently, we shall consider only the case of $l = 0$, or S-wave scattering, with a square well potential of radius a.

The function $U_i(r)$ (Eq. 15.52) in the region outside the potential well now becomes for a nucleon of energy E and mass M (note that $M \approx 2\mu$ where μ is the reduced mass for the neutron-proton system)

$$U_{\text{II}}(r) = \frac{B_0 \sin (k_2 r + \delta_0)}{k_2} \quad \text{with } k_2 = \sqrt{\frac{ME}{\hbar^2}}$$

and

$$\psi_{\text{II}} = \frac{B_0 \sin (k_2 r + \delta_0)}{k_2 r}$$

From Eq. (15.60) the total elastic cross section is

$$\sigma = 4\pi \lambda^2 \sin^2 \delta_0$$

Therefore, the problem is to evaluate δ_0. Inside the well ($r < r_0$) the equation satisfied by $U(r)$ is

$$\frac{d^2 U_{\text{I}}}{dr^2} + k_1^2 U_{\text{I}} = 0 \quad \text{where } k_1^2 = \frac{M}{\hbar^2}(E + V_0)$$

$$U_{\text{I}} = A \sin r \sqrt{k_1^2} \quad \text{for } \psi(r) \text{ to be finite as } r \longrightarrow 0$$

$$\psi_{\text{I}} = A \frac{\sin r \sqrt{k_1^2}}{r}$$

at $r = a_0$

$$\psi_I = \psi_{II} \quad \text{and} \quad \psi'_I = \psi'_{II}$$

so

$$A \frac{\sin k_1 a}{r_0} = B_0 \frac{\sin (k_2 a + \delta_0)}{k_2 a}$$

and

$$\frac{A k_1 \cos k_1 a}{a} - A \frac{\sin k_1 r_0}{a^2} = \frac{B_0 k_2 \cos (k_2 a + \delta_0)}{k_2 a} - \frac{B_0 \sin (k_2 a + \delta_0)}{k_2 a^2}$$

Dividing and rearranging yields

$$\tan (k_2 a + \delta_0) = \frac{k_2}{k_1} \tan k_1 a$$

or

$$\tan \left(\frac{\sqrt{ME}}{\hbar} a + \delta_0 \right) = \left(\frac{E}{E + V_0} \right)^{1/2} \tan \left(\frac{\sqrt{M(E + V_0)}}{\hbar} a \right)$$

This equation can be combined with the result for the deuteron bound state (Eq. (6.27)).

$$\tan \left(r_0 \frac{\sqrt{M(V_0 - W)}}{\hbar} \right) = - \frac{\sqrt{V_0 - W}}{W}$$

We now have two independent equations in the two unknowns V_0 and a, since W, δ_0, and E can be obtained from experiment. The equations can be combined to eliminate V_0 for instance, and the resulting agreement between δ_0 and E and a for n-p scattering is found to be good for a limited range of E. At high values of $E (> 10 \text{ Mev})$ the effect of P-wave scattering becomes important and at low energies we must consider the effect of the nucleon spins. As Wigner has pointed out, by choosing one value for the binding energy of the deuteron we have assumed that the triplet and singlet S-wave interactions are equal. A more realistic approach is to consider that an incoming unpolarized beam of neutrons could undergo both singlet and triplet interactions which would not necessarily make equal contributions. The reader should refer to the suggested references at the end of the chapter for further details.

15.11 EFFECTIVE RANGE THEORY

By a modification of Section 15.9 we may easily show that if the range of the interaction potential is very short compared to a wavelength, it is impossible to determine what the range or size of the potential is. This is an expected difficulty, since one can never determine spatial relationships to a precision better than the wavelength of the "radiation" one employs. By recognizing this difficulty, however, we can develop a general theory relating both a scattering cross section or scattering length and an effective potential range for any reasonable potential shape to the observed low-energy experimental phase shifts δ_0. This effective range theory was first considered by Breit, but we will use the subsequent pellucid development by Bethe.

We wish to compare the values of the phase shifts for two different energy values E_1 and E_2 occurring in the wave equations

$$\frac{d^2u_1}{dr^2} + [k_1^2 + W(r)]u_1 = 0$$

$$\frac{d^2u_2}{dr^2} + [k_2^2 + W(r)]u_2 = 0 \qquad (15.69)$$

where $k_1^2 = mE_1/\hbar^2$, $k_2^2 = mE_2/\hbar^2$, and $W(r) = -(m/\hbar^2)V(r)$. Multiplying the upper equation by u_2 and the lower by $-u_1$, and adding, we obtain

$$u^2\frac{d^2u_1}{dr^2} - u_1\frac{d^2u_2}{dr^2} + (k_1^2 - k_2^2)u_1u_2 = 0 \qquad (15.70)$$

Integrating over a dummy variable r' from 0 to r we obtain

$$\left[u_2\frac{du_1}{dr'} - u_1\frac{du_2}{dr'}\right]_0^r = (k_2^2 - k_1^2)\int_0^r u_1u_2\,dr' \qquad (15.71)$$

The lower limit of the integrated expression gives zero but the upper limit is nonzero. For large r,

$$u_1, u_2 \longrightarrow \sin(k_{1,2}r + \delta_{1,2}) \qquad (15.72)$$

where in Eq. (15.72) the $\delta_{1,2}$ denote the phase shifts for the $l = 0$ wave function for states 1 and 2 respectively. For the region beyond the potential range, let ψ_1 and ψ_2 be solutions to the asymptotic differential equations.

$$\frac{d^2\psi_1}{dr^2} + k_1^2\psi_1 = 0 \quad \text{and} \quad \frac{d^2\psi_2}{dr^2} + k_2^2\psi_2 = 0 \qquad (15.73)$$

For convenience, let $\psi_1(0) = \psi_2(0) = 1$; then

$$\psi_1 = \frac{\sin(k_1r + \delta_1)}{\sin\delta_1}, \quad \psi_2 = \frac{\sin(k_2r + \delta_2)}{\sin\delta_2} \qquad (15.74)$$

We want $\psi_{1,2}$ to be those solutions of Eq. (15.73) that agree with $u_{1,2}$ for larger r, that is, the solutions with the same phase shifts $\sin(k_{1,2}r + \delta_{1,2})$. ψ_1 and u_1 must be the same beyond the range of the potential; at $r = 0$, however, $\psi_1(0) \neq 0$, (unlike u) since ψ_1 is not the correct solution to the wave equation near $r = 0$. Proceeding with Eqs. (15.73) as we did with Eqs. (15.64) in obtaining Eq. (15.71), we find

$$\left[\psi_2\frac{d\psi_1}{dr'} - \psi_1\frac{d\psi_2}{dr'}\right]_0^r = (k_2^2 - k_1^2)\int_0^r \psi_1\psi_2\,dr' \qquad (15.75)$$

We know that $u_1 = \psi_1$ and $u_2 = \psi_2$ at large r beyond the range of the potential. Therefore, the left-hand sides of (15.71) and (15.75) are equal for the upper limit r. The lower limit of the left-hand side of (15.71) gives zero. By subtracting Eq. (15.75) from Eq. (15.71) we obtain a result independent of the complicated behavior of the left-hand sides of these equations at large r,

$$-\left[\psi_2\frac{d\psi_1}{dr'} - \psi_1\frac{d\psi_2}{dr'}\right]_{r'=0} = k_2 \cot \delta_2 - k_1 \cot \delta_1$$

$$= (k_2^2 - k_1^2) \int_0^r (\psi_1\psi_2 - u_1u_2)dr' \qquad (15.76)$$

where $k_2 \cot \delta_2 - k_1 \cot \delta_1$ is obtained by substituting (15.74) in the left-hand side of (15.76).

Outside the range of nuclear forces this integral vanishes. If we define the effective range $\rho(E_1, E_2)$ by the equation

$$\rho(E_1, E_2) = 2 \int_0^\infty (\psi_1\psi_2 - u_1u_2)dr \qquad (15.77)$$

Eq. (15.76) becomes

$$k_2 \cot \delta_2 - k_1 \cot \delta_1 = \tfrac{1}{2}(k_2^2 - k_1^2)\rho(E_1, E_2) \qquad (15.76a)$$

a very useful relation. The δ's may be derived from the measured cross sections; the effective range is a valuable theoretical quantity from which other cross sections may be obtained and to which potentials may be related. Contributions to ρ all come from ψ and u at small values of r; the two wave functions are illustrated in Figure 15.7. For low energy at small r, ψ and u are essentially independent of the kinetic energy, so that we may use the zero-energy wave functions in the effective range integral to obtain $\rho(0, 0) \equiv r_0$:

$$r_0 = 2 \int_0^\infty (\psi_0^2 - u_0^2)dr \qquad (15.78)$$

At low energies ψ_0 approaches a straight line,

$$\psi_0 \xrightarrow{r \sim 0} 1 + \left(\frac{\partial\psi}{\partial r}\right)_{r=0} r = 1 + (k \cot \delta)r$$

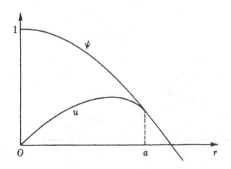

FIGURE 15.7 Graph of ψ and u illustrating that they differ only near the origin where the potential is different from zero.

We define α to be

$$\alpha \overset{\lim}{\underset{k \to 0}{=}} k \cot \delta$$

α is the reciprocal of the Fermi scattering length,

$$b = \frac{1}{\alpha} \overset{\lim}{\underset{\delta, k \to 0}{=}} \frac{1}{k} \frac{\sin \delta}{\cos \delta}$$

If we then take $k_1 = \delta_1 = 0$ and drop the subscripts on k_2 and δ_2, Eq. (15.76a) yields

$$k \cot \delta = k_1 \cot \delta_1 + \tfrac{1}{2}k^2 r_0 = \alpha + \tfrac{1}{2}k^2 r_0 \qquad (15.79)$$

a relation which remains remarkably accurate up to 20 Mev, for example, for scattering of neutrons by protons. Thus the phase shifts can be calculated from a knowledge of the two parameters, scattering length b and effective range r_0, which is the result we set out to obtain. A graph of u_0 and ψ_0 is presented in Figure 15.8.

An experimenter may interpret his scattering experiments by making a plot of $k \cot \delta$ as a function of k^2, which by Eq. (15.79) should yield the straight line illustrated in Figure 15.9. $k \cot \delta$ is obtained from the data by use of the relation between the observed scattering cross section σ and the known particle energy k^2. Eq. (15.60) with $l = 0$ yields

$$\sigma = \frac{4\pi \sin^2 \delta}{k^2}$$

Rearranging and noting that $\sin^{-2}\delta = 1 + \cot^2 \delta$ we obtain

$$\frac{4\pi}{\sigma} - k^2 = k^2 \cot^2 \delta$$

whence

$$k \cot \delta = \sqrt{\frac{4\pi}{\sigma} - k^2} \sim \sqrt{\frac{4\pi}{\sigma}} - \frac{1}{2}\sqrt{\frac{\sigma}{4\pi}}k^2 \qquad (15.80)$$

where $k \cot \delta$ has been put in terms of the measured quantities σ and k^2.

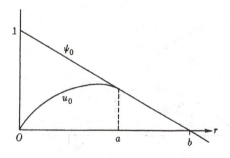

FIGURE 15.8 Graph of ψ_0 and u_0 near the origin.

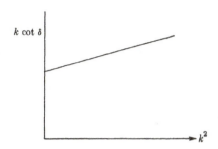

FIGURE 15.9   Graph of $k \cot \delta$ vs. k^2.

The intercept of the curve at $k^2 = 0$ in Figure 15.9 gives the reciprocal of the Fermi scattering length b. From Eq. (15.60) with $l = 0$ we have

$$\sigma = \frac{4\pi}{k^2} \sin^2 \delta_0 \underset{k, \delta_0 \to 0}{\lim} 4\pi b^2$$

in keeping with Eq. (15.61) so that b may be found from the experimentally determined cross section σ, i.e., $b = \sqrt{\sigma/4\pi}$); the slope gives the effective range. The range and depth parameters of any possible theoretical potential well must be such that their wave functions give the measured effective range in Eq. (15.79), and the measured scattering length.

BIBLIOGRAPHY

Bohr, Niels, "The Penetration of Atomic Particles Through Matter," *Proc. Danish Royal Society*, **18**, No. 8 (1948).

Fermi, E., *Notes on Quantum Mechanics*. Chicago: University of Chicago Press, 1961.

Schiff, L. I., *Quantum Mechanics*. New York: McGraw-Hill Book Co.: 1955.

Mott, N. F. and I. N. Snedden, *Wave Mechanics and Its Applications*. London: Oxford University Press, 1948.

Slater, J. C., *Quantum Theory of Atomic Structure*. New York: McGraw-Hill Book Co., 1960.

Pauling, L. and E. B. Wilson, *Introduction to Quantum Mechanics*. New York: McGraw-Hill Book Co., 1935.

XVI

NUCLEAR FORCES
AND MODELS

Nuclear physics is one of the most exciting chapters in contemporary physics exploration. The discovery of the neutron and neutrino, element transformation, energy of the stars, and nuclear fission have challenged man's previous understanding of his universe and ability to exploit scientific discovery to benefit human welfare. It is not possible to give a detailed and comprehensive discussion of the subject in a brief chapter. However, the brevity of a single chapter does permit a coherence in presenting the theoretical concepts which have proved so useful in developing our present understanding of the atomic nucleus.

When we consider the great and fundamental differences between the forces which bind atoms together and the forces that bind nuclei together, it is surprising that many of the concepts that have proved so useful in our discussion of atoms in earlier chapters should be of great utility in a discussion of the nucleus. In atoms, electrons are bound by long-range Coulombic forces to a massive, strongly attractive nucleus, whereas in nuclei, nucleons are bound by very short-range forces between equally massive, equally attractive particles. Furthermore, nuclear forces saturate, that is, the binding energy of a nucleon is relatively independent of the number of other nucleons for medium and heavy nuclei. This is in contrast to the atomic case, where the binding energy of the innermost electron increases rapidly with increasing nuclear charge. Consequently, whereas electrons are attracted to the effective center of charge density, nucleons would be expected to be attracted only to a few of their immediate neighbors. This expectation is changed when we consider the effect of the Pauli exclusion principle, according to which two particles can interact only if there are available unfilled states into which they can move and still conserve energy. Since in general there are no nearby unoccupied energy levels for a typical nucleon in the nucleus, the nucleons can move quite freely in the overall force field.

In this chapter we will first discuss what is known about nuclear forces experimentally, concentrating mainly on the two-body problem, that is, the interaction between two nucleons, which we have taken up several times earlier in the book in discussing the deuteron and the theory of the effective range. Then from what we know of the nuclear force we will be able to approach the problem of many nucleons interacting with one another. We introduce this problem with a simple atomlike model and then progress to methods of treating more complicated motion when the correlated behavior of several nucleons occurs. In this way we shall first treat the nucleus as consisting of individual nucleons moving in an overall smeared-out field of force provided by the attractive force fields of all the other nucleons acting collectively. We then consider the motion of energetic nucleons through nuclear matter. We conclude by considering the collective motions of some of these particles, in analogy to the motion of molecules within a drop of water, and to surface waves the collective motion brings about. Thus we are able to proceed after considering the peculiar nature of forces acting between two nucleons to more and more refined theoretical models systematizing our present understanding of nuclear structure.

16.1 NUCLEAR FORCES

Nuclear forces are for most of their behavior best observed by studying simple two-body systems. In discussing nuclear forces it is useful to reconsider the electromagnetic forces between charged particles. As is well known, the static electric field acting on a charged particle in electromagnetic theory is derived from Poisson's equation for the electric potential

$$\nabla^2 \psi = -4\pi\rho = -4\pi e \,\delta(\mathbf{r}) \tag{16.1}$$

where ρ is the electric charge density, e is the charge on a point charge and $\delta(\mathbf{r})$ is the Dirac delta function. The solution to this second-order differential equation is well known

$$\psi = -\frac{e^2}{r} \tag{16.2}$$

Here ψ is the electrostatic field potential for a point particle of charge e. This potential is long-range and the coupling (the electrostatic charge of the electron) is of intermediate strength between the coupling in gravitational fields and the coupling in nuclear fields. Now in contrast to this relatively weak but very long-range force which essentially extends to infinity, the nuclear force is observed to be stronger but of much shorter range. That is, the "nuclear charge" is far greater than the electric charge, and the range of nuclear forces is known to decrease very much more rapidly than simply inversely with r with increasing distance from the charges.

This electrostatic potential is invalid when the charges are in motion, for then the electric fields become magnetic fields, and vice versa, because moving charges or currents create additional fields. To treat the interaction between two charges relativistically one cannot use a static potential, for a static potential is not relativistically invariant. Instead, a formalism has been developed which, indeed, Feynman's original formulation of quantum mechanics was intended to clarify. In this formulation the electromagnetic interaction between two charges is represented by the exchange of a photon. When this interaction is calculated by the use of quantum electrodynamics, it is seen that this exchange of photons reduces in the nonrelativistic limit to the simple Coulomb force between two static charges. That is, the equation of the electric field potential in its simplest form for the static electric charges has the solution given in Eq. (16.2).

In this case, similar to Eq. (16.1) which is the equation for the photons of the electromagnetic field, we assume there is a field representing the forces between nucleons. The coupling is far stronger, but most important, the radial dependence has to be very different from the long-range Coulomb force. This field was invented by Yukawa, who made the enlightened, and since then well-confirmed, hypothesis that instead of exchanging massless photons or light quanta, nucleons exchange massive particles. If such particles are spinless, as it turns out they are, then the relativistic wave equation developed by Klein and Gordon, which governs the meson field, can be derived heuristically in the following way.

The relativistic energy obeys the equation, $E^2 = p^2c^2 + M^2c^4$. If we make the simple substitution that $p = (\hbar/i)\nabla$, the equation becomes $E^2 = -\hbar^2c^2\nabla^2 + M^2c^4$. This is not the complete Klein-Gordon equation, which includes the potential energy, because we assumed that the particle is a free particle; furthermore we can set $E = 0$ if we are just interested in considering the static case. As a result of this, we get the equation for the static meson field.

$$\nabla^2\psi - \frac{Mc^2}{\hbar}\psi = 4\pi\rho = 4\pi g_1\delta(\mathbf{r}) \tag{16.3}$$

which is the Klein-Gordon counterpart to Poisson's equation for the electrostatic potential. Similarly, if we introduce here a point nucleon at the origin, the "nucleon charge" g_1 times the Dirac delta function $\delta(\mathbf{r})$ replaces the "nucleon charge density" ρ. The solution to Eq. (16.3) is

$$\psi = -\frac{g_1}{r} \exp\left\{-\frac{Mc}{h}r\right\} \qquad (16.4)$$

The potential energy of a second nucleon in this field is, entirely analogous to electrostatic theory, $g_2\psi$ where g_2 is the effective "nucleonic charge" of the second nucleon. Of course this result reduces to the Coulomb case if the particles exchanged have zero mass, as photons do.

The nuclear force potential may be clarified if we note that h/Mc is the Compton wave length of a meson and is the length in which the exponential function decreases by $1/e$. This can be related to the Heisenberg uncertainty principle in energy and time. The total energy of the system is not conserved while the meson with rest energy Mc^2 is in flight. The time of flight permitted by the uncertainty principle is less than, or of the order of h/Mc^2. The distance the meson would travel at maximum possible velocity c is of the order h/Mc and, as we should expect, the heavier the meson, the shorter the range of the force for which the meson is responsible. If we had considered the more complicated case of the exchange of more than one meson, the potential acting between the mesons would be proportional instead to $\exp\{-n(Mc/h)r\}$ where n is the number of mesons exchanged. Since the meson coupling constant (nucleonic charge) is very much larger than the electron charge, nuclear forces have a much stronger coupling than electric forces. Hence we might anticipate further, as is indeed the case, that the exchange of many mesons is apt to occur and that this additional potential term $\exp\{n(Mc/h)r\}$ is an important part of the interaction.

Since Yukawa's invention of the theory in 1935, mesons have been experimentally discovered and there is little doubt that they are the agents responsible for forces between nucleons, just as photons are the agents responsible for forces between electric charges. There have been further refinements in the form of the coupling over the simple scalar form proposed by Yukawa. Furthermore, it is known that the deuteron has a small positive quadrupole moment, showing that the deuteron is not a sphere but is on the average cigar-shaped. This means that there must be more than just the $L = 0$ state in the ground state of the deuteron; there must also be some $L = 2$ admixture. However, the $L = 2$ contribution is small, being only about 4% of the $L = 0$ part. Since the quadrupole moment cannot be realized by a simple central force, there has to be a small contribution from a considerably more complicated interaction called a *tensor force*. Furthermore, since the mesons may be electrically charged, exchange forces are possible in which the nucleons are changed by the interaction. A neutron can emit a negative charged meson and change its charge to a proton. A proton which has been struck by a neutron could be changed into a neutron by the absorption of the negative meson which the neutron emitted in the interaction.

The exponential potential shape provided by the exchange of mesons gives us the characteristic short range of the nuclear force. As discussed earlier and in the next section, however, a simple central potential like this will not account satisfactorily for the properties of nuclei. In the next section we will discuss the behavior of nucleons moving in a sort of smeared-out overall potential quite similar to the atomic case treated earlier. Although the theory of the following section was first developed in 1935, a great stumbling-block to its acceptance was the fact that it did not order the levels in nuclei correctly. M. Mayer, and Haxel, Jensen, and Suess satisfactorily accounted for the properties of nuclear matter in such a potential by invoking an additional strong force among nuclei, the spin-orbit interaction. In atomic structure, a spin-orbit force appears from the well-known relativistic properties of the electromagnetic field, but in nuclear structure a strong spin-orbit force had to be hypothesized to account satisfactorily for a great many properties of nuclei.

Shortly after this force was hypothesized it was experimentally confirmed in the following classic experiment by Heusinkveld and Freier in 1952. Figure 16.1 illustrates schematically the experiment which they performed in scattering protons on helium. The results of the experiment were that many more protons registered in photographic plate 1 in the figure than in photographic plate 2. The interpretation of the experiment is the following: As labeled in the figure, particles coming in from the left and scattering from helium with trajectories below the helium atom are scattered to the right and have an angular momentum $r \times v$ pointing up out of the figure, whereas particles which are scattered to the left have an angular momentum pointing down into the page. The converse is true for particles with trajectories above the helium atom in the figure. Suppose that as a result of the spin-orbit force, the scattering cross section for protons were greater for protons whose spin and angular momentum are aligned in the same direction than for protons whose spin and angular momentum are oppositely aligned. Now, if this were the case, we should expect an equal number of protons scattered to the left and to the right from the helium in the first scattering interaction, assuming that the incoming beam of protons was unaligned. However, if the cross section with aligned spin and orbital momentum is greater, we should expect most of the particles scattered to the right to have a spin aligned with the orbital momentum, that is, pointing up out of the page, whereas most particles which are scattered to the left will have a spin aligned down predominantly into the page. If the beam scattered to the right in the first interaction is again scattered on helium, most of the particles will be scattered to the right in this second interaction and be detected by plate 1. Since the spins of these once-scattered particles are predominantly in the up

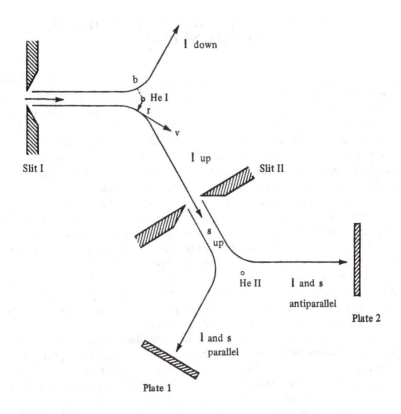

FIGURE 16.1 Schematic drawing of scattering of protons on He⁴ which give rise to polarized protons. (From Maria Goeppert Mayer and J. Hans D. Jensen, *Elementary Theory of Nuclear Shell Structure*, New York: John Wiley and Sons, Inc., 1955).

direction, the unequal number of counts in plates 1 and 2 shows that there is definitely a spin-orbit interaction from protons incident on helium. Just as in ordinary X-ray scattering, the first scattering process acts as a polarizer and the second scattering acts as an analyzer. The spin-orbit interaction turns out to be an extremely important additional force which must be taken into account in developing a theory of nuclear structure.

Other experimental evidence indicates that at low energies, the two-nucleon force is charge-independent; the nuclear forces between two protons, or the forces between two neutrons, or the forces between a proton and neutron, are all equal. In addition to being charge-independent, the two-nucleon force is also spatially symmetry-dependent, as demonstrated by the fact that two particles interacting in a spatially symmetric state interact much more strongly than they do in a spatially antisymmetric state.

16.2 THE INDIVIDUAL PARTICLE MODEL
OF THE NUCLEUS

Nuclei have their analogs to the noble gases in atomic physics, that is, there are nuclei which are relatively inert as far as nuclear reactions go and which seem to have especially stable binding energies. For example, it takes a great deal more energy to knock a proton or a neutron out of a calcium nucleus than it does out of nuclei near calcium in the chart of the nuclides. Furthermore, it is relatively difficult to get calcium to participate in a nuclear reaction by absorbing a low-energy neutron, that is, its neutron absorption cross section is relatively small compared to neighboring nuclei.

It was suspected as early as 1935 that nuclei behaved as though the nucleons had a level structure similar to the electrons in the central potential of an atom, but the levels could not be worked out by a simple central potential. As a result the theory was pretty much abandoned until the addition of the spin-orbit coupling with a suitable strength changed the level structure and ordered the levels to reproduce the results of experimental observations. With the individual particle model some surprising, although not necessarily contradictory, simplifications of the complicated nuclear system are possible. The mutual interaction of all the many nuclear particles may be imagined to be smoothed out so that an individual nucleon, neutron or proton, no longer feels a force from any one single neighboring particle but rather an aggregate force from all other particles acting together. Thus a single nucleon may be assumed to move in a limited region of force, completely independently of all the other individual particles in the nucleus.

As has already been mentioned, the assumption of independent motion of individual nucleons can be made plausible when the Pauli exclusion principle is included in the discussion. This situation, as we shall see later in discussing the optical model extension of the individual particle theory, also holds true for particles of low energy entering the nucleus during scattering experiments. In this case the incoming particle, if it has low energy, can interact with only those few particles having a high enough energy to be scattered and lifted up to unoccupied states of slightly higher energy. Particles buried deep below the unoccupied levels cannot contribute to the scattering. Therefore, the mean free path of incident particles in the core of the nucleus is very long if the incident particles have very low energies.

To give a more quantitative discussion of the individual particle model, or *nuclear shell model* as it is often called, we begin by listing two of the basic assumptions:

a) Each nucleon moves in a spherically symmetric potential which is independent of the exact instantaneous positions of the other nucleons.

b) The energy levels are successively filled by protons or neutrons in accordance with the Pauli exclusion principle. (The neutrons and protons will separately fill their respective levels.)

Although the exact form of the potential cannot be specified, reasonable guesses can be made on the basis of known nuclear properties. For instance, since the nuclear density is expected to be constant throughout the nucleus, we do not want the assumed spherically symmetric potential to be singular at the origin. Consequently, the choice for $V(r)$ should be such that

$$\left(\frac{dV}{dr}\right)_{r=0} = 0$$

Further, we know that the nuclear potential must go to zero rather abruptly at the nuclear surface.

It is easy to see that two different types of potentials with which we are already familiar are consistent with these criteria, namely the square well and harmonic oscillator. For ease of calculation the following two forms are chosen:

Square Well: $\qquad V(r) = -V_0 \quad$ for $\quad r < R$

$$= \infty \quad \text{for} \quad r > R$$

Harmonic Oscillator: $\quad V(r) = -V_0\left[1 - \left(\frac{r}{R}\right)^2\right]$

The value for V_0 is unknown, but there is experimental evidence to indicate that it should be reasonably large (of the order of 40 Mev) compared to the expected energy-level spacing.

In Section 8.7 the harmonic oscillator was discussed and it was shown that the energy levels for the three-dimensional oscillator are given by

$$E_{n_0} = \left(n_0 + \frac{3}{2}\right)\hbar\omega = \hbar\omega\left[2(n-1) + l + \frac{3}{2}\right] \qquad n_0 = 1, 2, 3, \cdots$$

Here l is the orbital momentum quantum number and n the radial quantum number. Each value of n_0 designates an energy shell. There is accidental degeneracy between levels of different n and l except for the two lowest levels where $n = 1, l = 0$ and $n = 1, l = 1$. For any given energy level $n_0\hbar\omega$ (we ignore the zero-point energy), the total degeneracy is obtained by considering the $2l + 1$ projections for each value of l as well as the spin degeneracy. This is indicated in Table 16.1 for the first few levels.

From Table 16.1 it is evident that any particular value of n_0 always has just even or odd l values associated with it, and in Section 8.2 we saw that the parity of the wave function for any case involving a spherically symmetric potential depends only on the angular part of the wave function, going as $(-1)^l$. Therefore, for the harmonic oscillator it follows that all the eigenfunctions belonging to one particular energy level have the same parity.

TABLE 16.1

n_0	n	l	Number of degenerate levels	Total number of states
0	1	0	2	2
1	1	1	6	8
2	2	0	2	
	1	2	10	20
3	2	1	6	
	1	3	14	40
4	3	0	2	
	2	2	10	
	4	4	18	70

Of course, we should not expect that either the harmonic oscillator potential or the square well with an infinite wall will actually represent the situation inside the nucleus. (Eckart, Bethe, and more recently Woods and Saxon have used a more realistic potential consisting of a constant core potential with exponentially flaring sides.) Nevertheless, since the square well and harmonic oscillator potentials represent somewhat opposite extremes, it is reasonable to anticipate that the real solution lies somewhere between them. This is illustrated in Figure 16.2, where the level sequences for both the isotropic harmonic oscillator and the square well are displayed. In the center of the figure is shown the approximate nuclear level sequence to be expected together with the total number of nucleons (either neutrons or protons) necessary to form closed shells. It is clear that these numbers do not coincide with the so-called *magic numbers* (2, 8, 20, 28, 50, 82, and 126) signifying the closure of shells for real nuclei. The situation is greatly improved, however, when a strong nucleon spin-orbit interaction is included.

In Section 11.5 dealing with the electron spin-orbit interaction we found that this interaction had the form $g(r)$ $(\mathbf{s} \cdot \mathbf{l})$, where $g(r)$ can be obtained from the known electromagnetic interaction. The same general form for the spin-orbit interaction is assumed for the nuclear case, only now the radial factor is unknown. In this situation one tries to make a reasonable assumption about the form of the radial factor which will give agreement with the experimental facts. As was the case for electrons, we have the expectation value of the spin-orbit interaction given as

$$\mathbf{s} \cdot \mathbf{l} = |\mathbf{j}|^2 - |\mathbf{l}|^2 - |\mathbf{s}|^2 = \hbar^2 \left[j(j+1) - l(l+1) - \frac{3}{4} \right]$$

For any given value of l, $j = l \pm 1/2$, and $\mathbf{s} \cdot \mathbf{l} = \hbar^2 l$ for $j = l + 1/2$ and $\mathbf{s} \cdot \mathbf{l} = -\hbar^2(l+1)$ for $j = l - 1/2$. From this we see that the spin-orbit splitting will be proportional to $2l + 1$ for any level specified by a given l and

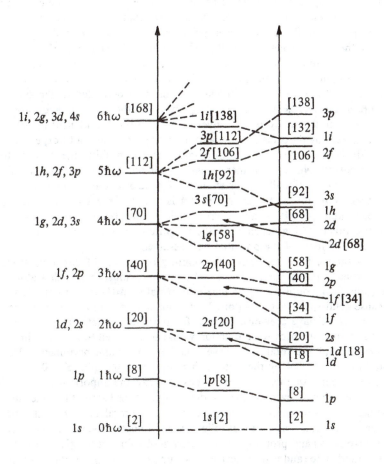

FIGURE 16.2 Energy-level system of the three-dimensional isotropic harmonic oscillator and the square well with infinitely high walls. The numbers in the brackets represent the total number of neutrons or protons in the nucleus when each particular level is filled. (From Maria Goeppert Mayer and J. Hans D. Jensen, op. cit.)

therefore we should reasonably expect that the spin-orbit interaction will increase with l. If in addition it is assumed that the radial term $g(r)$ decreases with increasing values of the radial quantum number n and increasing nuclear radius, then the order of the nuclear energy levels can be expected to appear as shown in Figure 16.3. Now we see that the inclusion of a spin-orbit dependent force with appropriate strength and radial dependence causes the splitting between a filled shell and the next group of levels of higher energy to be in agreement with the sequence of magic numbers.

To obtain predictions in reasonable agreement with properties found for real nuclei, various other refinements have to be added to the simple picture presented here. For instance, although the neutron and proton levels fill up separately in any given nucleus, proton and neutron levels characterized by the same set of quantum numbers should not be expected to coincide because the repulsive Coulomb interaction shifts the protons to higher energy. Nuclei with increasingly larger numbers of protons should then have proportionally larger excesses of neutrons compared to protons. Another limitation of the simple model is that the exact ordering of the various sublevels within any energy shell cannot be uniquely predicted because this ordering depends on the magnitude of the spin-orbit interaction and the form of the potential well assumed.

So far we have neglected the coupling or pairing of either neutrons or protons in partly filled subshells. We have previously noted that atomic electrons have coupled to give maximum total spin (Hund's rule). This is because total spin wave functions having maximum spin are entirely symmetric. As we have previously stated, for particles *attracted* to each other the total nucleon space wave function would be expected to have maximum symmetry and the total spin wave function, maximum asymmetry. This condition is satisfied by the state with minimum total spin, $S = 0$ or 1/2, and to account for this effect one needs to make the important additional assumption that an even number of identical nucleons having the same j and l values will always couple to give even parity and a total angular momentum of zero. From this assumption it follows that all nuclei having even numbers of neutrons (N) and protons (Z) must have zero spin and positive parity for the ground state (and a zero electric quadrupole moment). Furthermore, all nuclei comprised of a given odd Z, and even N, should have the same spin and parity in their ground state. A similar expectation holds for nuclei with a given odd N, and even Z. Generally speaking, these predictions of the shell model are in good agreement with experimental evidence. As an example, we note that all even N-even Z nuclei have been found to have a ground state spin and parity of $0+$. (Parity is discussed in Appendix D.)

By the very nature of the way in which the nuclear shell model has been formulated we should expect it to be most successful in regions near closed shells (magic numbers). In analogy to the atomic electron case, if there is

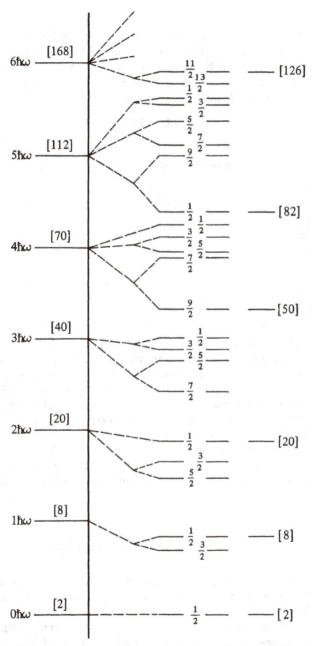

FIGURE 16.3 Schematic representation of the nuclear energy levels showing how the levels of Figure 16.2 are further split by the spin-orbit coupling leading to an energy shell structure in agreement with experimental results.

one nucleon outside a closed nuclear-energy shell, the total nuclear spin and parity will be determined by this particle or, if the shell lacks one particle, the net spin and parity will be the same as that of the hole. Table 16.2 gives examples of these situations. In each case there is excellent agreement between the model predictions and measured values.

TABLE 16.2

Nucleus	Predicted angular momentum of odd nucleon	Measured spin and parity
$_5B^{11}$	$p\ 3/2$	$3/2-$
$_7N^{15}$	$p\ 1/2$	$1/2-$
$_8O^{17}$	$d\ 5/2$	$5/2+$
$_{13}Al^{27}$	$d\ 5/2$	$5/2+$
$_{19}K^{39}$	$d\ 3/2$	$3/2+$
$_{21}Sc^{41}$	$f\ 7/2$	$7/2-$
$_{50}Sn^{119}$	$s\ 1/2$	$1/2+$
$_{81}Tl^{201}$	$s\ 1/2$	$1/2+$
$_{83}Bi^{209}$	$h\ 9/2$	$9/2-$

Although the nuclear shell model has had great success in predicting the spins and parities of ground states for nuclei having an odd number of nucleons, it by no means can be thought of as a complete model. Two of its more outstanding limitations are its failure to explain spins and parities of excited states and nuclear electric quadrupole moments in regions far from closed shells. According to the shell model, for an odd-A (A is the number of nucleons in the nucleus) nucleus the quadrupole moment should be just that resulting from the unpaired nucleon; in particular, an odd-A nucleus with an even number of protons should have zero electric quadrupole moment, a prediction which is in direct conflict with experiment for many nuclei of this type.

16.3 THE COMPOUND NUCLEUS

The shell model has subsequently been developed to a very high order of success, in explaining nuclear levels and many nuclear properties. The general idea of a nuclear well in which the nucleons move freely has also been used with great success to explain scattering experiments at rather high energy. In a three-dimensional harmonic oscillator potential, the level density obtained by using the WKB approximation, valid in the limit of high quantum numbers appropriate here, is given by

$$\rho(\epsilon)\, d\epsilon = c\sqrt{\epsilon}\, d\epsilon \qquad (16.5)$$

where ϵ is the energy of the particle in the well. Since the nucleon wavelength is long compared to the length in which the boundary is terminated (boundary thickness) this result is, as expected, the same as for particles in a square well (cf. Eq. (14.3)).

$$N(\epsilon)d\epsilon = \frac{V}{2\pi^2}\left(\frac{2m}{\hbar^2}\right)^{3/2}\sqrt{\epsilon}\,d\epsilon$$

where $N(\epsilon)\,d\epsilon$ is the number of particles having an energy between ϵ and $\epsilon + d\epsilon$ and V is the volume of the nucleus. If we integrate this equation over the particle energy up to the highest filled level, ϵ_F, we obtain the total number of particles in terms of ϵ_F.

$$N = \frac{V}{3\pi^2}\left(\frac{2m}{\hbar^2}\right)^{3/2}\epsilon_F^{3/2}$$

In this way we obtain the *Fermi energy* ϵ_F which is the energy of the particle in the highest filled energy level in the nucleus.

$$\epsilon_F = \frac{h^2}{8M}\left(\frac{3N}{\pi V}\right)^{2/3} = 54\,\frac{N}{A^{2/3}} \tag{16.6}$$

where M is the mass of a nucleon, A the atomic mass number, N the total number of either neutrons or protons, and N/V is the density of particles per unit volume in the nucleus. The Fermi energy can be quite large; ϵ_F is of the order of 30 Mev. Thus, we would expect that nucleon collisions will not often transfer small amounts of momentum to the nucleus, because the nucleon momentum states near the origin (in the core of the nucleus) are filled. When we are considering meson production on the other hand we might anticipate, as is indeed the case, that the threshold for meson production would be lower in heavy nuclei than it would be for free proton-proton or neutron-proton collisions. The nucleons in the highest filled levels will have a kinetic energy nearly equal to the Fermi energy and a momentum of $\sqrt{2m\epsilon_f}$. When neutrons hit a heavy target nucleus, some of the nucleons in the core have a kinetic energy of 30 Mev and a velocity one-fourth c in the direction of the incident neutrons, which lowers the threshold energy for meson production. The fact that experiments on meson production do show the expected decrease in the threshold energy corroborates the idea that there are nucleons in the target nucleus having kinetic energies of the order of 30 Mev as suggested by the consideration of the nucleus as a Fermi gas of individual particles.

Equation (16.5) gives us the total number of particle levels for an individual particle in the nucleus. However, when we consider the total nuclear level density, we must consider the overall level densities available from combining all of these individual particle excitations. In order to do this we will have to borrow some results from elementary statistical mechanics to obtain the thermodynamic function, the *Helmholtz free energy*. Then the total number of nucleons of one kind in the potential is equal to the integral over the particle energy of Eq. (16.5) times the probability of occupation of the states, given by statistical theory for Fermi-Dirac particles as $1/(1 + \exp\{\epsilon/kT - \xi/kT\})$, where ξ is a slowly varying function of the temperature, chosen to make the integral equal to the total number of particles, and T is the temperature of a nucleus excited by particle or gamma ray absorption. Therefore,

$$N = \int_0^\infty \frac{\rho(\epsilon)d\epsilon}{1 + \exp\{\epsilon/kT - \xi/kT\}} \tag{16.7}$$

The total energy of all nucleons of one kind in the potential is given by a related integral

$$U = \int_0^\infty \frac{\epsilon\rho(\epsilon)d\epsilon}{1 + \exp\{\epsilon/kT - \xi/kT\}} \tag{16.8}$$

These integrals can be solved approximately; they are the same integrals first developed by Summerfeld in his electron theory of metals. The result is

$$U = U_0 + \frac{\pi}{2} \frac{N}{\epsilon_F}(kT)^2 \tag{16.9}$$

where ξ has been determined from the integral for N. The total nuclear excitation energy can be written

$$Q = U - U_0 = \alpha\tau^2 \tag{16.10}$$

where $\tau = kT$, and $\alpha = (\pi/2)^2 N/\epsilon_F$.

The Helmholtz free energy of a gas is

$$F(\tau) = -2 \int_0^\tau \frac{\tau' Q(\tau')}{\tau'^2} = -\alpha\tau^2 \tag{16.11}$$

where $F(\tau)$ may be defined by

$$e^{-F(\tau)/\tau} = \int_0^\infty \omega(Q)e^{-Q/\tau} \, dQ \tag{16.12}$$

where $\omega(Q)$ is the total nuclear level density level we seek. The right-hand side of this equation is merely the Laplace transform of $\omega(Q)$. Hence, $\omega(Q)$ is merely the inverse Laplace transform of $e^{-F(\tau)/\tau}$ and is given by

$$\omega(Q) = \frac{1}{2\pi} \frac{\alpha^{1/4}}{Q^{3/4}} e^{2\sqrt{\alpha Q}} \tag{16.13}$$

This formula for the level density in heavy nuclei yields an excellent description of the level densities in nuclei observed in the evaporation of neutrons and protons from highly excited nuclei. For very high excitations, the formula has to be modified by a more exact description of the observed nuclear well shape.

16.4 THE OPTICAL MODEL

The model of individual particles moving in a nuclear well in which few of the particles can interact can be extended to consider the penetration of low-energy incoming particles into this nuclear well and is called the *optical model*. The optical model was developed primarily to explain scattering experiments. As was emphasized in the discussion of Eq. (4.25) in Chapter IV, particles move in a potential in a way analogous to light waves moving in a refractive medium. The analogy is inadequate to explain particle-scattering from nuclei in that a varying refractive index alone cannot account for particle absorption or inelastic scattering. The analogy can be made more complete by including a small imaginary potential which gives rise to an effective absorption coefficient. This has also been referred to as the cloudy-crystal-ball model. In discussing particle scattering in place of light scattering we consider a particle incident on a nuclear surface translucent to the particle waves. In other words, we are interested in the case where the incident particle may be either scattered or absorbed in passing through nuclear matter. In Section 4.5 we saw that the probability was unity that the particle would be either reflected (scattered) or transmitted for elastic scattering. Now we wish to include the possibility of the absorption of the particle by the nucleus, and this can be accomplished by choosing a complex potential for the region within the nuclear volume. The complex potential will give rise to an exponentially decaying wave function to account for the disappearance (absorption) of the incident particle.

For simplicity the discussion will be restricted to the one-dimensional case; the extension to three dimensions is straightforward and nothing essential is added to the discussion thereby. We choose a complex potential of the form shown in Figure 16.4. Outside the well in region I we have

$$\frac{d^2\psi_{\mathrm{I}}}{dx^2} + \frac{2m}{\hbar^2}E\psi_{\mathrm{I}} = 0$$

and

$$\psi_{\mathrm{I}} = Ae^{ikx} + Be^{-ikx}$$

where

$$k = \sqrt{\frac{2m}{\hbar^2}E}$$

ψ_{I} = incident wave + reflected wave

Inside the nucleus in region II the Schrödinger equation becomes

$$\frac{d^2\psi_{\mathrm{II}}}{dx^2} + \frac{2m}{\hbar^2}(E + V_0 + iW)\psi_{\mathrm{II}} = 0$$

and the solutions can be written as

$$\psi_{\mathrm{II}} = Ce^{\pm iKx}$$

where $K = (1/\hbar)\sqrt{2m(E + V_0 + iW)}$ = complex wave number. It is convenient to separate K into real and imaginary parts

$$K = k_1 + ik_2$$

Then, if W is assumed to be small compared to $E + V_0$, we have

$$K = \frac{1}{\hbar}\sqrt{2m(E + V_0)}\sqrt{1 + \frac{iW}{(E + V_0)}}$$

$$\cong \frac{1}{\hbar}\sqrt{2m(E + V_0)}\left(1 + \frac{iW}{2(E + V_0)}\right)$$

so that

$$k_1 \cong (1/\hbar)\sqrt{2m(E + V_0)}$$

$$k_2 \cong \frac{1}{\hbar}\frac{\sqrt{m}W}{\sqrt{2(E + V_0)}} = \frac{Wk_1}{2(E + V_0)}$$

For a wave moving in the plus x direction

$$\psi_{II} \cong Ce^{ik_1x - k_2x}$$

and the probability density $\psi_{II}^{*}\psi_{II} \rightarrow C^2e^{-2k_2x}$. It can readily be seen that inside the nucleus the probability is exponentially decaying. If the mean free path L of the particle is defined to be the distance in which the probability density decreases by a factor of $1/e$ then

$$L = \frac{1}{2k_2} = \frac{E + V_0}{Wk_1}$$

Therefore, we may regard the situation as one in which the incoming particle penetrates into the nucleus with a 63 percent probability of disappearing by the time it has travelled a distance L. The details of how it disappears are not considered here. We may imagine simply that the incoming particle loses all or part of its excess energy so that it is no longer identifiable as a particular incident particle with a given energy E. Even though this is admittedly a rather crude picture, nevertheless it has proved to be quite useful in describing various nuclear interactions. For actual calculations usually more sophisticated potentials have been assumed by various authors and the problem treated in three dimensions.

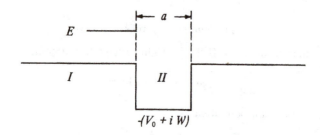

FIGURE 16.4 Schematic diagram illustrating the potential well seen by the incoming nucleon of energy E in the optical model.

It is interesting to see what a typical nuclear mean free path might be. If we assume an incident neutron to have a kinetic energy of 5 Mev, $W = 8$ Mev, and $V_0 = 45$ Mev then

$$L = \frac{1}{2k_2} = \frac{\hbar}{W} \sqrt{\frac{(E + V_0)}{2m}}$$

$$= \frac{1.05 \times 10^{-27} \text{ erg/sec}}{8 \text{ Mev}} \sqrt{\frac{50 \text{ Mev}}{2 \times 1.67 \times 10^{-24} \text{ gm} \times 1.6 \times 10^{-6} \text{ erg/Mev}}}$$

$$\cong 0.40 \times 10^{-13} \text{ cm}$$

This value is less than an average nuclear radius of approximately 10^{-12} cm.

The interaction of the incoming particle with the particles of the target nucleus is very much inhibited by the Pauli exclusion principle. Most of the nucleons are in low-lying energy levels and the incoming particle has insufficient energy to interact with them and raise one of them to an unoccupied level at higher energy. As a result the imaginary part of the potential is much less than the real part. Higher-energy incoming particles, however, can interact with a larger number of nucleons in the target nucleus since they have sufficient energy to exchange and raise a larger number of particles to unoccupied levels. Therefore, as we should expect, the imaginary part of the optical model potential is dependent on the energy of the incoming particle and increases in magnitude with this energy.

For the incoming particle having energies greater than the depth of the nuclear potential, the Pauli exclusion principle is relatively ineffective in restricting the number of interactions which can take place, and the mean free path of the incoming particle in nuclear matter becomes very short. In this case absorption coefficients are so high that the concept of a particle trajectory becomes complicated and of limited usefulness. Under these circumstances it is better just to consider the penetration of the particle into the potential with the resultant formation of an excited compound nucleus in which all trace of the initial incoming particle is lost.

For very high energies of the incoming particle (above 100 Mev or so) the mean free path again becomes quite long due to the small cross section at high energy for collisions with the particles in the nucleus. Therefore, at these energies the optical model is also a very good means of interpreting the scattering data.

16.5 THE COLLECTIVE OR UNIFIED MODEL

Many of the physical properties of the nuclei neighboring magic nuclei depend solely on the last few nucleons or holes apart from a filled nuclear shell. That is, the overall nuclear energy levels, spin, parity, and excitation properties are a function of the motion of the neutrons and protons corresponding to valence electrons in atomic physics. The physical properties of these nuclei are especially easy to calculate. The nuclear spin, magnetic moment, and quadrupole moment of near-magic nuclei are all estimated by considering the rotations about the nuclear core of the last few nucleons or the vacancies in the last shell. These holes or extra nucleons cause the nuclei to be aspherical in shape. If the outer nucleons or holes have an effective electrical charge, the result will be an aspherical charge distribution expressed as a nonzero quadrupole moment. The nuclear spin and magnetic moment arise of course from the circulation of mass and electric charge implicit in these rotations. Successful predictions for the nuclear spin, magnetic moment, neutron scattering, radioactive decay, unique energies possible for the nucleus, and the sign of the quadrupole moment of near-magic nuclei have all increased our confidence in the individual particle model.

The individual nucleons are essentially pictured as moving in a kind of box or hollow sphere whose walls limit the region of force in which these particles rattle around, but what would be the result if the walls should move? We might expect the most important effect to occur in the rough value for the quadrupole moment estimated from the motion of the extra nucleons outside the last closed spherical shell, since the estimate of the quadrupole moment was based on the motion of individual extra protons, or their absence, and did not include the effect of any ordered wall motion. Unfortunately, it is true that although the sign is given correctly, the magnitude of the quadrupole moment estimated from elementary nuclear-shell theory is indeed much smaller than the measurements indicate and when the outer nucleon or hole represents the presence or absence of a neutron, the quadrupole moment is incorrectly given as zero. Moreover, the individual particle theory is inadequate to describe nuclear fission.

An adequate theory of nuclear fission does exist, however. Shortly after the discovery of nuclear fission, N. Bohr and J. Wheeler were able to describe very nicely most of the gross features of fission with an extreme collective model. They pictured the nucleus as liquid drop which shook and wobbled when excited so that when sufficiently excited it would break apart into two or more smaller drops. The liquid-drop model had even earlier proved to be a very useful guide for predicting the approximate dependence of nuclear binding energy on the nuclear surface area. In the liquid-drop model the nucleus is thought to have a surface tension, which means that the nuclear surface helps hold the nucleus together in the same way water drops and mercury drops are kept from fragmenting by their surface tension. Therefore, the smallest possible surface area results in the most binding energy. This idea agrees very well with measurements of nuclear binding energies.

The theory is quite elementary. A liquid drop has a uniform central density independent of its size and outside of this a surface layer. Thus we would expect that the total binding energy of a nucleus of mass number A should depend linearly upon the atomic mass number. This is of course because of the saturation of nuclear forces. The binding energy of a piece of matter to the earth, for example, depends on the total mass so that the total binding energy of matter in the earth would go as the square of the mass of the earth. However, because of the saturation of nuclear forces, the binding energy of the entire nucleus goes linearly with the atomic number.

We also expect a surface energy. This is a negative energy and is due to the unsaturated nuclear bonds which particles on the nuclear surface experience. That is, some of the bonds that a nucleus would make with neighboring nuclei are absent on the nuclear surface. We would also anticipate a Coulomb repulsion because we have a sphere of positive charge which is mutually repulsive and there would be a negative energy due to that. The positive potential energy of a uniformly charged sphere is $(3/5)(Z^2e^2/r_0 A^{1/3})$ where Z is the number of protons in the nucleus, e is the charge of a proton, and $r_0 A^{1/3}$ is the nuclear radius.

Finally, there is a fourth physical effect to be accounted for and it comes about from consideration of the antisymmetry which must exist in the total wave functions of any two identical particles in the nucleus. As we have said before, the space-symmetric state is the one of most binding force between any two nucleons, but this is not a possible state if the two particles are identical and are in a symmetric spin state. For this reason unlike particles have more binding energy in the nucleus, and the nucleus has maximum binding energy for a maximum number of pairs of unlike nucleons. On the other hand, the interaction of any two pairs would depend inversely upon the nuclear volume, proportional to the atomic mass number. We may take this symmetry effect into account by adding a term inversely proportional to the nuclear volume and proportional to the square of the maximum number of symmetric pairs, $(N - Z)^2/A$.

As a result of all these considerations, we find that the total binding energy in Mev of a nucleus is given by $B = +a_1 A - a_2 A^{2/3} - a_3 Z^2/A^{1/3} - a_4(N - Z)^2/A$, where a_3 is the coefficient given above for the Coulomb repulsive term and the other three terms are fitted to the normal binding energies of nuclei as determined by mass spectroscopy and other means. For instance, a typical set of coefficients might be given by $a_1 = 14.0$, $a_2 = 13.1$, $a_3 = 0.659$, $a_4 = 19.4$.

$$B = 14.0A - 13.1A^{2/3} - 0.659Z^2/A^{1/3} - 19.4(N - Z)^2/A$$

We have yet to include an important additional term due to a pairing energy which occurs between two identical nucleons in the same space state. If, for example, a neutron is added to a stable nucleus having a core consisting of an even number of protons and an even number of neutrons and with one additional neutron in the highest-ground-state energy level, the added neutron will go into the remaining vacant lowest-energy space state shared by the odd neutron. There will be an additional binding energy from the attractive interaction between the two neutrons in the same space state. If, instead, a proton had been added to the initial nucleus, there would be two unpaired nucleons in the resultant nucleus and its binding energy would be less by the amount of the pairing energy in two nucleon pairs. This effect may be represented an by additional term of $+34/A^{2/3}$ for even-even nuclei, 0 for nuclei of odd A, and $-34/A^{2/3}$ for odd-odd nuclei. The coefficient of this additional term has been evaluated by a least-squares fitting to energies of beta rays emitted in the decay of odd-odd nuclei to even-even nuclei.

This familiar *semiempirical mass formula*, first developed by Weissacker, reproduces nuclear binding energies quite well. Better mass formulas have since been developed from a least-squares fitting to a power expansion in A, Z, and N-Z, but they have less physical content.

J. Rainwater, A. Bohr, and B. Mottelson, among others, successfully invoked some features of the liquid-drop model so useful in the past to predict many additional features of nuclei, particularly the magnitude of the quadrupole moment. Their model is particularly important in considering nuclei which have several particles in an unfilled shell, or several holes in an unfilled shell. Several additional particles above a filled shell, or holes in a nearly filled shell, will move collectively and because of this distort the spheroidal shape of the nuclear well. The collective motion of these particles on the surface of the nucleus becomes quite difficult to handle by means of the individual particle model because instead of considering just one particle outside the spheroidal nuclear core one must consider the mutual interaction of four or five particles or more. It becomes much easier to revert to the nuclear liquid drop and just consider the wall motion.

We imagine the nucleus as a sort of quivering drop of liquid on which surface waves rotate. The amplitude of these surface waves (the number of nuclear particles moving in the waves) depends on the number of nucleons or vacancies present in an unfilled outer shell of the individual particle model. Only a very small fraction of the nuclear matter participates in this wall motion. Although most of the nuclear matter remains stationary, the wall motion in further distorting the nucleus from a spherical shape clearly will add to its quadrupole moment. Besides successfully predicting measured quadrupole moments, the energy in these surface waves has been predicted and the theory confirmed by observations of gamma rays resulting from the disappearance of such waves. Furthermore, observations of vibrational surface waves have also been made, and S. G. Nilson, in particular, has shown theoretically that the motion of a single particle is altered as the box in which it is confined changes shape.

The underlying assumption in the *collective*, or *unified*, model is that particles moving within the nucleus can exert a centrifugal pressure on the surface. For nuclei having filled shells the pressure will be isotropic and the nuclear shape will be spherical in agreement with the shell model. However, in a partially filled shell the outer nucleons will exert anisotropic pressure, thereby causing the nucleus to be deformed into a nonspherical shape. This will in turn cause the effective potential seen by nucleons in the filled inner core of the nucleus to become nonspherical also. The nucleus can then be thought of as having its internal energy shared by three separate kinds of motion (similar to the diatomic molecule case discussed in Chapter XIII), the single-particle motion, and rotation and vibration of the nuclear surface. In regions far from closed shells the nonspherical shape can explain the observed large electric quadrupole moments and at the same time oscillations in the shape can be used to explain collective effects such as fission and the ordering of excited states.

Figure 16.5 shows the coupling scheme for a symmetrical deformed nucleus. As usual, I is the total nuclear angular momentum or spin, M is the component of I along the axis z fixed in an arbitrary direction, K is the projection of I on the symmetry axis of the nucleus, and R is the component of I perpendicular to K and results from the rotation of the nucleus as a whole.

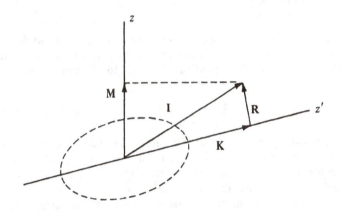

FIGURE 16.5 Coupling scheme for angular momenta for rotationally symmetric deformed nuclei. I is the total nuclear angular momentum with a component M along the space fixed axis z and component K along the nuclear symmetry axis z'.

It is constructive to consider the case for deformed even-even nuclei. From the previous discussion of the shell model we know the ground state of these nuclei will be $0+$. Consequently, $K = 0$ and the lowest excited states should be just the rotational energy levels given in analogy with the results in Section 8.4 by

$$E_{\text{rot}} = \frac{\hbar^2}{2d}\langle R^2 \rangle = \frac{\hbar^2}{2d}\langle I^2 \rangle = \frac{\hbar^2}{2d}I(I + 1) \qquad \textbf{(16.14)}$$

However, it has been found from experiments that here the moment of inertia d is considerably less than the rigid body moment of inertia obtained by considering the nucleus as whole to rotate. Consequently, we should not consider that the nucleus as a whole is actually rotating, but rather that the spheroidal surface of the nucleus is undergoing a rotation.

Also, for the particular case chosen of even-even nuclei, not all values of I for low-lying excited rotational levels are allowed. This can be seen in several ways; perhaps the easiest approach is to compare the situation here with that for homonuclear diatomic molecules (Section 13.7). In this case if we consider the nucleus to be made of two separate halves, each half will have an even number of nucleons and therefore have an integral spin. The nucleus then obeys Bose-Einstein statistics and the total nuclear wave function must be symmetric. Since the total spin is just due to the excited rotational levels, only even rotational levels can be excited so that we will have

	spin	parity
ground state	0	+
1st excited state	2	+
2nd excited state	4	+
	etc.	

Just such a series of low-lying levels have been identified for many even-even deformed nuclei. In addition, the ratio of the spacing between the various excited levels is predicted by Eq. (16.14). For instance, for a given nucleus

$$\frac{E_4}{E_2} = \frac{4(4+1)}{2(2+1)} = \frac{10}{3}$$

Figure 16.6 shows the ratios actually obtained for some even-even deformed nuclei. It can be seen that the agreement for E^4/E^2 is generally very good, but that for higher levels the experimental data deviate considerably from the predicted ratios. The differences can be reduced if one considers that there will be an increase in the moment of inertia with increasing rotational angular momentum because of the action of the centrifugal force. This is entirely similar to situations of the rotating diatomic molecule, and a similar correction term is added so that Eq. (16.14) becomes

$$E_{\text{rot}} = \frac{\hbar^2 I(I+1)}{2d} - BI^2(I+1)^2 \quad \text{with } I = 0, 2, 4, 6 \cdots$$

This equation has two adjustable parameters (d and B).

Nuclei which are spherical, or nearly spherical, do not have low-lying rotational states. Since they can undergo a collective vibrational oscillation about the equilibrium shape, the lowest-lying states would ideally be given by solutions of the harmonic oscillator problem. The situation is similar to

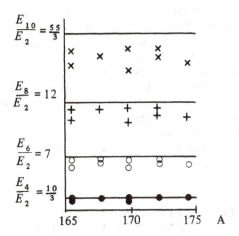

$$\frac{E_{10}}{E_2} = \frac{55}{3}$$

$$\frac{E_8}{E_2} = 12$$

$$\frac{E_6}{E_2} = 7$$

$$\frac{E_4}{E_2} = \frac{10}{3}$$

165 170 175 A

FIGURE 16.6 Ratios of rotational energy levels for deformed even-even nuclei. The solid lines represent the ratios predicted by Eq. (16.14).

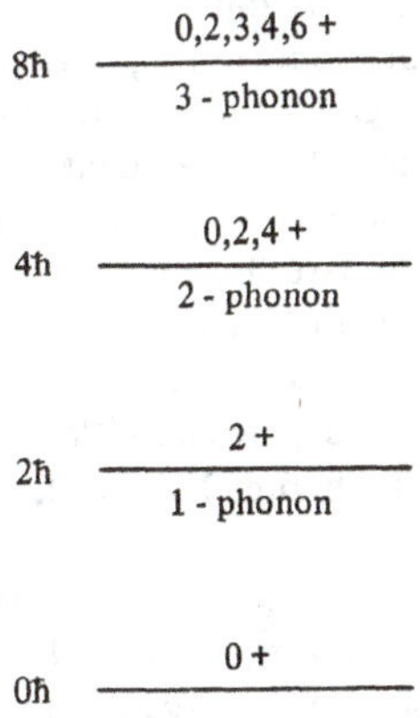

8ħ   0,2,3,4,6 +
      ─────────────
      3 - phonon

4ħ   0,2,4 +
      ─────────────
      2 - phonon

2ħ   2 +
      ─────────────
      1 - phonon

0ħ   0 +
      ─────────────

FIGURE 16.7 Energy-level diagram for even-even nuclei assuming the ideal conditions of pure vibrational motion.

that for diatomic molecules except that in nuclei the level spectra resemble those of a simple quadrupole oscillator. A dipole oscillation requires a separation of the center-of-charge density and the center-of-mass, which occurs only at high excitation. Hence only quadrupole and higher-order excitations can occur for transitions among low-lying collective states. The excited states would then correspond to one, two, or three vibrational quanta called *phonons*. For an even-even nucleus with a 0+ ground state, equally spaced excited pure vibrational levels of 2+; 0+, 2+, 4+; and 0+, 2+, 3+, 4+, 6+ should be expected, as is shown in Figure 16.7. However, in real nuclei it is found that the degeneracies of the 2— and 3— phonon states are removed and the levels are split. The existence of nuclear vibrational states has been experimentally verified for a variety of nuclei.

A nucleus undergoing vibration can also rotate. In this case, however, the simple rotational spectra is changed because the moment of inertia (Mr^2) is changed due to the change in r from the vibrations. This effect causes a small coupling between rotational and vibrational states which increases with the energy of these states.

16.6 THE EFFECTIVE MASS OF NUCLEONS BOUND IN A POTENTIAL

We have seen how the Pauli exclusion principle prohibits interactions between nucleons in the nucleus unless the exchange of momentum and energy can be such that the nucleons end up in previously unoccupied states. Therefore we should expect that the interaction of a particle incident on the nucleus will depend on the energy of the incoming particle because the more energy the incident particle has the greater will be the number of nucleons in filled states with which it can interact. Consequently, we should expect that the potential the incident particle experiences should be energy-dependent, particularly in the absorptive part. This may be interpreted as implying that the potential produced by all the nucleons in the nuclear well changes the mass of the particle, or in other words, the effective mass of a bound particle is different from that of a free particle. (We have already seen one such example in the discussion of Chapter XIV on the effective electron mass in metals.)

To be more specific, we assume that the energy (velocity) dependence is quadratic in v and may be represented by $V = -cv^2$, where c is an appropriate constant. Then the total energy of a nucleon would be given by

$$E = V_0 - cv^2 + \frac{1}{2}mv^2$$

where V_0 is the constant potential part of the interaction. The terms $1/2\ mv^2 - cv^2$ can be combined to give $1/2\ m^*v^2$, with the effective mass m^* being equal to $m - 2c$. Thus the velocity or energy-dependent part of the potential can be included as a change in the effective mass of the kinetic energy term.

The theory of nuclear matter as developed by Brueckner, Bethe, and others yields an effective mass of nucleons in a nucleus in its ground state of about one-half the mass of free nucleons, whereas for the unfilled levels above the Fermi level a nucleon has an effective mass very close to the free nucleon mass. Therefore, we should expect that an effective nuclear mass of $(1/2)m$ would be the appropriate mass to use in writing equations of motion for nuclei in their ground state. On the other hand, for the problem of neutron and proton evaporation from highly excited nuclei (involving level densities of highly excited nuclei) the mass term to be used in the equations would be essentially the mass of a free nucleon.

BIBLIOGRAPHY

Enge, Harold A., *Introduction to Nuclear Physics*. Reading, Mass.: Addison-Wesley Publishing Company, Inc., 1966.

Mayer, M. G. and J. H. D. Jensen, *Elementary Theory of Nuclear Shell Structure*. New York: John Wiley & Sons, Inc., 1955.

Siegbahn, K., ed., *Alpha- Beta- and Gamma-Ray Spectroscopy*. Amsterdam: North-Holland Publishing Company, 1965

APPENDIX A

THE VIBRATING STRING

As an analogy to the more complicated three-dimensional problem of electromagnetic waves in a box we take up the problem of waves in a string. If a string fastened at both ends (at $x = 0$ and L) is displaced slightly in the y direction, the restoring force of an infinitesimal portion of the string dx will be given by the change in the y component of the tension on the string F.

$$d\left[F\frac{dy}{\sqrt{(dx)^2 + (dy)^2}}\right] \sim d\left[F\frac{\partial y}{\partial x}\right] = F\frac{\partial^2 y}{\partial x^2}\,dx$$

The equation of motion for the infinitesimal string element is then given by

$$m\,dx\frac{\partial^2 y}{\partial t^2} = d\left[F\frac{\partial y}{\partial x}\right]$$

or

$$m\frac{\partial^2 y}{\partial t^2} = F\frac{\partial^2 y}{\partial x^2}$$

where m is the mass per unit length of the string.

We may attempt to solve this differential equation by assuming that y may be written as a product of a function of t times a function of x. The assumption is justified if a solution is obtained.

$$y = T(t)X(x); \quad mX(x)\frac{\partial^2 T(t)}{\partial t^2} = FT(t)\frac{\partial^2 X(x)}{\partial x^2}$$

which may be rewritten as

$$\frac{m}{F}\frac{1}{T(t)}\frac{\partial^2 T(t)}{\partial t^2} = \frac{1}{X(x)}\frac{\partial^2 X(x)}{\partial x^2}$$

The left-hand side of this equation is independent of x, the right-hand side is independent of t, and since they are equal, each must be independent of both x and t, and therefore constant. Let the constant be written as $-\omega^2$. Thus

$$\frac{m}{F}\frac{1}{T(t)}\frac{\partial^2 T(t)}{\partial t^2} = -\omega^2$$

or

$$\frac{\partial^2 T(t)}{\partial t^2} + \frac{F}{m}\omega^2 T(t) = 0$$

whose solution is

$$T(t) = A_1 \cos\sqrt{\frac{F}{m}}\,\omega t + A_2 \sin\sqrt{\frac{F}{m}}\,\omega t$$

where A_1 and A_2 are undetermined constants. We also have the equation for $X(x)$

$$\frac{\partial^2 X(x)}{\partial x^2} + \omega^2 X(x) = 0$$

whence

$$X(x) = B_1 \cos\omega x + B_2 \sin\omega x$$

where B_1 and B_2 are undetermined constants.

Since the string cannot vibrate at $x = 0$ and L, $X(0) = X(L) = 0$. Therefore, $B_1 = 0$ and ωL must be some integral multiple of π:

$$X(x) = B_{2_n} \sin\frac{n\pi}{L}x$$

where n is an integer and there is a solution for every value of n.

The time dependence of the solution is determined by the initial conditions—the initial values of y and $\partial y/\partial t$ between $x = 0$ and $x = L$. If we assume the string is vibrating at maximum amplitude at $t = 0$, then $A_2 = 0$, and we have ($y \neq 0$, $\partial y/\partial t = 0$)

$$T(t) = A_{1_n} \cos\sqrt{\frac{F}{m}}\frac{n\pi}{L}t$$

Let $A_{1_n}B_{2_n} = C_n$; then the final complete solution may be written

$$y(x, t) = \sum_n C_n \cos\sqrt{\frac{F}{m}}\frac{n\pi}{L}t \sin\frac{n\pi}{L}x$$

The C_n depend on the shape in which the string is made to vibrate (whether it is plucked in the middle or near one end, etc.) and on the amplitude of vibration, and need not concern us for the present. What is important is that the frequencies (inverse of the period of the time dependence) are some discrete multiple of $\sqrt{(F/m)}(\pi/L)$; only these frequencies are permitted, as a result of the boundary conditions on a string of finite length.

It is also worthy of note that we have considered only one independent degree of freedom for the vibration of the string. An identical, completely independent solution is obtained if we consider the vibration of the string in the z direction.

APPENDIX B

MATRICES

Matrix representation is an especially useful means to represent operators belonging to various physical observables. In perturbation theory we have used matrix elements representing the perturbation energy. In discussing orbital momentum and spin, it is particularly useful to represent their operators in matrix notation.

We may explain matrix notation in the following way. Let $\varphi_1, \varphi_2 \ldots$ form a complete orthonormal set of wave functions, say the eigenfunctions of some operator A. As we already explained, any arbitrary wave function ψ can then be represented as a series

$$\psi = \sum a_n \varphi_n, \text{ where } a_n = \int d\tau \, \varphi_n^* \psi$$

Indeed we can specify the function by giving only the set of coefficients $(a_1 \, a_2 \ldots)$, since the series can always be reconstructed. There is a one-one correspondence, in other words, between wave functions ψ and ordered sets of numbers (i.e., between a wave function and a vector $a_1, a_2 \ldots$, or $\{a_n\}$ for short). One can also set up a correspondence between operators and doubly ordered sets of numbers (i.e., between operators and matrices) as follows:

$$H_{mn} = \int d\tau \, \varphi_m^* H \varphi_n$$

Now let $x = H\psi$ and compute the coefficients $\{b_n\}$ of the new function x:

$$b_m = \int d\tau \, \varphi_m^* x$$

$$= \int d\tau \, \varphi_m^* H\psi$$

$$= \sum_n \left[\int d\tau \, \varphi_m^* H\varphi_n \right] a_n = \sum_n H_{mn} a_n$$

which is the familiar rule for the multiplication of a vector by a matrix. We can think of a wave function ψ and an operator H, therefore, as a vector $\{a_n\}$ and a matrix H_{mn}; and if we do, the result of letting the operator act on the wave function will be the same as the result of multiplying the vector by the matrix. (Note, by the way, that the vectors and matrices are generally infinite-dimensional.)

The next thing to show is that the product of two operators corresponds in the proper way to the product of the two matrices. Let $C = AB$, where A, B, and C are operators. Then

$$C_{mn} = \int d\tau \, \varphi_m^* C\varphi_n$$

$$= \int d\tau \, \varphi_m^* AB\varphi_n$$

Now the series for $B\varphi_n$ is

$$B\varphi_n = \sum_p \left(\int d\tau \, \varphi_p^* B\varphi_n \right) \varphi_p$$

$$= \sum_p B_{pn} \varphi_p$$

Hence

$$C_{mn} = \int d\tau \, \varphi_m^* A \left(\sum_p B_{pn}\varphi_p \right)$$

$$= \sum_p \int d\tau \, \varphi_m^* A\varphi_p B_{pn}$$

$$= \sum_p A_{mp} B_{pn}$$

But this is simply the product of the two matrices, which is what we set out to prove.

It should be emphasized that the matrix elements depend on the choice of the orthonormal set φ_1, φ_2 . . . In other words, there is more than one matrix representation. If the φ_n's are the eigenfunctions of some operator A, one speaks of the A representation. For example, if they are the simultaneous eigenfunctions of the operators H, L^2, and L_z of Chapter VIII, we have the so-called energy-angular momentum representation. An operator is said to be *diagonal* in a certain representation if its matrix elements vanish when the indices are different.

$$A_{mn} = 0 \text{ whenever } m \neq n.$$

(*Exercise:* Show that the operator A is diagonal in the A representation, that is the representation in which $A\varphi_n = a_n\varphi_n$.) It is very often important to *find* a representation in which a given operator (usually the Hamiltonian) is diagonal, and the process of doing so is called *diagonalization*. This, in fact, is what we were doing in the second-order perturbation theory; we were finding an orthonormal set of wave functions (eigenfunctions of H_0) in which the operator V was diagonal.

We might review the material presented in the text with respect to their matrix representation. In this regard, degenerate perturbation theory provides a good illustrative introduction to the significance of and requisite conditions for diagonality of an operator matrix. For example, when $H^0\psi_{ik}^0 = E_i^0\psi_{ik}^0$, we say that the Hamiltonian or energy operator is diagonal in this matrix representation, for H^0 operating on any of these ψ^0's always gives us back the same wave function. If the ψ^0's are orthogonal, multiplying by $\psi_{ik'}^{0*}$ on the left and integrating over all space would yield

$$\langle\psi_{ik'}^{0*}H^0\psi_{ik}^0\rangle = E_i^0\delta_{k'k} \qquad (1)$$

The subscripts k and k' may have any values from 1 to n for an n-fold degenerate system. These n^2 numbers may be ordered in the form of a matrix where the first index refers to the row and the second index refers to the column. Thus Eq. (1) is written

$$
\begin{bmatrix}
\langle\psi_{i1}^{0*}H^0\psi_{i1}^0\rangle & \langle\psi_{i1}^{0*}H^0\psi_{i2}^0\rangle & \cdots & \langle\psi_{i1}^{0*}H^0\psi_{in}^0\rangle \\
\langle\psi_{i2}^{0}H^0\psi_{i1}^0\rangle & \langle\psi_{i2}^{0*}H^0\psi_{i2}^0\rangle & \cdots & \langle\psi_{i2}^{0*}H^0\psi_{in}^0\rangle \\
\cdot & \cdot & & \cdot \\
\cdot & \cdot & & \cdot \\
\cdot & \cdot & & \cdot \\
\langle\psi_{in}^{0}H^0\psi_{i1}^0\rangle & \langle\psi_{in}^{0*}H^0\psi_{i2}^0\rangle & \cdots & \langle\psi_{in}^{0*}H^0\psi_{in}^0\rangle
\end{bmatrix}
= E_i^0
\begin{bmatrix}
1 & 0 & 0 & \cdots & 0 \\
0 & 1 & 0 & \cdots & 0 \\
0 & 0 & 1 & \cdots & 0 \\
& \cdots & & & \cdot \\
& \cdots & & & \cdot \\
& \cdots & & & \cdot \\
0 & 0 & 0 & \cdots & 1
\end{bmatrix}
\qquad (2)
$$

If an additional term V' is added to the Hamiltonian operator, $\langle\psi_{ik'}^{0*}H\psi_{ik}^0\rangle$ is no longer necessarily zero if $k \neq k'$, as we know. Hence we obtain a nondiagonal matrix, because V' operating on ψ_{ik}^0 yields a wave function different from ψ_{ik}^0. If

$$\int \psi_{ik'}^{0*}V'\psi_{ik}^0\,d\tau \neq V'\delta_{k'k}$$

it is because $V'\psi_{ik}^0$ yields some $\psi_{ik'}^0$ (plus some other ψ^0's, in general), for $V'\psi_{ik}^0$ has to yield some off-diagonal ψ's in order that this equation be true since $\int \psi_{ik'}^{0*}\psi_{ij}^0\,d\tau = 0$ unless $k' = j$. In several of the simple examples treated already, we discovered other representations in which the energy was again diagonal; that is, we *diagonalized* the Hamiltonian operator.

B.1 ANGULAR MOMENTUM MATRICES

Besides the Hamiltonian or energy operator, another vital operator representing a physical observable is the square of the total angular momentum operator or of operators representing any of its components. While one almost always chooses a representation in which the energy is diagonal, in accordance with the central importance of the energy of a physical state, a given component of the angular momentum may not be diagonal in the representation chosen for a given system. Indeed, only one component of the angular momentum may be diagonal in any one representation, the other two components being necessarily nondiagonal. Thus L_x and L_y are nondiagonal in a representation in which L_z is diagonal. Let us for the moment confine our attention just to the orbital momentum.

Note that if L_x, for example, operates on a wave function such as $R(r)Y_l^m$ which is represented in spherical coordinates with polar axis along the z axis, new wave functions result (cf. Sections 8.1, 8.2). L_z operating on these particular spherical harmonic wave functions yields the original wave function, and therefore $R(r)Y_l^m$ is an eigenfunction of L_z but not of L_x,

$$L_z R(r)Y_l^m(\theta, \varphi) = m\hbar R(r)Y_l^m(\theta, \varphi) \tag{3}$$

whereas

$$L_x R_{n,l}(r)Y_l^m(\theta, \varphi)$$

$$= \frac{\hbar}{i}\left\{-\sin\varphi\frac{\partial}{\partial\theta} - \cot\theta\cos\varphi\frac{\partial}{\partial\varphi}\right\}R_{n,l}(r)\frac{(-1)^{(l+|m|)/2}}{\sqrt{4\pi}}$$

$$\times\sqrt{\frac{(2l+1)(l-|m|)!}{(l+|m|)!}}\,\Theta_l^{|m|}(\cos\theta)e^{im\varphi}$$

$$= \frac{\hbar}{2}\{\sqrt{(l-m)(l+m+1)}R_{n,l}(r)Y_l^{m+1}(\theta, \varphi)$$

$$+ \sqrt{(l-m+1)(l+m)}R_{n,l}(r)Y_l^{m-1}(\theta, \varphi)\} \tag{4}$$

where the Y_l^m have been written explicitly once as a product of a function of θ and $e^{im\varphi}$ to illustrate the functions the derivatives operate on. (The Y_l^m have been suitably normalized by the complicated function of l and m occurring in (4).) Hence use of the operator L_x in this representation results in a linear combination of new wave functions.

We will now put the angular momentum operator into a matrix representation. (It will be finite-dimensional representation, since we are going to consider only one value of l at a time. There will be one dimension for each value of m, and therefore $2l + 1$ in all.) An eigenfunction of this state may be written as a linear combination of spherical harmonics having allowed m (i.e., $-l \leqslant m \leqslant l$):

$$\psi = c_l Y_l^l(\theta, \varphi) + c_{l-1} Y_l^{l-1}(\theta, \varphi) + \ldots + c^{-l} Y_l^{-l}(\theta, \varphi)$$

where the c's are constant coefficients giving the contribution of each particular Y_l^m to the eigenfunction. A different ψ would have different c's since it would be made up of a different combination of the $(2l + 1)Y_l^m$'s. In short, if each Y_l^m is compared to a unit vector parallel to the x, y, or z axis in coordinate space, each constant coefficient acts as a component of any given eigenfunction in the same way as x, y, or z is a component of a vector. These component c's thus act likè coordinates in a $(2l + 1)$-dimension vector space, and ψ could be represented as a column vector

$$\psi = \begin{bmatrix} c_l \\ c_{l-1} \\ \cdot \\ \cdot \\ \cdot \\ c_{-l} \end{bmatrix} \tag{5}$$

The operator L_z operating on Y_l^m must yield the same Y_l^m times a constant $(m\hbar)$, depending on the component of ψ it operates on. Hence in a representation with the z axis parallel to the polar axis, the operator L_z would be represented by a diagonal matrix

$$L_z = \hbar \begin{bmatrix} l & 0 & 0 & \cdots & 0 \\ 0 & (l-1) & 0 & \cdots & 0 \\ 0 & 0 & (l-2) & \cdots & 0 \\ \cdot & \cdot & \cdot & & \cdot \\ \cdot & \cdot & \cdot & & \cdot \\ \cdot & \cdot & \cdot & & \cdot \\ 0 & 0 & 0 & \cdots & -l \end{bmatrix} \tag{6}$$

while the operator L_x would have to be represented by a nondiagonal matrix since this operator introduces a different combination of Y_l^m's, Eq. (4):

$$L_x = \frac{\hbar}{2} \begin{bmatrix} 0 & a_l & 0 & 0 & \cdots & 0 & 0 \\ a_l & 0 & a_{l-1} & 0 & \cdots & 0 & 0 \\ 0 & a_{l-1} & 0 & a_{l-2} & \cdots & 0 & 0 \\ 0 & 0 & a_{l-2} & 0 & \cdots & 0 & 0 \\ \cdot & \cdot & \cdot & \cdot & & \cdot & \cdot \\ \cdot & \cdot & \cdot & \cdot & & \cdot & \cdot \\ \cdot & \cdot & \cdot & \cdot & & \cdot & \cdot \\ 0 & 0 & 0 & 0 & \cdots & 0 & a_{-l+1} \\ 0 & 0 & 0 & 0 & \cdots & a_{-l+1} & 0 \end{bmatrix} \tag{7}$$

where by Eq. (4)

$$a_{l-n} = \sqrt{(2l - n)(n + 1)}$$

Hence the operator L_x is said to be nondiagonal in the spherical coordinate representation with the z axis parallel to the polar axis. This is because (see below) L_z is diagonal in this representation and L_x and L_y do not commute with L_z. In general, only operators that commute can both be diagonal in the same representation; conversely if a representation exists in which two operators are diagonal these two operators commute. Since the energy of a complete system is constant and physically observable, a representation can always be found in which it is diagonal. In Chapter X it was pointed out that the representation in which the energy operator was diagonal was called the *true* or *correct representation* and the wave functions in this representation were called the *true* or *correct functions* of the system.

We can easily demonstrate that two noncommuting operators P and Q cannot both be diagonal in the same representation. We ask, what is $Q\psi$ if $P\psi = p\psi$ and $[P, Q] \equiv PQ - QP = R$, the capital letters denoting operators and p denoting the eigenvalue? To determine the answer to this question we operate on $Q\psi$ with P:

$$P(Q\psi) = PQ\psi = (QP + R)\psi = pQ\psi + R\psi$$

Thus P operating on $Q\psi$ does not give the same eigenvalue and unchanged eigenfunction as are obtained with P operating on ψ. Q changes the wave function (and hence must be nondiagonal) so that $P(Q\psi)$ gives a different result, even within a factor of the eigenvalue of Q, from that given by $P\psi$. The converse of this statement, incidentally, is not true. It does not follow, if P commutes with Q, and if P is diagonal, that Q is also diagonal. This is shown by a simple example. The operator L^2 is diagonal in the present representation (with diagonal elements $l(l + 1)$), and it commutes with L_x, which is not diagonal.

B.2 SPIN MATRICES

Here we present operators corresponding to the orbital angular momentum operators L_x, L_y, and L_z which will give us the spin of a complete electron state. It is true that we have used an operator of this kind in calculating the spin-orbit energy of an electron (Eq. 11.18), but in our previous discussion we managed to avoid specifying just what it was.

The spatial angular momentum is a property of the state of a system; if the state is degenerate in the projection of its angular momentum on any axis, so that the values of l do not specify all the levels of equal energy, then the state may be represented by a unit-rank tensor (a column tensor or vector) with $2l + 1$ components made up of the $2l + 1$ degenerate wave functions (Eq. 5). In the case of half-integral spin only two independent spin wave functions are possible, one representing a projected spin up, the other spin down. The spin eigenfunction of a single electron is either α or β. If we treat the spin eigenfunction as a two-component vector, we have

$$\alpha = \begin{vmatrix} 1 \\ 0 \end{vmatrix} \equiv \text{spin up}, \qquad \beta = \begin{vmatrix} 0 \\ 1 \end{vmatrix} \equiv \text{spin down} \qquad (8)$$

The operator representing the spin variable would then be a 2×2 matrix

$$\sigma_i = \begin{bmatrix} a_{11} & a_{12} \\ a_{21} & a_{22} \end{bmatrix} \qquad (9)$$

where $i = x, y,$ or z and the a's are yet to be evaluated constants. We normalize σ_i to give an eigenvalue of ± 1 so that

$$\sigma_x^2 = \sigma_y^2 = \sigma_z^2 = \begin{vmatrix} 1 & 0 \\ 0 & 1 \end{vmatrix} \equiv 1 \tag{10}$$

where 1 is the unit matrix. σ is a vector of constant length and has unit projection on any axis. For the component of σ along an arbitrary axis with direction cosines $\gamma_x, \gamma_y, \gamma_z,$ we have

$$(\gamma_x \sigma_x + \gamma_y \sigma_y + \gamma_z \sigma_z)^2 = 1$$

which means that

$$\sigma_x \sigma_y + \sigma_y \sigma_x = \sigma_x \sigma_z + \sigma_z \sigma_x = \sigma_y \sigma_z + \sigma_z \sigma_y = 0 \tag{11}$$

that is, the σ_i's *anticommute*. We might note here that the operator σ^2 is *not* a unit operator even though it has unit projection on any axis (see Problem 1). Arbitrarily we decide on a representation in which σ_z is diagonal and is different from the unit matrix. Then

$$\sigma_z = \begin{vmatrix} a & 0 \\ 0 & b \end{vmatrix}$$

We have $\sigma_z \alpha = \alpha, \sigma_z \beta = -\beta.$ Therefore $a = 1, b = -1,$ and

$$\sigma_z = \begin{vmatrix} 1 & 0 \\ 0 & -1 \end{vmatrix} \tag{12}$$

which satisfies Eq. (10).

In Chapter 5 we showed that real physical observables can be represented only by Hermitian operators. The corresponding property for matrices is $a_{nk}^* = a_{kn},$ as is readily verified by writing

$$a_{nk}^* = \left[\int \varphi_n^* A\varphi_k \, d\tau \right]^* = \int \varphi_n (A\varphi_k)^* \, d\tau = \int \varphi_k^* A\varphi_n \, d\tau = a_{kn}$$

That is, the complex conjugate of the transposed matrix must be equal to the original matrix, $a_{nk}^* = a_{kn}.$ Hence, the diagonal elements must be real and

$$\sigma_x = \begin{bmatrix} a_{11} & a_{12} \\ a_{12}^* & a_{22} \end{bmatrix}$$

From Eq. (11), $\sigma_x \sigma_z + \sigma_z \sigma_x = 0,$ so that multiplying matrices gives us

$$\begin{bmatrix} a_{11} & -a_{12} \\ a_{12}^* & -a_{22} \end{bmatrix} + \begin{bmatrix} a_{11} & a_{12} \\ -a_{12}^* & -a_{22} \end{bmatrix} = \begin{bmatrix} 2a_{11} & 0 \\ 0 & -2a_{22} \end{bmatrix} = 0$$

Therefore $a_{11} = a_{22} = 0.$ From Eq. (10),

$$\sigma_x^2 = \begin{bmatrix} |a_{12}|^2 & 0 \\ 0 & |a_{12}|^2 \end{bmatrix} = \begin{bmatrix} 1 & 0 \\ 0 & 1 \end{bmatrix}$$

Thus $|a_{12}|^2 = 1,$ which means that the magnitude (or modulus) of a_{12} is unity, but since a_{12} is a complex number its phase μ is as yet undetermined;

therefore $a_{12} = e^{i\mu}$. If we choose $\mu = 0$, we obtain for this component of the spin operator

$$\sigma_x = \begin{bmatrix} 0 & 1 \\ 1 & 0 \end{bmatrix} \tag{13}$$

The same relations hold for σ_y. In order for σ_y to anticommute with σ_x, we must have $\mu = \pm\pi/2$. We choose $-\pi/2$ and obtain

$$\sigma_y = \begin{bmatrix} 0 & -i \\ i & 0 \end{bmatrix} \tag{14}$$

The spin of the electron is given by the operator **S**,

$$\mathbf{S} = (\hbar/2)\boldsymbol{\sigma} \tag{15}$$

so that

$$S_z\alpha = S_z \begin{vmatrix} 1 \\ 0 \end{vmatrix} = \frac{\hbar}{2} \begin{vmatrix} 1 \\ 0 \end{vmatrix} = \frac{\hbar}{2}\alpha \tag{16}$$

since by straight matrix multiplication $\sigma_z \begin{vmatrix} 1 \\ 0 \end{vmatrix} = \begin{vmatrix} 1 \\ 0 \end{vmatrix}$. That is, the eigenvalue of the spin operator S_z is $\hbar/2$ when S_z operates on α. Similarly

$$S_z\beta = -\frac{\hbar}{2}\beta \tag{17}$$

and

$$S_x\alpha = \frac{\hbar}{2}\beta \quad \text{and} \quad S_x\beta = \frac{\hbar}{2}\alpha \tag{18}$$

Note, however, that

$$S_x(S_x\alpha) = S_x\left(\frac{\hbar}{2}\beta\right) = \frac{\hbar^2}{4}\alpha$$

as it should since the operator σ_x^2 equals the unit matrix. **S** is an operator representing the spin or intrinsic angular momentum of the electron, and it operates on the spin eigenfunctions (not on space!). The operator S^2 has the eigenvalue $\frac{3}{4}\hbar^2$. If we include the **L·S** term in the Hamiltonian (the H operator) for atomic electrons, then if we use central field potential wave functions, H is not diagonal, for **L·S** does not commute with L_z or S_z, which are diagonal in this representation. Operation with the **L·S** term will mix states of different L_z, S_z (since the **L·S** operator has nondiagonal matrix elements) having the same $m = m_l + m_s$, as we have discussed in the section on fine structure in Chapter XI (Eqs. 11.19, 11.20). m_l and m_s are the z-axis projections of the orbital and spin angular momentum vectors respectively. The operator **L·S** was found to have the eigenvalues $(\hbar^2/2)l$ for $j = l + s = l + \frac{1}{2}$ and $(-\hbar^2/2)(l + 1)$ for $j = l - \frac{1}{2}$ in this representation. Table 1 and Eq. (36) give the transformation matrix from the $|m_l, m_s\rangle$ to the $|j, m\rangle$ wave functions in the diagonal $(|j, m\rangle)$ representation where in this instance the subscript 1 in the table is l and 2 is s.

Problem 1: Prove that

$$\sigma^2 = 3, \quad \sigma_x\sigma_y = i\sigma_z, \quad \sigma \times \sigma = 2i\sigma$$

Problem 2: Show that the operator

$$e^{\sigma_y} = \cosh 1 + \sigma_y \sinh 1$$

Problem 3: Show that $\mathbf{L}\cdot\mathbf{S}$ does not commute with L_z or S_z but that it does commute with J_z, where $J_z = L_z + S_z$.

B3 VECTOR ADDITION MATRICES

Even though the total spin and orbital momentum of a state may be given, the total angular momentum is not specified thereby. Any two vectors such as the spin and orbital momentum add up vectorially to a vector, in this case the total angular momentum, whose magnitude ranges from the difference of the two vectors to their sum. To illustrate, we will consider a total eigenfunction with total angular momentum $\mathbf{j}$, made up of two component eigenfunctions having angular momenta $\mathbf{j}_1$ and $\mathbf{j}_2$, as illustrated in Figure 1. This would represent, for example, the total angular momentum of a state composed of a target nucleus with initial angular momentum $\mathbf{j}_1$ which absorbed a nucleus with angular momentum $\mathbf{j}_2$.

First we note that in spherical coordinates a wave function has two angular degrees of freedom; hence there are two eigenvalues associated with the angular momentum. Two angular momentum vectors such as $\mathbf{l}$ and $\mathbf{s}$ or, more generally, $\mathbf{j}_1$ and $\mathbf{j}_2$ will thus have four eigenvalues characterizing their wave functions, j_1, m_1 and j_2, m_2 (from which the angular eigenfunctions may be deduced.) We wish to know how these two vectors add. That is, given a wave function ψ which is a simultaneous eigenfunction of the commuting operators j_1^2, j_2^2, j^2, and m, we wish to express it as a linear combination of eigenfunctions of the commuting operators j_1^2, j_2^2, m_1, and m_2. This is important because the latter functions are simpler (and in fact already known to us in the case of $\mathbf{l}$ and $\mathbf{s}$ for a single electron), whereas the former are the only ones that can also be energy eigenfunctions (as explained in the section on spin-orbit coupling).

We might express our task as finding out how to go from a representation in which the operators $j_1^2, j_2^2, j_{1z} = m_1$ and $j_{2z} = m_2$ are diagonal to a representation in which j_1^2, j_2^2, j^2, and $j_z = m$ are diagonal, where $[\mathbf{j} = \mathbf{j}_1 + \mathbf{j}_2$ and $j_z = j_{1z} + j_{2z}]$. Since j_1^2 and j_2^2 are the same in both representations, we will denote the wave functions in the two representations by $|m_1, m_2\rangle$ and $|j, m\rangle$ respectively. We will also frequently write $|m_1, m_2\rangle$ as $|j_1, m_1\rangle|j_2, m_2\rangle$. Our task is to find out how to go from the first representation to the second, that is to find out how the vectors $\mathbf{j}_1$ and $\mathbf{j}_2$ add.

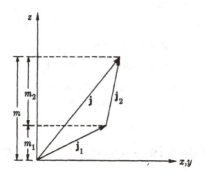

FIGURE 1 Illustration of vector model of angular momenta, showing how two vectors j_1 and j_2 add to a total vector j.

For example, a nucleus with known spin may absorb a thermal neutron. The state of the resultant compound nucleus has a total spin j and z projection m and is comprised of various combinations of states of the target nucleus and of the incident neutron state with identical j and m. It is our purpose to determine what these combinations are. The $|j, m\rangle$ may be given as a superposition of the complete set of $|m_1, m_2\rangle$:

$$|j, m\rangle = \sum_{m_1, m_2} a_{jmm_1m_2} |m_1, m_2\rangle$$

Multiplying by an arbitrary state vector $\langle r_1, r_2|$ on the left of both sides we obtain $\langle r_1, r_2|j, m\rangle = a_{jmr_1r_2}$, owing to the orthonormality of the base wave functions $|m_1, m_2\rangle$. Thus, writing m_1 for r_1 and m_2 for r_2 we have

$$|j, m\rangle = \sum_{m_1, m_2} \langle m_1, m_2|j, m\rangle |m_1, m_2\rangle \qquad (19)$$

where $m_1 + m_2 = m$ for each term (since the matrix element vanishes when $m \neq m_1 + m_2$).

The normalized wave function $|j, m\rangle$ is given in the first representation by a linear combination of different $|m_1, m_2\rangle$ whose coefficients are given by the $\langle m_1, m_2|j, m\rangle$. (Of course, $\langle m_1, m_2|j, m\rangle = 0$ if $m \neq m_1 + m_2$.)

If the two vectors j_1 and j_2 are parallel and their projection on the z axis is a maximum, then only one compound state is possible:

$$|j, m\rangle = |j, j\rangle = |j_1, j_1\rangle|j_2, j_2\rangle \equiv |j_1, j_2\rangle \qquad (20)$$

where $j = j_1 + j_2$ and the third and fourth terms are merely different notations for the same state. If, on the other hand, $m = j_1 + j_2 - 1$, two states are possible,

$$|j_1, j_1\rangle|j_2, j_2 - 1\rangle \quad \text{and} \quad |j_1, j_1 - 1\rangle|j_2, j_2\rangle$$

which may also be written as $|j_1, j_2 - 1\rangle$ and $|j_1 - 1, j_2\rangle$. The complete state $|j_1 + j_2, j_1 + j_2 - 1\rangle$ is a linear combination of these two with appropriate normalized coefficients.

To find these coefficients we first make use of the commutation relations for angular momenta discussed in Chapter VIII to obtain operators which will raise or lower the j_z projection operator eigenvalue m.

$$j_z j_x = j_x j_z + i\hbar j_y$$
$$j_z j_y = j_y j_z - i\hbar j_x$$

Hence

$$j_z(j_x \pm ij_y) = (j_x \pm ij_y)(j_z \pm \hbar) \tag{21}$$

by virtue of the commutation relations for angular momenta.

Since

$$j_z |j, m\rangle = m\hbar |j, m\rangle$$

Eq. (21) yields

$$j_z(j_x \pm ij_y)|j, m\rangle = (m \pm 1)\hbar(j_x \pm ij_y)|j, m\rangle \tag{22}$$

(Note that eigenvalues, being just numbers, can be written on either side of an operator.) Therefore, $(m \pm 1)\hbar$ is the eigenvalue of j_z (that is, $(m \pm 1)\hbar$ is the projection of the angular momentum on the z axis) when j_z operates on the wave function resulting from the operation of $(j_x \pm ij_y)$ on $|j, m\rangle$. Thus, we conclude that operating on $|j, m\rangle$ with $(j_x \pm ij_y)$ gives us a new eigenfunction $|j, m \pm 1\rangle$ times an as yet undetermined coefficient.

The operator $j_x + ij_y$ is called a *raising* operator

The operator $j_x - ij_y$ is called a *lowering* operator.

The use of raising and lowering operators (also known as *creation* and *annihilation* operators), of which this is an example, is an especially elegant method for obtaining the properties of the harmonic oscillator.

The coefficient of the eigenfunction $|j, m - 1\rangle$ obtained by operating on $|j, m\rangle$ with $(j_x - ij_y)$ can be determined by finding the normalization constant N which goes with the operator $j_x - ij_y$ operating on $|j, m\rangle$ so that both $\langle j, m - 1 | j, m - 1\rangle = 1$ and $\langle j, m | j, m\rangle = 1$ (i.e., so that both eigenfunctions are normalized). Since

$$N(j_x - ij_y)|j, m\rangle = |j, m - 1\rangle \tag{23}$$

and

$$\langle j, m - 1 | j, m - 1\rangle = 1$$

therefore

$$\langle j, m - 1 | j, m - 1\rangle = |N|^2 \langle j, m |(j_x + ij_y)(j_x - ij_y)| j, m\rangle = 1 \tag{24}$$

where the complex conjugate of the state $(j_x - ij_y)|j, m\rangle$ has been written on the left, and N is the normalization constant we seek. Further manipulation utilizing commutativity gives

$$|N|^2 \langle j, m | j_x^2 + j_y^2 - i(j_x j_y - j_y j_x) | j, m \rangle$$
$$= |N|^2 \langle j, m | j^2 - j_z^2 + \hbar j_z | j, m \rangle$$
$$= |N|^2 \hbar^2 [j(j+1) - m(m-1)] = 1 \qquad (25)$$

Hence,

$$|N| = \frac{1}{\hbar \sqrt{(j+m)(j-m+1)}} \qquad (26)$$

(The phase factor in N is arbitrary, since $e^{i\varphi} \psi$ represents the same physical state as ψ.)

Therefore, from Eqs. (23, 26) we conclude that for $m_1 = j_1$,

$$(j_{1x} - ij_{1y})|j_1, j_1 \rangle = \hbar \sqrt{(j_1 + j_1)(j_1 - j_1 + 1)}|j_1, j_1 - 1 \rangle \qquad (27)$$
$$= \hbar \sqrt{2j_1} |j_1, j_1 - 1 \rangle$$

Similarly, we have

$$(j_x - ij_y)|j_1 + j_2, j_1 + j_2 \rangle = \hbar \sqrt{2(j_1 + j_2)}|j_1 + j_2, j_1 + j_2 - 1 \rangle \qquad (28)$$

where the same argument has been applied to the total angular momentum $\mathbf{j} = \mathbf{j}_1 + \mathbf{j}_2$. Now there is only one way in which the eigenfunction $|j_1 + j_2, j_1 + j_2 \rangle$ can be formed from eigenfunctions $|j_1, m_1 \rangle$, $|j_2, m_2 \rangle$, namely

$$|j_1 + j_2, j_1 + j_2 \rangle = |j_1, j_1 \rangle |j_2, j_2 \rangle \qquad (29)$$

since any other product would necessarily have $m = m_1 + m_2 < j_1 + j_2$. Therefore the left-hand side of Eq. (28) can be written as

$$(j_{1x} - ij_{1y} + j_{2x} - ij_{2y})|j_1, j_1 \rangle |j_2, j_2 \rangle$$
$$= \hbar \sqrt{2j_1} |j_1, j_1 - 1 \rangle |j_2, j_2 \rangle + \hbar \sqrt{2j_2} |j_1, j_1 \rangle |j_2, j_2 - 1 \rangle$$

Hence

$$|j_1 + j_2, j_1 + j_2 - 1 \rangle = \sqrt{\frac{j_1}{j_1 + j_2}} |j_1, j_1 - 1 \rangle |j_2, j_2 \rangle$$
$$+ \sqrt{\frac{j_2}{j_1 + j_2}} |j_1, j_1 \rangle |j_2 j_2 - 1 \rangle \qquad (30)$$

which gives two of the coefficients $\langle m_1, m_2 | j, m \rangle$, namely

$$\langle j_1 - 1, j_2 | j_1 + j_2, j_1 + j_2 - 1 \rangle = \sqrt{\frac{j_1}{j_1 + j_2}} \qquad (31)$$

$$\langle j_1, j_2 - 1 | j_1 + j_2, j_1 + j_2 - 1 \rangle = \sqrt{\frac{j_2}{j_1 + j_2}} \qquad (32)$$

Similarly, from (29), we have

$$\langle m_1, m_2 | j_1 + j_2, j_1 + j_2 \rangle = \begin{cases} 0, \text{ unless } j_1 = m_1 \text{ and } j_2 = m_2 \\ 1, \text{ if } j_1 = m_1 \text{ and } j_2 = m_2 \end{cases}$$

All of the other coefficients $< m_1, m_2 \,|\, j, m >$ are obtained by continuing this process, but the calculations get to be very tedious. They are known as the *vector-addition coefficients*, or the *Clebsch-Gordon coefficients*.

In general,

$$|j, m\rangle = \sum_{m_1, m_2} \langle m_1, m_2 \,|\, j, m \rangle |m_1, m_2\rangle \tag{19}$$

where the summation must be carried out over all m_1 and m_2 consistent with

$$m = m_1 + m_2$$
$$\mathbf{j} = \mathbf{j}_1 + \mathbf{j}_2 \tag{33}$$

in order to obtain the total wave function $|j, m\rangle$. The coefficients on the right-hand side of Eq. (19) are frequently written as

$$C_{m_1 m_2 m}^{j_1 j_2 j} = \langle m_1, m_2 \,|\, j, m \rangle \tag{34}$$

and in many other ways by various authors.

The $C_{m_1 m_2 m}^{j_1 j_2 j}$ are complicated functions of the j_1, j_2, m_1, and m_2, as we have seen, which must be evaluated for each possible combination of $\mathbf{j}_1$ and $\mathbf{j}_2$, m_1, and m_2 occurring in the sum. The C coefficients (Clebsch–Gordon coefficients) occur frequently in atomic and nuclear physics, and extensive tables of possible combinations of m_1, m_2, j_1, and j_2 have been published,* as well as tables for very much more extensive and complicated three-vector addition coefficients for which

$$\mathbf{j} = \mathbf{j}_1 + \mathbf{j}_2 + \mathbf{j}_3$$
$$m = m_1 + m_2 + m_3$$

Table 1 is a table of vector addition coefficients for the important special case $j_2 = \frac{1}{2}$. With these coefficients one may determine the way the orbital momentum of an electron ($j_1 = l$) may add to its spin ($j_2 = s$) to yield a total angular momentum of j and a z projection of m. If two electrons having a total spin of unity (spin-symmetric state) and a total orbital momentum of l were treated in terms of their total angular momentum and its projection, a 3×3 table having nine coefficients would result.†

What Table 1 means in l, s notation is that the spin-space projection wave functions $|m_l, m_s\rangle$ are related to the total angular momentum wave functions $|l \pm \frac{1}{2}, m\rangle$ in the following way: ($m = m_l + m_s$)

$$|l + \tfrac{1}{2}, m\rangle = \sqrt{\tfrac{1}{2} + \frac{m}{2l+1}}\, |m - \tfrac{1}{2}, \tfrac{1}{2}\rangle + \sqrt{\tfrac{1}{2} - \frac{m}{2l+1}}\, |m + \tfrac{1}{2}, -\tfrac{1}{2}\rangle$$

$$|l - \tfrac{1}{2}, m\rangle = -\sqrt{\tfrac{1}{2} - \frac{m}{2l+1}}\, |m - \tfrac{1}{2}, \tfrac{1}{2}\rangle + \sqrt{\tfrac{1}{2} + \frac{m}{2l+1}}\, |m + \tfrac{1}{2}, -\tfrac{1}{2}\rangle$$

$$\tag{35}$$

---

* See, for example, Condon, E. V. and G. H. Shortley, cited in bibliography of this appendix.

† Given in Condon and Shortley, *op cit.*, p. 76.

In matrix notation these wave functions may be written as two-component first-rank tensors or column vectors, and the coefficients in the form of a 2×2 second-rank tensor called a *transformation matrix*, enabling us to go from one representation (the $|m_l, m_s\rangle$) to the other representation ($|j, m\rangle$):

$$\begin{bmatrix} |l + \tfrac{1}{2}, m\rangle \\ |l - \tfrac{1}{2}, m\rangle \end{bmatrix} = \begin{bmatrix} \sqrt{\tfrac{1}{2} + \dfrac{m}{2l+1}} & \sqrt{\tfrac{1}{2} + \dfrac{m}{2l+1}} \\ -\sqrt{\tfrac{1}{2} - \dfrac{m}{2l+1}} & \sqrt{\tfrac{1}{2} + \dfrac{m}{2l+1}} \end{bmatrix} \begin{bmatrix} |m - \tfrac{1}{2}, \tfrac{1}{2}\rangle \\ |m + \tfrac{1}{2}, -\tfrac{1}{2}\rangle \end{bmatrix}$$

(36)

which gives Eq. (35) by the rules of matrix multiplication.

TABLE 1

Vector addition coefficients for the special case of $j_2 = \tfrac{1}{2}$. Only four different combinations of m_1, m_2, j_1, and j_2 are possible satisfying Eq. (33).

	$m_1 = m - \tfrac{1}{2}$ $m_2 = \tfrac{1}{2}$	$m_1 = m + \tfrac{1}{2}$ $m_2 = -\tfrac{1}{2}$
$j = j_1 + \tfrac{1}{2}$	$\sqrt{\dfrac{1}{2} + \dfrac{m}{2j_1 + 1}}$	$\sqrt{\dfrac{1}{2} - \dfrac{m}{2j_1 + 1}}$
$j = j_1 - \tfrac{1}{2}$	$-\sqrt{\dfrac{1}{2} - \dfrac{m}{2j_1 + 1}}$	$\sqrt{\dfrac{1}{2} + \dfrac{m}{2j_1 + 1}}$

If a spin-1 system were added to the total orbital momentum, three $|j, m\rangle$ wave functions would result, $|l + 1, m\rangle$, $|l, m\rangle$, and $|l - 1, m\rangle$, from the three $|m_l, m_s\rangle$ wave functions $|m - 1, 1\rangle$, $|m, 0\rangle$, and $|m + 1, -1\rangle$. Therefore a 3×3 transformation matrix would be required.

Problem 4: Express a total two-electron spin-symmetric state with total orbital momentum l and total angular momentum projected on a given axis of m in terms of the three product wave functions of the two separate electrons, $|m - 1, 1\rangle$, $|m, 0\rangle$, and $|m + 1, -1\rangle$.

BIBLIOGRAPHY

Condon, E. V. and G. H. Shortley, *The Theory of Atomic Spectra*. New York: Cambridge University Press, 1957. Difficult and advanced, this has been a physicist's bible for the material presented in this appendix since publication of the first edition in 1935.

Slater, J. C., *Quantum Theory of Atomic Structure*, Vol. 1. New York: McGraw-Hill Book Co., 1960. This contains very extensive and especially lucid discussions of angular momenta and the rest of the material presented in this appendix.

Fermi, E., *Notes on Quantum Mechanics*. Chicago: University of Chicago Press, 1961. Fragmentary and unpolished; Fermi never arranged these notes for publication, but they are outstanding for their concise and physically illuminating treatment.

Mayer, M. G. and J. H. D. Jensen, *Elementary Theory of Nuclear Shell Structure*. New York: John Wiley & Sons, Inc., 1955.

APPENDIX C

REDUCTION OF
INFINITE SERIES IN
HIGHER-ORDER
PERTURBATION
THEORY

One of the cumbersome difficulties of second-order and higher-order perturbation theory is that the infinite series called for in Eqs. (9.29, 9.30) require evaluation of an infinite number of integrals. Bethe and Salpeter have mentioned a method by which these infinite numbers of integrals may be replaced by only one or two integrals, which Dalgarno and his collaborators have done for the hydrogen atom system. The development used here follows Dalgarno and Lewis as described by Schwartz, who has generalized and extended the method to higher order than second-order perturbation theory.

Let us define an operator F such that

$$[F, H_0]\psi_i^0 = (V' - E_i')\psi_i^0 \tag{1}$$

Substitute this equation into (9.29) and neglect V_{ii}''

$$E_i'' = \sum_{\substack{m \neq i \\ m=0}}^{\infty} \frac{V_{im}' \int \psi_m^{0*}(FH_0 - H_0F + E_i')\psi_i^0 \, d\tau}{E_i^0 - E_m^0}$$

$$= \sum_{\substack{m \neq i \\ m=0}}^{\infty} \frac{V_{im}' \int \psi_m^{0*}(FE_i^0 - E_m^0 F)\psi_i^0 \, d\tau + E_i' \int \psi_m^{0*}\psi_i^0 \, d\tau}{E_i^0 - E_m^0}$$

$$= \sum_{\substack{m \neq i \\ m=0}}^{\infty} V_{im}' \int \psi_m^{0*} F \psi_i^0 \, d\tau \equiv \sum_{\substack{m \neq i \\ m=0}}^{\infty} V_{im}' F_{mi}$$

$$= \sum_{m=0}^{\infty} V_{im}' F_{mi} - V_{ii}' F_{ii}$$

$$= (V'F)_{ii} - E_i' F_{ii}$$

Thus we have obtained our second-order perturbation result in terms of two integrals which replace the infinite sum of integrals in Eq. (9.29). The problem is to find F, which is given as the solution of Eq. (1). One way of doing this is to write out Eq. (1) explicitly

$$(\nabla^2 F)\psi_i^0 + 2(\nabla F)\cdot\nabla\psi_i^0 = \frac{2m}{\hbar^2}(V' - E_i')\psi_i^0 \tag{2}$$

As a simple example of the method we take Problem 9.7 letting $V' = -eGr\cos\theta$. The ground state of the hydrogen-like atom, a $(1s)$ state, is

$$\psi_0^0 = (1/\sqrt{\pi a^3})\, e^{-r/a}$$

where $a = \hbar^2/zme^2$, and $E_0' = 0$. Hence, after dividing out ψ_0^0, Eq. (2) may be written

$$\nabla^2 F - \frac{2}{a}\frac{\partial F}{\partial r} = -\frac{2meG}{\hbar^2} r\cos\theta$$

where

$$\nabla^2 F = \frac{1}{r^2}\frac{\partial}{\partial r} r^2 \frac{\partial F}{\partial r} + \frac{1}{r^2 \sin\theta}\frac{\partial}{\partial \theta}\sin\theta\frac{\partial F}{\partial \theta}$$

and the solution is

$$F = \frac{meG}{\hbar^2}\cos\theta\left(a^2 r + \frac{1}{2}ar^2\right)$$

Therefore, after an elementary integration over the ground state hydrogen-like wave functions,

$$E_0' = (V'F)_{00} = -\frac{9}{4}\frac{a^3 G^2}{Z}$$

APPENDIX D

PARITY

Parity refers to whether or not the wave function changes sign when the coordinate system is reversed. ψ itself is not a physical observable although $\psi^*\psi$ is, as we know. The change of sign of ψ when the coordinate system is reversed is possible and, as we shall see, is a very useful property of the eigenfunction. It is also convenient to note that the parity of a state may be inferred from the mathematical properties of the function representing the state. We shall also see that operators representing physical observables can be constructed which will tell us whether the parity of a system's eigenfunction *changes* even though the parity itself is not directly observed. In most physical processes this does not happen (i.e., parity is usually *conserved*), but when it does happen that the parity of the system's wave function changes during a physical process, parity is said to be *not conserved*.

The parity operation is equivalent to the exchange operation for systems composed of two identical particles; its application, however, includes also systems of nonidentical particles. The parity operator changes the signs of directions of all coordinates. That is, all vectors $\mathbf{r}_1$, $\mathbf{r}_2$ ... become $-\mathbf{r}_1$, $-\mathbf{r}_2$ ... by a change in the directions of the coordinate system, $x \longrightarrow -x$, $y \longrightarrow -y$, $z \longrightarrow -z$. (This amounts to going from a right-handed to a left-handed coordinate system, which reflection in any plane would also accomplish.) The behavior of the wave function under the parity operation is an important property of the wave function. Since only $|\psi|^2$ is measured experimentally, not ψ itself, the phase of ψ may be affected by the parity operation and the parity operator P may have an eigenvalue different from $+1$. As for the exchange operation, to which the parity operation is equivalent for systems composed of two identical particles, we have

$$P\psi(\mathbf{r}) = \psi(-\mathbf{r}) = k\psi(\mathbf{r}) \tag{1}$$

When the parity operation is taken twice, the original eigenfunction must result:

$$P^2\psi(\mathbf{r}) = k^2\psi(\mathbf{r}) = \psi(\mathbf{r}) \tag{2}$$

Therefore $k = \pm 1$; if k is $+1$, the parity of the state is said to be even; if -1, it is said to be odd. Note that the spin coordinate is unaffected by the parity operation. For the *orbital* angular momentum we have

$$
\begin{aligned}
L_x &= yP_z - zP_y \\
&\rightarrow (-y)(-P_z) - (-z)(-P_y) \\
&= L_x, \text{ not } -L_x
\end{aligned}
\tag{3}
$$

and the x component of the spin, S, must behave in the same way.

If any operator does not depend on time explicitly, its time dependence is determined from its commutation with the Hamiltonian (Eq. 5.17):

$$[P, H] = PH - HP \tag{4}$$

Note that this result shows that parity is conserved (unchanged) for all space-symmetric Hamiltonians, for H being space-symmetric does not change when operated on by P. Since $(PH) = H$, $PH = HP$, and therefore $PH - HP = 0$. ((PH) means P operates only on H while PH means P operates on *everything* to the right of P.) The consequence, $dP/dT = 0$, means that if the initial state of a system is even (or odd), it will remain so in the course of time. The parity of the state will never change. Formerly it was thought that all systems have Hamiltonians that commute with the parity operator, but as we shall presently see this is now known to be false.

The parity of a theoretical wave function is readily determined by reversing the coordinate system and calculating the new wave function. Through such calculations it becomes clear that in the case of a spherical harmonic wave function, the parity is equal to $(-1)^l$, where l is the total orbital momentum quantum number. Therefore, s, d, and g states ($l = 0$, 2, 4) have even parity ($k = 1$), while p, f, and h states ($l = 1, 3, 5$) have odd parity ($k = 1$).

The parity of the total state of an entire system is not directly determinable experimentally; it must be inferred from a measurement of the total orbital momentum of the state. Let us consider in detail what we can and cannot learn about the parity of a state through physical measurement.

An even-parity operator representing an observable, O_+, would yield the result

$$\langle \psi^* | O_+ | \psi \rangle = \langle \psi_+^* | O_+ | \psi_+ \rangle + \langle \psi_-^* | O_+ | \psi_- \rangle$$
$$+ \langle \psi_+^* | O_+ | \psi_- \rangle + \langle \psi_-^* | O_+ | \psi_+ \rangle \qquad (5)$$

for a ψ impure in parity, being composed of ψ_+ an even-parity and ψ_- an odd-parity wave function. That is,

$$P\psi_+(\mathbf{r}) = \psi_+(\mathbf{r}) \quad \text{and} \quad P\psi_-(\mathbf{r}) = -\psi_-(\mathbf{r}) \qquad (6)$$

When the integrals over all space are taken, the cross-product terms $\langle \psi_+^* | O_+ | \psi_- \rangle$ and $\langle \psi_-^* | O_+ | \psi_+ \rangle$ give zero. The other two terms are positive definite. A measurement of the physical quantity represented by O_+ would therefore not yield any information about the parity of ψ. However, the fact that the parity of the state is *not pure* could be determined with an odd-parity operator of a different observable, O_-. An odd-parity operator, which changes sign if the coordinate system is reversed, gives a finite result only for the cross-product terms $\langle \psi_+^* | O_- | \psi_- \rangle$ and $\langle \psi_-^* | O_- | \psi_+ \rangle$. Hence, if the physical observable represented by O_- is measured experimentally and a nonzero result is obtained, we have definite evidence of an impure parity state representing the system on which the measurement is made. On the other hand, if the parity of the state is pure, but changes during some transition to another state and if O_- is somehow measured between the initial state on the right and the final state on the left and gives a nonzero result, we know that the parity of the system changed during the transition.

If the parity of a state were to change during some physical process, say particle emission, a partial change in parity of the system state function during the particle-emission process might be observed as an asymmetry in the spatial distribution of the emitted particles. In one famous example, beta decay as initially commented on by C. N. Yang and T. D. Lee, one might observe the following: the correlations or scalar products of electron momentum **p** and the spin (sum of intrinsic nucleon spins plus the nuclear orbital momentum) of the emitting nucleus **J**, (**p·J**); electron momentum and the circular polarization $\boldsymbol{\Sigma}$ (angular momentum) of an emitted gamma ray associated with the beta decay, (**p·$\boldsymbol{\Sigma}$**); circular polarization of a gamma ray and its propagation vector; or electron momentum and its intrinsic spin (**p·$\boldsymbol{\sigma}$**). Each of these operators includes an angular momentum vector **J**, $\boldsymbol{\Sigma}$, or $\boldsymbol{\sigma}$ which is not a true vector. If the coordinate system is reversed (that is, $x \longrightarrow -x$, $y \longrightarrow -y$, $z \longrightarrow -z$) so that one goes from a so-called right-handed to a left-handed coordinate system, all these scalar products change sign. The axis of this asymmetry would clearly have to have some physical meaning, for all directions in space would be equivalent unless some property of the system had a direction one could choose as a coordinate axis for analyzing the emission event. Such directions or axes with physical meaning are furnished by pseudovectors representing physical variables, e.g., $\mathbf{J}$, $\boldsymbol{\sigma}$, or $\boldsymbol{\Sigma}$. Ordinary scalar quantities such as the mass of a sack of potatoes or the temperature of a room never depend on a coordinate system in this way; therefore, the scalar product of angular momentum and ordinary momentum is called a *pseudoscalar*. A pseudoscalar operator such as $\mathbf{p} \cdot \boldsymbol{\sigma}$ is, therefore, an odd-parity operator which represents a physical observable. A positive measurement of correlation between electron momentum $\mathbf{p}$ and electron spin $\boldsymbol{\sigma}$ is proof that an impure parity state exists.

For beta decay, an initial atomic state is represented as transformed to a final state by use of the beta decay operator:

$$\psi_f = H_\beta \psi_i$$

where ψ_f is composed of product wave functions of the residual nucleus, the atomic electrons, the emitted beta ray, and neutrinos, and ψ_i represents the initial atom and H_β the beta decay operator.

The expectation of a pseudoscalar operator, for example $\mathbf{J} \cdot \mathbf{p}$ in beta decay, is given by

$$\langle \mathbf{J} \cdot \mathbf{p} \rangle = \langle \psi_f^* | \mathbf{J} \cdot \mathbf{p} | \psi_f \rangle = \langle \psi_i^* H_\beta^* | \mathbf{J} \cdot \mathbf{p} | H_\beta \psi_i \rangle \tag{7}$$

In integration over all space an odd number of odd-parity factors in the integrand will cause the integral to be zero.

Ordinarily, without use of the $\mathbf{J} \cdot \mathbf{p}$ operator, no interference can be observed between the parity-conserving and parity-nonconserving terms in the beta decay operator; the calculated result is proportional to the *square* of these terms and hence their parity cannot be ascertained. The $\mathbf{J} \cdot \mathbf{p}$ operator, however, is an odd-parity operator, i.e., it is odd with respect to reversing the coordinates and when integrated gives a finite result only if multiplied by an odd-parity function on one side and an even-parity function on the other side. Thus $\mathbf{J} \cdot \mathbf{p}$ or any pseudoscalar operator would have an expected value of zero unless the beta decay interaction operator contained both even and odd terms. The presence of both odd and even parity terms was first determined experimentally by C. S. Wu and co-workers at Columbia and the N.B.S., by observing the beta decay electron momentum $\mathbf{p}$ upon emission from nuclei whose nuclear spins $\mathbf{J}$ were aligned all in one direction. The electrons were not emitted isotropically as expected; more electrons were measured in one direction along the axis of nuclear spin than in the opposite direction. The observed correlation of emitted beta rays with nuclear spin was incontrovertible evidence of partial parity nonconservation in the beta decay process. This and μ-meson decay were the physical interactions or processes first observed not to conserve parity. For nuclear or electromagnetic interactions, such as give rise to alpha particle decay or different nuclear states, operators such as $\mathbf{J} \cdot \mathbf{p}$ are observed to give a zero eigenvalue; the rule appears to be that strong interactions conserve parity and weak interactions such as beta decay do not.

INDEX

Triplet state, 187

Uhlenbeck and Goudsmit, intrinsic electron
  spin, 162
Uncertainty, 12
  frequency band width and duration of a
    wave, 19-21
Uncertainty principle, Heisenberg's, 33-34
Unified model of the nucleus, 296-300
Unit matrix, 311

Variational method, 130-132
Vector addition:
  with j-j coupling, 206
  with R-S coupling, 199
Vector addition matrices, 313-318
Velocity of a wave:
  phase, 22, 23
  group, 22, 23
Vibrating string, 302-304

Wave equation:
  electromagnetic, 2
  Schrödinger, 49-51, 95
Wave function:
  antisymmetric, 184
  diatomic molecule, 228-230
  harmonic oscillator:
    linear, 81
    spherical, 115
  normalized, 56, 81, 238

one-electron atom, 112, 154
  particle, 44-47
  particle in a square well, 108
  photons, 43
  rigid rotator, 105
  symmetric, 184
Wave packets, 21-24
  particle, 52, 53
Waves:
  in a box, 1-4 (*see also* Radiation)
  particle, 51-53
Wave velocity:
  group, 22, 23
  phase, 22, 23
Wien's law, 7
Wentzel-Kramers-Brillouin approximation,
  (WKB), 84-90
  barrier penetration, 90-92
  connection formulas, 87-89
  energy levels in potential wells, 89
Weissaker atomic mass formula (*see* Semiempirical mass formula)

X-rays, 207-210
  bremsstrahlung, 208
  characteristic, 208

Yukawa field, 278-279

Zeeman effect, 35
  multi-electron atom, 209-210
  one-electron atom, 171-178, 180-181